T0340712

Advances in Ceramic Armor VII

Advances in Ceramic Armor VII

A Collection of Papers Presented at the 35th International Conference on Advanced Ceramics and Composites
January 23–28, 2011
Daytona Beach, Florida

Edited by
Jeffrey J. Swab

Volume Editors
Sujanto Widjaja
Dileep Singh

A John Wiley & Sons, Inc., Publication

Published by John Wiley & Sons, Inc., Hoboken, New Jersey.
Published simultaneously in Canada.

Library of Congress Cataloging-in-Publication Data is available.

ISBN 978-1-118-05990-6

oBook ISBN: 978-1-118-09525-6
ePDF ISBN: 978-1-118-17309-1

ISSN: 0196-6219

Printed in the United States of America.

10 9 8 7 6 5 4 3 2 1

Contents

NONDESTRUCTIVE CHARACTERIZATION

PHENOMENOLOGY AND MECHANICS OF CERAMICS SUBJECTED TO BALLISTIC IMPACT

Preface

Over the past nine years the Armor Ceramics Symposium at the International Conference & Exposition on Advanced Ceramics and Composites in Daytona Beach, FL has grown to become one of the premier international conferences for the latest developments in the fabrication and application of ceramic materials to meet the needs of the armor community. The symposium continues to foster discussion and collaboration between academic, government and industry personnel from around the globe. The 9th edition focused on subjects such as Phenomenology and Mechanics of Ceramics Subjected to Ballistic Impact, High-Rate Real-Time Characterization, Microstructural Design, Nondestructive Characterization, Multi-Scale Modeling and Manufacturing. The manuscripts in these proceedings are a snapshot of some of the 70+ presentations that comprised the symposium.

On behalf of the organizing committee I would like to thank all of the presenters, authors, session chairs and manuscript reviewers for their efforts in making the symposium and the associated proceedings a success. As always a special thank you must go out to Marilyn Stoltz and Greg Geiger of The American Ceramic Society for their tireless efforts and for always being there to answer my questions and provide the guidance and administrative support necessary to ensure that symposium is successful.

JEFFREY J. SWAB
Symposium Chair

Introduction

This CESP issue represents papers that were submitted and approved for the proceedings of the 35th International Conference on Advanced Ceramics and Composites (ICACC), held January 23–28, 2011 in Daytona Beach, Florida. ICACC is the most prominent international meeting in the area of advanced structural, functional, and nanoscopic ceramics, composites, and other emerging ceramic materials and technologies. This prestigious conference has been organized by The American Ceramic Society's (ACerS) Engineering Ceramics Division (ECD) since 1977.

The conference was organized into the following symposia and focused sessions:

Symposium 1	Mechanical Behavior and Performance of Ceramics and Composites
Symposium 2	Advanced Ceramic Coatings for Structural, Environmental, and Functional Applications
Symposium 3	8th International Symposium on Solid Oxide Fuel Cells (SOFC): Materials, Science, and Technology
Symposium 4	Armor Ceramics
Symposium 5	Next Generation Bioceramics
Symposium 6	International Symposium on Ceramics for Electric Energy Generation, Storage, and Distribution
Symposium 7	5th International Symposium on Nanostructured Materials and Nanocomposites: Development and Applications
Symposium 8	5th International Symposium on Advanced Processing & Manufacturing Technologies (APMT) for Structural & Multifunctional Materials and Systems
Symposium 9	Porous Ceramics: Novel Developments and Applications

Symposium 10	Thermal Management Materials and Technologies
Symposium 11	Advanced Sensor Technology, Developments and Applications
Symposium 12	Materials for Extreme Environments: Ultrahigh Temperature Ceramics (UHTCs) and Nanolaminated Ternary Carbides and Nitrides (MAX Phases)
Symposium 13	Advanced Ceramics and Composites for Nuclear and Fusion Applications
Symposium 14	Advanced Materials and Technologies for Rechargeable Batteries
Focused Session 1	Geopolymers and other Inorganic Polymers
Focused Session 2	Computational Design, Modeling, Simulation and Characterization of Ceramics and Composites
Special Session	Pacific Rim Engineering Ceramics Summit

The conference proceedings are published into 9 issues of the 2011 Ceramic Engineering & Science Proceedings (CESP); Volume 32, Issues 2-10, 2011 as outlined below:

- Mechanical Properties and Performance of Engineering Ceramics and Composites VI, CESP Volume 32, Issue 2 (includes papers from Symposium 1)
- Advanced Ceramic Coatings and Materials for Extreme Environments, Volume 32, Issue 3 (includes papers from Symposia 2 and 12)
- Advances in Solid Oxide Fuel Cells VI, CESP Volume 32, Issue 4 (includes papers from Symposium 3)
- Advances in Ceramic Armor VII, CESP Volume 32, Issue 5 (includes papers from Symposium 4)
- Advances in Bioceramics and Porous Ceramics IV, CESP Volume 32, Issue 6 (includes papers from Symposia 5 and 9)
- Nanostructured Materials and Nanotechnology V, CESP Volume 32, Issue 7 (includes papers from Symposium 7)
- Advanced Processing and Manufacturing Technologies for Structural and Multifunctional Materials V, CESP Volume 32, Issue 8 (includes papers from Symposium 8)
- Ceramic Materials for Energy Applications, CESP Volume 32, Issue 9 (includes papers from Symposia 6, 13, and 14)
- Developments in Strategic Materials and Computational Design II, CESP Volume 32, Issue 10 (includes papers from Symposium 10 and 11 and from Focused Sessions 1, and 2)

The organization of the Daytona Beach meeting and the publication of these proceedings were possible thanks to the professional staff of ACerS and the tireless dedication of many ECD members. We would especially like to express our sincere thanks to the symposia organizers, session chairs, presenters and conference attendees, for their efforts and enthusiastic participation in the vibrant and cutting-edge conference.

ACerS and the ECD invite you to attend the 36th International Conference on Advanced Ceramics and Composites (http://www.ceramics.org/daytona2012) January 22–27, 2012 in Daytona Beach, Florida.

SUJANTO WIDJAJA AND DILEEP SINGH
Volume Editors
June 2011

High-Rate Real-Time Characterization

THE INFLUENCE OF TEMPERATURE AND CONFINEMENT PRESSURE ON THE DYNAMIC RESPONSE OF DAMAGED BOROSILICATE GLASS

Xu Nie, Weinong W. Chen
School of Aeronautics and Astronautics
Purdue University
West Lafayette, IN, USA

ABSTRACT

Comminution and pulverization of glass is frequently encountered in the process of projectile penetration on vehicle and building glass windows. During such events, the glass material in front of the penetrator nose will be severely damaged and counteract with further penetration. Both high confinement pressures and high temperatures are expected to take place under these circumstances, and subsequently influence the dynamic mechanical response of damaged glass. The purpose of this study is to investigate such influences on a damaged borosilicate glass with a modified high temperature Kolsky bar setup. Intact glass samples were confined by metal sleeves and loaded to fracture by the primary incident pulse. A secondary incident pulse, which arrived several hundred microseconds later, continued to load on the damaged glass samples under different confinement levels. Non-contact heating technique was also used to investigate the possible temperature effects. The results showed that under dynamic loading conditions, the shear strength of damaged glass increases linearly with confinement pressures. However, the temperature itself does not show any noticeable effect on the shear strength.

INTRODUCTION

In a lightweight armored vehicle, direct blast and projectile impact from Improvised Explosive Devices (IEDs) and/or small armor piercing bullets has posed a significant challenge in armor design especially for transparent window armor panels, since they are one of the most critical, but yet vulnerable parts in the armor system. In the past decades, numerous experimental and analytical work [1-10] has indicated that in the event of ballistic penetration, the material that is in direct contact with the projectile is comminuted and forms the so called "Mescall zone" [4] ahead of the penetrator due to the extremely high shear stress. It is those failed materials, but not the intact materials that counteract further penetration by eroding the projectile nose. In the mean while, efforts on numerical modeling have also been put forward to take into account the damage development and the associated evolution in material responses [1-15]. As the penetration proceeds, the damaged material flows by the nose of the projectile and are ejected backwards along the shank. Considering the nature of such a process, both high pressure (resulted from the confinement of nearby materials) and high temperature (resulted from friction, severe plastic deformation of the projectile in a short period of time, etc.) are expected to present together with high rate of deformation. It is therefore desired to incorporate the dynamic mechanical response of damaged armor materials under these loading conditions into the constitutive and finite elementary models for penetration simulation. Unfortunately, most of the current modeling work does not take all these parameters into account due to the lack of reliable experimental results available to the community. A recent modification on Kolsky bar testing technique has targeted on characterization of dynamic response of damaged ceramic materials under different confinement levels [16]. In this newly developed technique, double pulse loading was utilized with the first pulse crushing the intact ceramic specimen, and the second pulse immediately loading on the damaged material. An improved version of this loading technique will be introduced to study the damaged glass. In addition, the pneumatic high temperature system is incorporated here to investigate the possible temperature effects.

EXPERIMENTAL TECHNIQUES

In this study, a high temperature Kolsky bar setup which has similar design with Apostol et. al. [17] was used for the dynamic characterization on damaged borosilicate glass over a wide range of temperatures. Special modifications were made to avoid possible thermal shock while at the same time ensure dynamic force equilibrium during the loading process. A picture of this system is shown in Fig. 1. Initially, the specimen was held by the sample slider and was kept in the furnace during the heating stage. Both incident and transmission bars were kept a little distance away from the gage section. Once the desired temperature was reached, a signal was sent by the computer to retract the specimen back to the loading line. This retraction motion will be detected by the position sensor on the sample slider, and another signal will be sent to the bar manipulator such that the transmission bar will be pushed forward to engage the specimen and the incident bar. In the mean while, the electromagnetic valve on the gas gun is triggered to facilitate the impact loading pulse. As opposed to the room temperature tests on glass materials where the specimens are in direct contact with the Kolsky bars upon loading, the current high temperature tests require the implementation of "thermal cushions" in the gage section between the hot sample and the cold incident and transmission bars. This is achieved through placing a pair of polished steel platens on both sides of the specimen and heating them up together. The platens are made of the same material as the Kolsky bars and have the same diameter to ensure wave impedance is matched. By this means, the thermal gradient in the glass sample will be greatly attenuated and thus eliminate the possibility of thermal shock induced pre-mature failure. In addition, since the specimen is no longer in direct engagement with the cold bars, the temperature loss is further minimized. Besides the built-in thermal couple in the furnace which is used to control the heating profile, another thermal couple is placed close to the glass specimen to monitor the real time sample temperature.

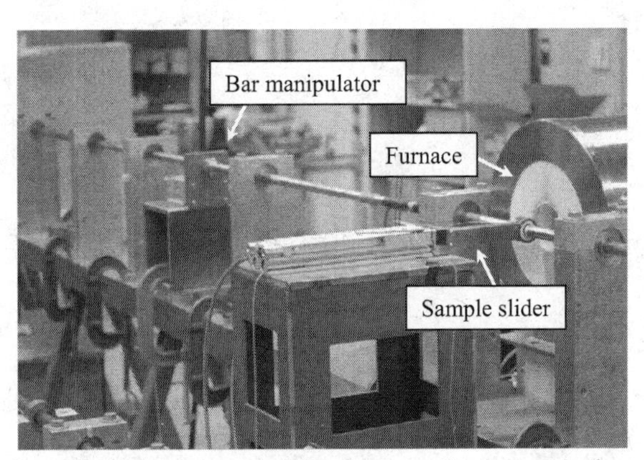

Figure 1: Illustration of the high temperature Kolsky bar setup for characterization of damaged glass

A double pulse loading technique was previously established by Luo et. al. [18] for the testing of dynamic response of comminuted ceramic materials under confinement pressure. In this technique, 2 strikers were connected head to tail by an elastic spring of certain length and stiffness. Figure 2 shows a schematic of this double striker configuration. The springs provide initial acceleration to the first striker while the compressed air is released from the gas reservoir and pushes on the second striker. While the first striker hits on the incident bar end, an impact pulse is then facilitated and propagating both ways—the one goes forward along the incident bar is called incident pulse which loads and breaks the intact specimen, and the one goes backward into the striker slows down the striker as the wave front swipes through. At the time the first striker is completely stopped, momentum will keep the second striker moving forward to close the preset gap and finally hit on the end of the first striker. By this means, a second impact pulse is then initiated to load the damaged specimen. The loading duration of each pulse is decided by the striker length, while the time interval between the pulses is determined by the preset gap and the striker velocity. An obvious drawback of this double pulse technique is the precise control of the preset gap between the strikers, which consequently affects the time interval between the two incident pulses. Stiffness of the springs and the striker velocity are both designing variables. Once the loading rate needs to be changed, i.e. the striker velocity needs to be changed; a different set of springs with different length and stiffness may be needed to achieve similar loading conditions.

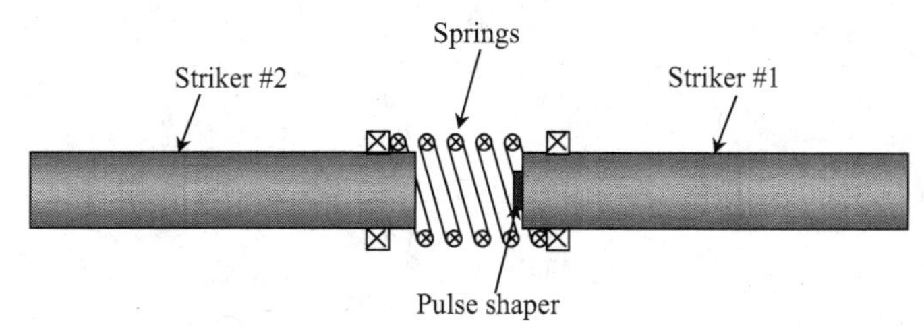

Figure 2: The schematic of a two striker assembly for previous ceramics testing.

In this study, we propose an alternative method to facilitate the dynamic double pulse loading. Instead of using elastic springs, an aluminum tube will be used for the connection of the two strikers. The inner diameter of this aluminum tube is slightly bigger than the diameter of the striker bars so it will easily slide on. Both sides of this aluminum tube are glued on the strikers with a preset gap in between. A schematic illustration of this assembly is given in Fig. 3. The shear strength of this glue should be good enough to withstand the inertia force of the assembly while it is accelerated by the compressed air, but would not survive the impact of the first striker on the incident bar. By this means, the gap will be closed upon shear failure of the adhesive layer. Thus the second striker, after traveling the preset distance at a certain velocity, will impact on the first striker and initiate the secondary incident pulse. The time interval between the two incident pulses is determined by the preset gap distance between the two strikers.

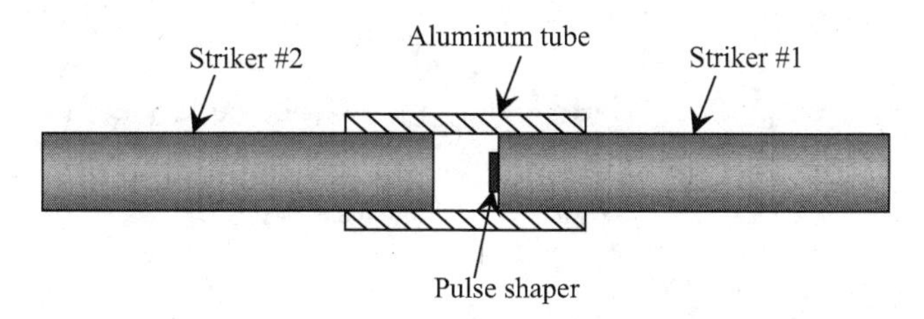

Figure 3: The improved double striker technique.

RESULTS AND DISCUSSION

The glass specimens used in this study are 0.25" diameter by 0.375" thickness cylinders. The end surfaces were mechanically polished after cutting and grinding to ensure smooth contact with the Kolsky bar ends, and the circumference was fire polished so as to eliminate possible edge defects introduced by the fabrication process. The specimens were then tightly fitted into the confinement collar before being tested. Two different wall thicknesses were chosen for the 304 stainless steel collars to achieve different pressure levels.

Figure 4 shows a set of original experimental record from oscilloscope. The primary triangular incident pulse (indicated by "1") was generated when the first striker impacted the incident bar through proper pulse shaping. The purpose of facilitating this linear ramp was to load the intact glass specimen to fracture under constant strain rate considering borosilicate glass typically has linear elastic response (indicated by "3") before fracture. Although the dynamic behavior of intact glass is not the focus of current research, to maintain a constant strain rate deformation of the specimen would be beneficial to ensure relatively homogeneous stress field up to failure. About 100 μs after the unloading of the primary incident pulse, the second striker hit on the end of the first striker to facilitate the secondary incident pulse (indicated by "2"). As the damaged glass exhibited plastic flow-like behavior

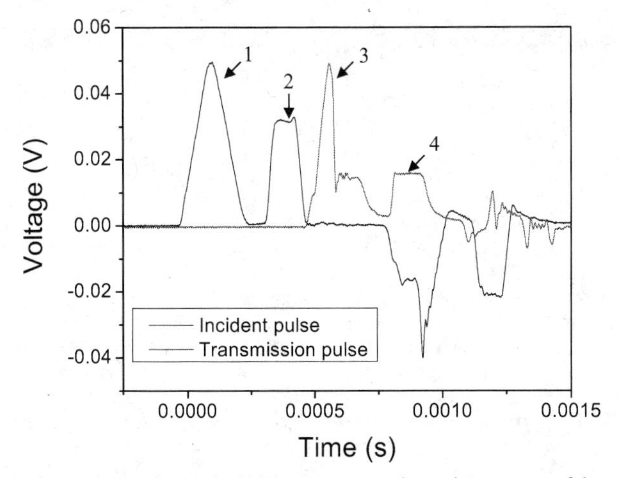

Figure 4: Original experimental record of high temperature dynamic response of damaged glass under confinement.

while loaded under confinement (indicated by "4"), the second pulse was designed in a way that it has a plateau region after initial rise time to achieve constant strain rate deformation. The experiments were carried out at four different temperatures (25°C, 200°C, 400°C, 600°C) on two confining sleeves which have the wall thicknesses of 0.38 mm and 0.71 mm, respectively, while the strain rates on intact and damaged specimens are ~670s^{-1} and 1000s^{-1}, respectively. Five to eight samples were tested under each loading and temperature condition to ensure the results are representative, and a typical stress-strain curve from one test at 200°C is shown in Fig. 5. It is evident that the glass specimen failed while the axial stress reached 2000 MPa, which is also the compressive strength of the intact borosilicate glass. A previous study revealed that the dynamic compressive failure process of this same glass was dominated by axial splitting, and then followed by buckling of the splitted columns [19]. Without any confinement, the glass fragments would shatter upon further progressing of the incident bar. If confinement is provided by using metal sleeves, all the fragments will be constrained after fracture. For this reason, the stress did not drop to zero after failure. Instead, it dropped to a certain value and immediately jumped up as the damaged glass was compacted in the sleeve. This reflects on the stress-strain curve as the first plateau region (indicated by "1" in Fig. 5). After the primary incident pulse was unloaded, the stress dropped back to zero until the arrival of the secondary incident pulse loading on the damaged sample at constant strain rate. The second plateau region (indicated by "2" in Fig. 5), which is referred to as the flow response of the damaged glass, was therefore created on the stress-strain curve.

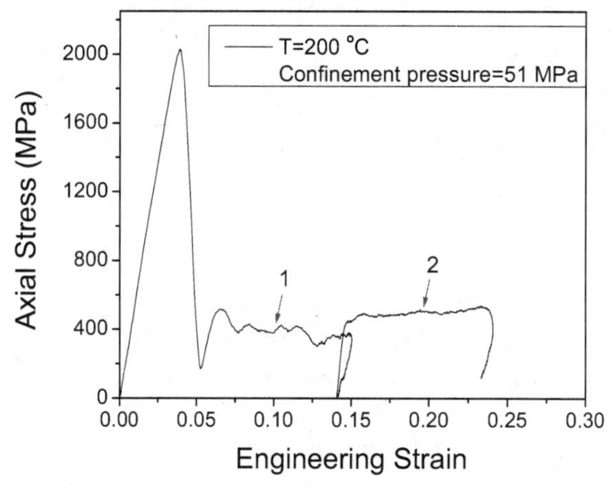

Figure 5: A typical stress-strain curve on damaged glass under confinement at the temperature of 200°C.

The damaged borosilicate glass exhibited a "plastic flow"-like behavior at all the temperatures and confinement levels. This "flow" stress was plotted against confinement pressure and the results are shown in Fig. 6. Surprisingly, the flow strength obtained from different temperatures all fall on the same fitted curve. This fitted curve starts from the origin, which is based on the assumption that the damaged glass could not handle any axial stress without radial confinement, and then passes through the crowd of all the data points as the confinement pressure increases. It is reasonable to draw the conclusion that the flow behavior of the damaged borosilicate glass is independent of temperature in the tested temperature range. The differences in the flow strength exhibited at different temperatures are solely because of the change in yield strength of the stainless steel confinement sleeves, and thus the change in the confinement pressure levels. A good example of this is given in Fig. 6 by comparing the flow strength of glass samples tested at 600°C with 0.71 mm wall thickness steel sleeves and those tested at room temperature with 0.38 mm wall thickness steel sleeves. The confining pressures in these two loading conditions are close to each other and both are around 60 MPa, as is indicated by an arrow. It is also evident that the flow strength in these two cases are similar despite of the large differences in the testing temperatures. The data points shown in Fig. 6 can be converted to equivalent yield strength versus hydrodynamic pressure diagram which is commonly used to describe the pressure sensitivity on the failure behavior of ceramic materials, and the results are presented in Fig. 7. The equivalent shear strength increases linearly with the hydrodynamic pressure. Again, the data points at different temperatures all fall on the same fitted line. As the borosilicate glass rubbles exhibits no axial compressive strength without any confining pressure, the extension of this line should go through the origin. However, while the particles are subjected to high pressure during axial loading, interlocking between particles may have taken place which would eventually result in compaction of

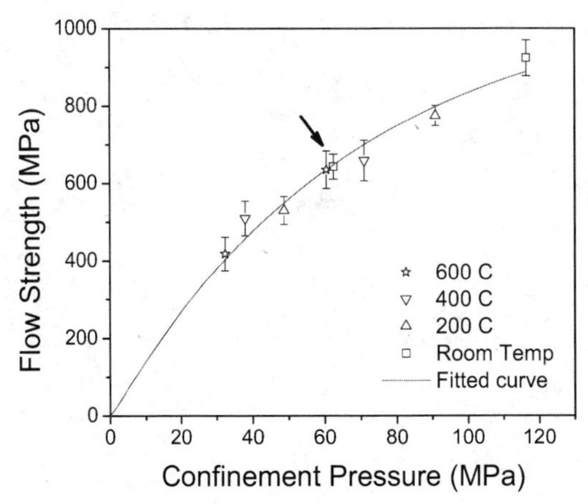

Figure 6: The flow strength of glass rubbles as a function of confining pressure at different temperatures.

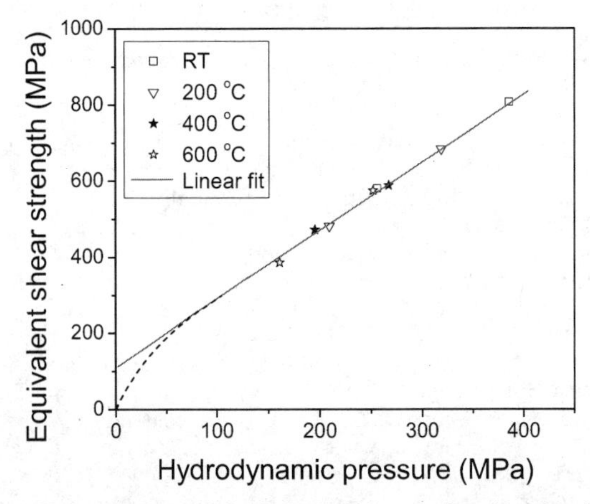

Figure 7: The equivalent shear strength of glass rubbles as a function of applied hydrodynamic pressure, the imaginary dash curve represents possible transition at low pressure.

the damaged glass. The mechanism of this process is similar to that in ceramic powder compaction investigated by Zeuch et. al. [20]. Therefore, these heavily compacted glass rubbles may possess certain shear strength without confining pressure. This is confirmed in our current study by extrapolating the fitted line in Fig. 7 to low pressure and intersecting with the vertical axis at a positive value. Hydrodynamic pressure in the range of 150-400 MPa was achieved in this research. The pressure lower than 150 MPa was not investigated due to the limited availability of glass specimens. But it is reasonable to assume that the particle compaction effect should diminish with decreasing pressure. This means that below a certain pressure level the shear strength of damaged glass may deviate from its linear dependency with pressure, and gradually curve down to the origin. The dashed curve in Fig. 7 describes such a possible transition for which further experimental evidences are still needed.

CONCLUSION

The high temperature dynamic compressive behavior of damaged borosilicate glass was investigated on a modified Kolsky bar setup. An improved double pulse loading technique was developed in this study to precisely control the profile of two consecutive incident pulses, for which the first pulse was designed to fracture the intact glass sample and the second pulse to load the as-damaged specimen under confinement. The glass specimens were tightly fitted into stainless steel confinement sleeves and then dynamically loaded at different target temperatures. The change in the confining pressure was achieved both by changing the wall thickness of the metal sleeves and by varying the testing temperature such that the yield strength of the sleeves are different. With the aid of single pulse loading technique, specimen recovery after a set of precisely controlled dual incident loading pulses was realized to maintain the morphology of deformed samples. The experimental results showed that as compared to the temperature, the confining pressure has profound effect on the flow strength of damaged borosilicate glass. Other than affecting the yield strength of confinement sleeves, the change in temperature does not have significant impact on the dynamic response of damaged glass. In the range of experimental conditions tested in this study, the glass rubbles showed a linear dependency between flow strength and hydrodynamic pressure, and the mechanical behavior that resembles those of compacted powders. Further investigations are needed in the low hydrodynamic pressure range where possible deviation from linear response may be revealed.

REFERENCES

[1] D. A. Shockey, A. H. Marchand, S. R. Skaggs, G. E. Cort, M. W. Burkett and R. Parker, “Failure Phenomenology of Confined Ceramic Targets and Impacting Rods”, International Journal of Impact Engineering, 9, 263-275, 1990

[2] J. Mescall and V. Weiss, “Materials Behavior under High Stress and Ultrahigh Loading Rates—Part II”, Proceeding of 29th Sagamore Army Conference, Army Materials and Mechanics Center, Watertown, MA, 1984

[3] J. Sternberg, “Material Properties Determining the Resistance of Ceramics to High Velocity Penetration”, Journal of Applied Physics, 65, 3417-3424, 1989

[4] D. R. Curran, L. Seaman, T. Cooper and D. A. Shockey, “Micromechanical Model for Comminution and Granular Flow of Brittle Material under High Strain Rate Application to Penetration of Ceramic Targets”, International Journal of Impact Engineering, 13, 53-83, 1993

[5] R. W. Klopp and D. A. Shockey, “ The Strength Behavior of Granulated Silicon Carbide at High Strain Rate and Confining Pressure”, Journal of Applied Physics, 70, 7318-7326, 1991

[6] W. A. Gooch, W. J. Perciballi, R. G. O’Donnell, R. L. Woodward and B. J. Baxter, “Effects of Ceramic Type on Fragmentation Behavior During Ballistic Impact”, 93-100, Proceeding of the 13th International Symposium on Ballistics, Stockholm, 1-3 June, 1992

[7] A. M. Rajendran, "Modeling the Impact Behavior of AD85 Ceramic under Multiaxial Loading", International Journal of Impact Engineering, 15, 749-768, 1994
[8] R. J. Clifton, "Response of Materials under Dynamic Loading", International Journal of Solids and Structures, 37, 105-113, 2000
[9] N. K. Bourne, J. C. F. Millett and J. E. Field, "On the Strength of Shocked Glasses", Proceeding of the Royal Society A, 455, 1275-1282, 1999
[10] Th. Behner, C. E. Anderson, D. L. Orphal, V. Hohler, M. Moll and D. W. Templeton, "Penetration and Failure of Lead and Borosilicate Glass Against Rod Impact", International Journal of Impact Engineering, 35, 447-456, 2008
[11] X. Sun and M.A. Khaleel and R.W. Davies, "Modeling of Stone-Impact Resistance of Monolithic Glass Ply Using Continuum Damage Mechanics", International Journal of Damage Mechanics, 14, 165-178, 2005.
[12] T. J. Holmquist and G. R. Johnson, "Response of Silicon Carbide to High Velocity Impact", Journal of Applied Physics, 91, 5858-5866, 2002.
[13] T. J. Holmquist, D. W. Templeton and K. D. Bishnoi, "Constitutive modeling of aluminum nitride for large strain, high-strain rate, and high-pressure applications", International Journal of Impact Engineering, 25, 211-231, 2001
[14] G.R. Johnson, T.J. Holmquist and S.R. Beissel, "Response of Aluminum Nitride to Large Strains, High Strain Rates and High Pressures", Journal of Applied Physics, 94, 1639-1646, 2003.
[15] S. Chocron, K. A. Dannemann, J. D. Walker, A. E. Nicholls and C. E. Anderson, "Constitutive Model for Damaged Borosilicate Glass under Confinement", Journal of the American Ceramic Society, 90, 2549-2555, 2007
[16] Chen, W. and Luo, H., "Dynamic Compressive Responses of Intact and Damaged Ceramics from a Single Split Hopkinson Pressure Bar Experiment", Experimental Mechanics, vol. 44, No. 3, pp. 295-299, 2004.
[17] M. Apostol, T. Vuoristo, V. Kuokkala, "High temperature high strain rate testing with a compressive SHPB", Journal de Physique IV 110:459-464, 2003
[18] H. Y. Luo, "Experimental and Analytical Investigation of Dynamic Compressive Behavior of Intact and Damaged Ceramics", Ph.D thesis, University of Arizona, 2005.
[19] X. Nie, W. Chen, X. Sun, and D. Templeton, "Dynamic failure of borosilicate glass under compression/shear loading", Journal of the American Ceramics Society, 90, 8, 2556-2562, 2007
[20] D. H. Zeuch, J. M. Grazier, K. G. Ewsuk and J. G. Arguello, "Comparison of the Mechanical Properties of Micron- and Submicron Size Alumina Powders", 100th Annual meeting of the American Ceramic Society. 3-6 May 1998, Cincinnati, Ohio.

THE STRAIN-RATE DEPENDENCE OF THE HARDNESS OF AlN DOPED SiC

N. Ur-rehman*, P. Brown°,*, L.J. Vandeperre*

* Centre for Advanced Structural Ceramics & Department of Materials, Imperial College London, London SW7 2AZ, UK
° Defence Science and Technology Laboratory, Porton Down, Salisbury, SP4 0JQ, UK

ABSTRACT

Tailored nanoindentation experiments were used to determine the strain rate dependence of the hardness of AlN doped SiC. It is shown that in the nano-indentation regime, where cracking is limited, the hardness of SiC reduces by 0.8 GPa per decade reduction in strain rate at room temperature for both a coarse and fine grained material. The coarse grained material was tested at higher temperatures as well and the strain rate sensitivity increases strongly with temperature becoming 1.6 GPa and 2.9 GPa per decade of strain rate at 373 K and 473 K respectively. Estimates for the strain rate dependence of the lattice resistance (Peierls' stress) of SiC derived from these measurements agree quite well with high temperature data for SiC. These experiments therefore open up the possibility to determine real constitutive equations for plasticity in SiC and other armour ceramics.

INTRODUCTION

A high hardness is generally considered a desirable property for armour ceramics[1], probably aided by the fact that hardness experiments are relatively straightforward to carry out. However, the resulting hardness number is effectively a system property influenced by the properties of the material (Young modulus, Poisson ratio), the damage mechanisms that are activated during indentation (plastic flow[2], densification[3], phase transformations[4-7], cracking[8], etc.), and the properties and shape of the indenter[9-14]. It is therefore not always easy to determine what a hardness value means. Thus, it is also difficult to decide which of a range of hardness values is the most relevant for ranking ceramics in terms of their ability to resist impact, although some authors have argued that the Knoop hardness is the better measure[8].

The fact that different indentation experiments excite different responses of the material can also be considered as an opportunity. It is well known that there is a relationship between hardness and yield strength[9, 15-18], but indentation techniques have also been used to determine the mechanical reliability (Weibull modulus) of ceramics[19], to measure stress-strain curve of a material[20, 21], to quantify the relaxation of elastic properties in visco-elastic materials[22, 23], and to measure the toughness of brittle materials[24-27]. With the introduction of depth-sensing indentation instruments, which record load and displacement during indentation, it has also become possible to investigate the time dependence of the hardness[28-31]. Recently, it was shown that quite reasonable estimates for the constitutive behaviour of dislocation flow in SiC could be obtained from the strain rate sensitivity of the room temperature hardness of SiC[32]. However, room temperature data alone does not suffice to determine the parameters in full. The aim of this paper is therefore to investigate the strain rate sensitivity of the hardness of SiC at different temperatures so that all parameters for the model can be determined experimentally. By also comparing the strain rate sensitivity of the hardness for materials with a distinctly different grain size it was hoped to clarify also whether grain size is important for plastic flow in SiC. Hence, based on initial processing trials[33, 34], two types of material were selected for this study. A fine grained material obtained by spark plasma sintering 20 mm discs using a blend with 2.50 wt% AlN and 3 wt% C leading to the microstructure shown in Figure 1a, and a coarse grained material obtained by hot pressing 30 mm discs using a blend containing 3.75 wt% AlN and 3 wt% C resulting in the microstructure shown in Figure 1b. Most of the work was carried out on the coarse grained material as it was anticipated that it would yield information on the resistance to plastic flow of SiC without interference of grain boundaries or grain boundary structures: the grain boundaries in AlN doped SiC are known to be rich in Al[33, 35, 36] and it has been suggested that this influences the hardness[37]. The measurements on the fine grained material were therefore made to clarify how important grain size and grain boundary structures are for the hardness as obtained from small scale hardness imprints.

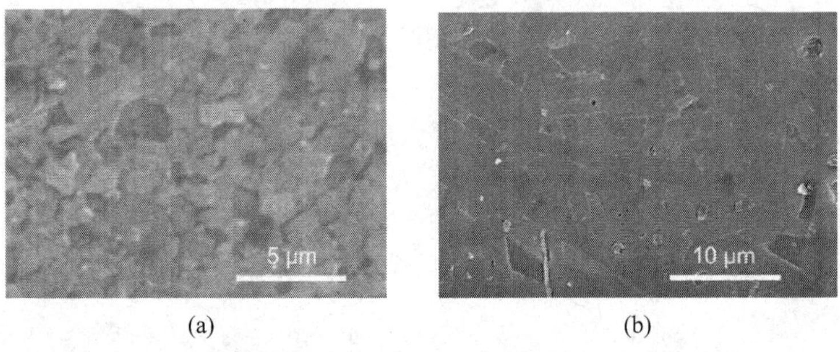

Figure 1 Scanning electron micrographs of the microstructures of the materials. Grains were made visible by etching in boiling Murakami's reagent for 1-2 minutes.

EXPERIMENTAL

Material production

50 g blends of commercially available powders of -SiC (H.C. Starck, Grade UF-15), AlN (H.C. Starck, Grade C) and a phenolic resin (CR-96 Novolak, Crios Resinas, Brazil) as a source of carbon were prepared by 24 h ball milling in methyl ethyl ketone using Si_3N_4 milling balls (Union Process Inc., USA). The mixures were dried using a rotary evaporator (Rotavapor R-210/215, Büchi, Germany). The powder blends were then heated to 600 °C under flowing Argon gas to pyrolise the resin and subsequently hot pressed or spark plasma sintered. Details of the sintering schedules for the two types of materials are reported in Table I and II.

Table I: Sintering schedule used for producing the large grain size material by hot pressing

Step	Heating rate (K min^{-1})	Dwell (min)	Pressure (MPa)
298 K – 660 K	n/a		8.5
660 K – 1903 K	10	30	8.5
1903 K – 2093 K	10	30	14.2
2093 K – 2168 K	5	30	21.2
2168 K – 2318 K	5	30	21.2

Table II: Sintering schedule used for producing the fine grain size materials by spark plasma sintering.

Step	Heating rate (K min^{-1})	Dwell (minutes)	Pressure (MPa)
< 723 K	(n/a)		16
723 K – 1973 K	100	0	16
1973 K – 2373 K	50	5	60

Hardness measurements

Indentation experiments were carried out at 298 K, 373 K and 473 K on an instrumented nano-indenter (Nanotest 600, Micromaterials, Wrexham, UK) equipped with a hot stage capable of heating samples up to 750 °C and a heated diamond Berkovich indenter. The indentation schedule consisted of the following sequence: the thermal stability of the capacitive displacement sensor was measured for 1 minute before the experiment whith the indenter was held in contact with the surface of the sample. Apart from testing the thermal stability of the displacement sensor, this also ensures that thermal equilibrium is reached between indenter and sample. This was followed by loading at a fixed rate of load increase to a predetermined maximum load. The maximum load was maintained for 60 s before unloading at the same rate as during loading. A second stability measurement was carried out during unloading at 20% of the maximum load. The measured drift of the displacement sensor was typically 0.02 nm s^{-1} for room temperature experiments. Because the independent heating of the tip and sample allows balancing the heat flow between the sample and the indenter, the drift of the displacement sensor increased only moderately to 0.1 nm s^{-1} and 0.2 nm s^{-1} for measurements made at 100°C and 200°C respectively. At each temperature, 10 measurements were made at each of 2 maximum loads (200 and 400 mN), and at each maximum load tests were conducted with 3 loading rates. The maximum loads were selected to be small enough to avoid the influence of cracking on the hardness, while being large enough to reduce a possible influence of the uncertainty on the calibration of the tip shape. The loading rates were set to nominally achieve 3 apparent strain rates, $\varepsilon^{\bullet}$, at the end of loading: 0.125 s^{-1}, 0.025 s^{-1} and 0.05 s^{-1} as calculated using a standard expression for the characteristic strain rate for an indentation experiment[28]:

$$\varepsilon^{\bullet} = \frac{1}{2F}\frac{dF}{dt} \tag{1}$$

where F is the applied force and t is time. Because strain rate effects relax rapidly during indentation experiments[38], the hardness was determined from the data collected during loading using a variant of the Oliver and Pharr method[39] in which the contact depth is determined assuming the reduced modulus remains constant during the experiment as described in full in ref.32. These values were then used to calculate the average hardness and average strain rate during loading. To widen the range of strain rates covered by the experiments, the data collected during the dwell at maximum load was analyzed with the same technique and related to apparent strain rate through:

$$\varepsilon^{\bullet} = \frac{1}{h}\frac{dh}{dt} \tag{2}$$

where h is the indentation depth. Again the average strain rate and average hardness for all data collected during dwelling was determined from all data.

RESULTS

The hardness obtained from experiments at 200 and 400 mN did not differ significantly. Therefore, all data was taken together. The average hardness versus average strain rate is plotted in Figure 2. At room temperature, the strain rate sensitivity of the hardness is limited to 0.8 GPa per decade of strain rate. However, as the temperature is increased, the low strain rate hardness decreases markedly and the sensitivity to strain rate increases to about 1.6 GPa per decade of strain rate at 100 °C and 2.9 GPa per decade of strain rate at 200 °C. Figure 2b, which shows the variation of hardness against temperature, illistrates that the results obtained here are consistent with data obtained elsewhere[40]. It is clear from Figure 2a also that despite the significant difference in grain size, the room hardness and the room temperature strain rate sensitivity of the two types of materials is not very different.

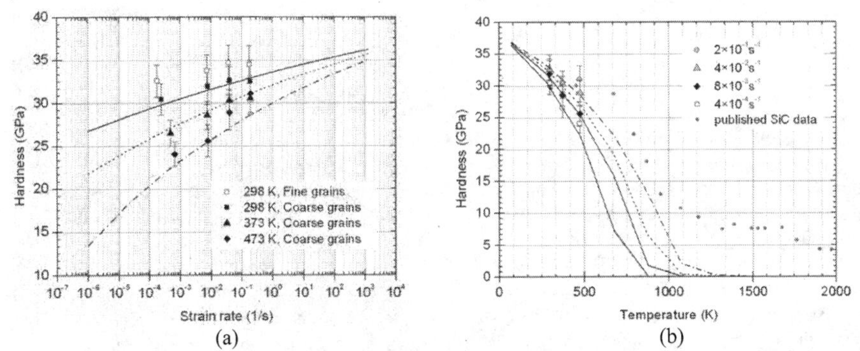

Figure 2 (a) Hardness versus strain rate and (b) hardness versus temperature compared with data taken from ref.[40]. The lines shown were calculated with the model discussed in the text.

DISCUSSION

The hardness as measured here is higher than the typical hardness values reported for armour SiC, which normally are of the order of 24 GPa[41]. However, such measurements are made using much higher loads and extensive cracking occurs. In fact a similar decrease in hardness with applied load was observed for the materials studied here also[34]. The absence of cracking in the low load experiments makes it reasonable to assume that the strain rate sensitivity is the result of the strain rate dependence of the dislocation slip, which caused the indentation to form.

Therefore, the hardness values were converted into shear flow stresses using the generalised relationship between hardness, P, and yield strength, Y, derived by Vandeperre et al.[18] by modifying the expanding cavity solution of Hill[2, 42]:

$$\frac{P}{Y}=\frac{2}{3}\left\{1+\frac{3}{3-6\lambda}\ln\left(\frac{(3+2\mu)(2\lambda(1-\zeta)-1)}{2\lambda\mu-3\mu-6\lambda}\times\frac{1}{\zeta}\right)\right\}$$

with

$$\zeta=\frac{E_r}{E_r-2H\tan\alpha},\quad \lambda=\frac{(1-2\nu)Y}{E},\quad \mu=\frac{(1-\nu)Y}{E} \tag{3}$$

and Y is the yield stress, E is the Young's modulus (470 GPa), ν is the Poisson's ratio (0.17), P is hardness, E_r is the reduced modulus and = 70.3° is the semi-angle of a conical indenter with the same area to depth characteristic as a Berkovich indenter. In deriving these relationships the Tresca yield criterion is used and therefore the uni-axial yield stress obtained is twice the shear flow stress. The variation of the shear flow stress with strain rate and temperature as derived from the hardness is shown in Figure 3.

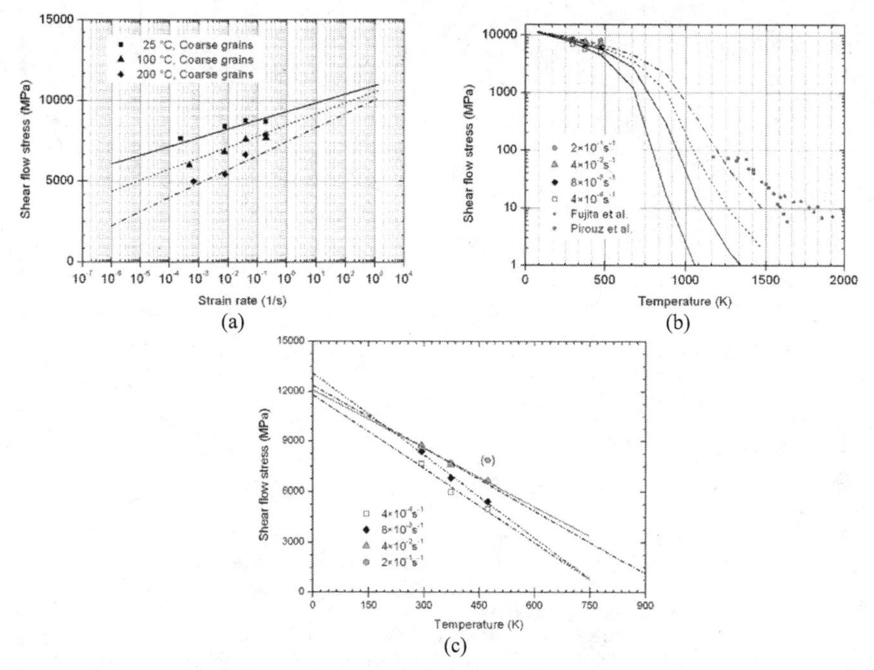

Figure 3 (a) Shear flow stress versus strain rate, and (b) shear flow stress versus temperature. Literature data[43, 44] from high temperature compression tests on SiC is shown for comparison. The lines were calculated with the model discussed in the text. (c) Shear flow stress versus temperature and linear regression lines used to derive the model parameters. The brackets round 1 data point indicate that it was considered an outlier and not included in the fit.

The lattice resistance or Peierls' stress of SiC is high. Hence, it can be expected that the plastic behaviour will be dominated by the lattice resistance. In fact, the limited influence of grain size observed here is consistent with this assumption: when moving dislocations through the lattice is difficult, the effect of grain size on the stress needed to move the dislocations will be limited.

To derive constitutive data from the experimental results, a simple model for the relation between the strain rate and shear flow stress can be developed starting from the well established relation[44] between the strain rate, the mobile dislocation density, $_m$, and dislocation velocity, v,:

$$\frac{d\gamma}{dt} = \rho_m \cdot b \cdot \mathrm{v} \tag{4}$$

where the velocity of the dislocations can be estimated using a standard approach for a stress activated process[45]:

$$\mathrm{v} = \nu \cdot b \cdot \left[\exp\left(-\frac{(\tau_p - \tau)V}{kT} \right) - \exp\left(-\frac{(\tau_p + \tau)V}{kT} \right) \right] \tag{5}$$

where is the attempt frequency, $_p$ the Peierls' stress or the stress needed to make the dislocations move in the absence of any thermal energy, the applied shear stress, V the activation volume, T the temperature and k the Boltzman constant. Combining these equations and solving for the applied shear stress, yields the following relationship:

$$\tau = \frac{kT}{V} \sinh^{-1}\left[\frac{\gamma^{\bullet}}{2 \cdot \rho_m \cdot \nu \cdot b^2} \exp\left(\frac{\tau_p V}{kT} \right) \right] \tag{6}$$

For large applied stresses, where the second term in the square brackets of eq.(5) vanishes relative to the first, and assuming the activation volume can be treated as constant, a simpler relation is obtained:

$$\tau = \tau_p + \frac{kT}{V} \ln\left[\frac{d\gamma}{dt} \right] - \frac{kT}{V} \ln\left[\rho_m \cdot b^2 \cdot \upsilon \right] \tag{7}$$

Equation 7 shows that the activation volume can be determined by linear regression of the shear flow stress against the natural logarithm of the strain rate, whereas the mobile dislocation density and the Peierls' stress can be found by plotting the flow stress versus temperature on linear axis as in Figure 3c. For the burgers' vector a value of 0.154 nm was used as flow in SiC is assumed to be by partial dislocations[44, 46, 47]. The attempt frequency was taken to be 10^{11} s^{-1} in line with the work of Frost and Ashby[48]. The values obtained from each of the datasets are summarised in Table III together with the mean values.

Table III. Raw estimates for the model parameters and mean values (* geometric mean, ° arithmetic mean).

Data set	Activation volume (m^3)	Peierls' stress (GPa)	Mobile dislocation density (m^{-2})
293 K	2.40×10^{-29}	(n/a)	(n/a)
373 K	1.70×10^{-29}	(n/a)	(n/a)
473 K	1.26×10^{-29}	(n/a)	(n/a)
2×10^{-1} s^{-1}	(n/a)	12.4	5.10×10^{14}
4×10^{-2} s^{-1}	(n/a)	12.1	3.41×10^{13}
8×10^{-3} s^{-1}	(n/a)	13.1	2.61×10^{15}
4×10^{-4} s^{-1}	(n/a)	11.8	1.78×10^{13}
Mean value	$1.73 \times 10^{-29(*)}$	$12.3 \pm 0.5^{(o)}$	$1.69 \times 10^{14(*)}$

The estimated dislocation densities are high but quite reasonable for heavily deformed material and this agrees with the high dislocation densities that have been observed when indentations were observed by TEM[49-51]. The value of the Peierls' stress, 12.3 GPa, is 5% of the shear modulus, which is quite close to what is expected from the Peierls model (7% of the shear modulus)[45, 47]. The activation energy (Q = $_p$×V) is estimated to be 2.1×10^{-19} J (1.33 eV), which is markedly lower than the result of Fujita et al. (3.4 eV)[43] but close to the valued obtained by Pirouz, Demenet and Hong of

1.67±0.16 eV [44]. Moreover, the model lines in Figures 2 and 3, which were obtained with the mean values for the parameters and eq.(6), do not only show good agreement with the data obtained here, but agree reasonable with measurements for the changes in hardness with temperature and of the flow stress at high temperature obtained elsewhere. This suggests that quite reliable parameters can be obtained from indentation. Given that these experiments are much easier than high temperature compression tests on single crystals, indentation could be used to rapidly establish whether different variants of armour ceramics show differences in the plasticity component of their response.

CONCLUSIONS

The hardness and room temperature strain rate sensitivity of the hardness of two types of SiC with significantly different grain sizes was determined to be of the order of 0.8 GPa per decade of strain rate. Grain size has a limited effect on small scale hardness. Additional measurements at higher temperatures (373 K & 473 K) for the coarse grained material showed that the strain rate sensitivity increased with temperature to 1.6 GPa per decade of strain rate at 373 K and 2.9 GPa per decade of strain rate at 473 K. The measured hardness values were converted in shear flow stresses and used to determine the parameters of a constitutive equation for plastic flow in SiC assuming the lattice resistance (Peierls' stress) controls the flow. The parameters obtained were a Peierls' stress of 12.3±0.5 GPa, an activation volume of 1.7×10^{-29} m^3, an activation energy for flow of 2.1×10^{-19} J (1.33 eV), and a mobile dislocation density of 1.7×10^{14} m^{-2} assuming an attempt frequency of 10^{11} s^{-1}. This limited set of parameters is capable of capturing the variation of hardness with temperature and strain rate quite accurately and yields fair agreement with hardness and shear flow stress measurements made elsewhere. Hence, relatively straightforward indentation experiments, which can be carried out on actual armour ceramics rather than single crystals, could be used to rapidly assess whether the plasticity contribution of ranges of armour ceramics differ or not.

ACKNOWLEDGEMENTS

Naeem Ur-rehman and Luc Vandeperre thank the Defence Science and Technology Laboratory (Dstl) for providing the financial support for this work under contract No Dstlx-1000015398.

REFERENCES

1. Swab, J., *Recommendations for determining the hardness of armor ceramics.* International Journal of Applied Ceramic Technology, 2004. **1**(3): p. 219-225.
2. Marsh, D.M., *Plastic Flow in Glass.* Proc. R. Soc. A, 1963. **279**: p. 420-435.
3. Suzuki, K., et al., *Densification Energy during Nanoindentation of Silica Glass.* J. Am. Ceram. Soc., 2002. **85**(12): p. 3102-3104.
4. Gerk, A.P. and D. Tabor, *Indentation Hardness and Semiconductor-metal transition of Germanium and Silicon.* Nature, 1978. **271**: p. 732-733.
5. Vandeperre, L.J., et al., *The hardness of silicon and germanium.* Acta Materialia, 2007. **55**(18): p. 6307-6315.
6. Weppelman, E.R., J.S. Field, and M.V. Swain, *Observation, analysis, and simulation of the hysteresis of silicon using ultra-micro-indentation with spherical indenters.* J. Mater. Res., 1993. **8**(4): p. 830-840.
7. Bradby, J.E., et al., *Spherical Indentation of Compound Semiconductors.* Phil. Mag. A, 2002. **82**(10): p. 1931-1939.
8. Niesz, D.E. and J.W. McCauley, *Advanced Metals and ceramics for Armor and Anti-Armor Applications High-Fidelity Design and Processing of Advanced Armor Ceramics.* 2007, Army Research Laboratory. p. 124.

9. Tabor, D., *The physical meaning of indentation and scratch tests.* Br. J. Appl. Phys., 1956. **7**: p. 159-166.
10. Cheng, Y.T. and C.M. Cheng, *What is Indentation Hardness ?* Surface Coatings and Technology, 2000. **133-134**: p. 417-424.
11. Cheng, Y.T. and C.M. Cheng, *Can Stress-Strain Relationships be Obtained from Indentation Curves using Conical and Pyramidal Indenters.* J. Mat. Res., 1999. **14**(9): p. 3493-3496.
12. Hill, R., E.H. Lee, and S.J. Tupper, *The Theory of Wedge Indentation of Ductile Materials.* Proc. R. Soc. A, 1947. **188**: p. 273-289.
13. Dugdale, D.S., *Wedge Indentation Experiments with Worked Metals.* J. Mech. Phys. Solids, 1953. **2**: p. 14-26.
14. Hirst, W. and G.J.W. Howse, *The Indentation of Materials by Wedges.* Proc. Roy. Soc., 1969. **A311**: p. 429-444.
15. Tabor, D., *Hardness of Metals.* 1951, Oxford: Clarendon Press.
16. Atkins, A.G. and D. Tabor, *Plastic Indentation in Metals with Cones.* J. Mech. Phys. Solids, 1965. **13**: p. 149-164.
17. Cheng, Y.T. and Z. Li, *Hardness obtained from conical indenters with various cone angles.* J. Mat. Res., 2000. **15**(12): p. 2830-2835.
18. Vandeperre, L.J., F. Giuliani, and W.J. Clegg, *Effect of elastic surface deformation on the relation between hardness and yield strength.* Journal of Materials Research, 2004. **19**(12): p. 3704-3714.
19. Warren, P.D., *Fracture of brittle materials: effects of test method and threshold stress on the Weibull modulus.* Journal of the European Ceramic Society, 2001. **21**(3): p. 335-342.
20. Basu, S., A. Moseson, and M.W. Barsoum, *On the determination of spherical nanoindentation stress-strain curves.* Journal of Materials Research, 2006. **21**(10): p. 2628-37.
21. Fischer-Cripps, A.C., *Elastic-plastic behaviour in materials loaded with a spherical indenter.* Journal of Materials Science, 1997. **32**: p. 727-736.
22. Tweedie, C.A. and K.J. Van Vliet, *Contact creep compliance of viscoelastic materials via nanoindentation.* Journal of materials research, 2006. **21**(6): p. 1576-1589.
23. Oyen, M.L. and R.F. Cook, *Load-displacement behavior during sharp indentation of viscous-elastic-plastic materials.* Journal of Materials Research, 2003. **18**(1): p. 139-150.
24. Chiang, S.S., D.B. Marshall, and A.G. Evans, *The Response of Solids to Elastic/Plastic Indentation : II. Fracture Initiation.* J. Appl. Phys., 1982. **53**(1): p. 312-317.
25. Lawn, B.R. and D.B. Marshall, *Hardness, toughness, and brittleness: an indentation analysis.* Journal of the American Ceramic Society, 1979. **62**(7-8): p. 347-350.
26. Chantikul, P., et al., *A Critical Evaluated of Indentation Techniques for Measuring Fracture Toughness: II, Strength Method.* Journal of the American Ceramic Society, 1981. **64**(9): p. 539-543.
27. Cook, R.F. and G.M. Pharr, *Direct Observation and Analysis of Indentation Cracking in Glasses and Ceramics.* Journal of the American Ceramic Society, 1990. **73**(4): p. 787-817.
28. Grau, P., et al., *Strain rate dependence of the hardness of glass and Meyer's law.* Journal of the American Ceramic Society, 1998. **81**(6): p. 1557-1564.
29. Keulen, N.M., *Indentation Creep of Hydrated Sóda-Lime Silicate Glass Determined by Nanoindentation.* J. Am. Ceram. Soc., 1933. **76**(4): p. 1904-1912.
30. Ngan, A.H.W., J.B. Pethica, and H.P. Ng, *Strain-rate sensitivity of hardness of nanocrystalline Ni75at.%Al25at.% alloy film.* Journal of Materials Research, 2003. **18**(2): p. 382-386.
31. Golovin, Y.I., Y.L. Iunin, and A.I. Tyurin, *Strain-rate Sensitivity of the Hardness of Crystalline Materials under Dynamic Nanoindentation.* Doklady Physics, 2003. **48**(9): p. 505-508.
32. Vandeperre, L.J., N. Ur-rehman, and P. Brown, *Strain rate dependence of hardness of AlN doped SiC.* Advances in Applied Ceramics, 2010. **109**(8): p. 493-497.

33. Ur-rehman, N., P. Brown, and L.J. Vandeperre, *Evolution of the AlN distribution during sintering of AlN doped SiC.* Ceramic Engineering and Science Proceedings, 2010. **31**(5): p. 231-238.
34. Ur-rehman, N., P. Brown, and L.J. Vandeperre, *The role of carbon in processing hot pressed aluminium nitride doped silicon carbide.* Ceram. Eng. Sci. Proc., 2010. **31**(2): p. 27-36.
35. Hu, J., et al., *Core-shell structure from the solution-reprecipitation process in hot-pressed AlN-doped SiC ceramics.* acta materialia, 2007. **55**: p. 5666-5673.
36. Huang, R., et al., *The sintering mechanism and microstructure evolution in SiC-AlN ceramics studied by EFTEM.* Int. J. Mat. Res., 2006. **97**(5): p. 1-7.
37. Balog, M., et al., *Nano- versus macro-hardness of liquid phase sintered SiC.* Journal of the European Ceramic Society, 2005. **25**(4): p. 529-534.
38. Quin, G.D., P.J. Patel, and I. Lloyd, *Effect of Loading Rate Upon Conventional Ceramic Microindentation Hardness.* J. Res. Natl. Inst. Stand. Technol., 2002. **107**: p. 299-306.
39. Oliver, W.C. and G.M. Pharr, *An Improved Technique for Determining Hardness and Elastic Modulus using Load and Displacement Sensing Indentation Experiments.* J. Mater. Res., 1992. **7**(6): p. 1564-1583.
40. Milman, Y., S.I. Chugunova, and I.I. Timofeeva, *The resistance of silicon carbide to static and impact local loading.* International journal of impact engineering, 2001. **26**(1-10): p. 533-542.
41. Chen, W.W., et al., *Dynamic fracture of ceramics in armor applications.* Journal of the American Ceramic Society, 2007. **90**(4): p. 1005-1018.
42. Hill, R., *The Mathematical Theory of Plasticity.* 1950, Oxford: Clarendon Press.
43. Fujita, S., K. Maeda, and S. Hyodo, *Dislocation Glide Motion in 6H SiC Single Crystals subjected to High-Temperature Deformation.* Phil. Mag. A, 1987. **55**(2): p. 203-215.
44. Pirouz, P., J.L. Demenet, and M.H. Hong, *On transition temperatures in the plasticity and fracture of semiconductors.* Philosophical Magazine A, 2001. **81**(5): p. 1207-1227.
45. Pitchford, J.E., *Effects of Structure on Mechanisms of High Temperature Plastic Deformation in Oxide Ceramics*, in *Department of Materials Science and Metallurgy.* 1999, University of Cambridge: Cambridge. p. 246.
46. Ohsawa, K., et al., *The Critical stress in a discrete Peierls-Nabarro Model.* Phil. Mag. A, 1994. **69**(1): p. 171-181.
47. Clegg, W.J., L.J. Vandeperre, and J.E. Pitchford, *Energy changes and the lattice resistance.* Key Engineering Materials, 2006. **317-318**: p. 271-76.
48. Frost, H.J. and M.F. Ashby, *Deformation mechanism maps: the plasticity and creep of metals and ceramics.* 1982, Oxford: Pergamon Press.
49. Lloyd, S.J., J.M. Molina-Aldareguia, and W.J. Clegg, *Deformation under Nanoindents in Sapphire, Spinel and Magnesia Examined Using Transmission Electron Microscopy.* Philosophical Magazine A, 2002. **82**(10): p. 1963-1969.
50. Lloyd, S.J., J.M. Molina-Aldareguia, and W.J. Clegg, *Deformation under Nanoindents in Si, Ge, and GaAs Examined through Transmission Electron Microscopy.* J. Mater. Res., 2001. **16**(12): p. 3347-3350.
51. Lloyd, S.J., et al., *Observations of nanoindents via cross-sectional transmission electron microscopy: a survey of deformation mechanisms.* Proc. R. Soc. Lond. A, 2005. **461**(2060): p. 2521-2543.

STATIC AND DYNAMIC INDENTATION RESPONSE OF ION-ARMOR™ GLASS

Phillip Jannotti and Ghatu Subhash
Department of Mechanical and Aerospace Engineering, University of Florida
Gainesville, FL, USA

Arun Varshneya
Saxon Glass Technologies, Inc.
Alfred, New York, USA

ABSTRACT

The mechanical properties of Ion-Armor™, a chemically strengthened glass, were investigated using static and dynamic micro-Vickers indentation. The strengthened glass was reported to have high surface residual compressive stress approaching 1 GPa.[1] Static indentations were performed using a standard Vickers hardness tester with a loading duration of 15 seconds. Dynamic indentations were generated using a custom-made Dynamic Indentation Hardness Tester (DIHT) with a loading duration of 60 microseconds. It was shown that the strengthened glass exhibited a 22% greater static hardness and a 12% greater dynamic hardness compared to the virgin glass. Furthermore, the strengthened glass showed an increase in Vickers hardness of 23% under dynamic loading compared to static loading as the result of apparent strain rate-sensitivity. Using Meyer's Law, or power law analysis, it was also determined that both the strengthened and virgin glass revealed increased load-dependence (i.e., indentation size effect) under dynamic loading.

INTRODUCTION

Chemically strengthened glasses have emerged as promising candidates for a diverse range of advanced civilian and military applications requiring high strength and optimal transparency. The glasses utilized in these applications have been engineered using ion-exchange to promote a significant resistance to impact-induced cracking. Civilian applications include autoinjector cartridges, screens in personal electronic communication devices (cell phones and MP3 players),[2] hurricane- and earthquake-resistant architectural windows for buildings and homes, windows for airplanes and automobiles,[1] etc. Examples of military applications, often requiring more stringent safety and security standards, include bullet- and blast-resistant windows for armored defense vehicles[1] and combat eye wear.[3]

Understanding the mechanical properties of ion-exchanged glass is crucial to the success of a given application. A convenient and cost-effective means of evaluating the mechanical properties of brittle materials, such as chemically strengthened glass, is by conducting Vickers indentation hardness studies.[4] Indentation hardness is an important material property for armor ceramics and is often regarded as a measure of the material's impact resistance.[5]

In this study, the utility of chemically strengthened glass, trade-named Ion-Armor™, in dynamic applications is evaluated in terms of its dynamic indentation hardness. This technique allows for indentations on the order of tens of microseconds[6-8] rather than tens of seconds as in standard static hardness tests. In addition, the dependence of indentation hardness on indentation load and strain rate for Ion-Armor™ glass is evaluated. The findings will be used to quantify the increased hardness resultant from the chemical strengthening treatment.

MATERIALS

Rectangular bars of chemically strengthened and virgin ("untreated") lithium aluminosilicate glass were provided by Saxon Glass Technologies, Inc., Alfred, NY. The chemically strengthened samples received concurrent ion-exchanges of potassium ions (K^+) and sodium ions (Na^+) ions. This treatment resulted in surface compression of approximately 1 GPa to a depth of 1 mm.[1] The sample dimensions were 23 mm x 9.9 mm x 7.6 mm.

PROCEDURE

Static Indentation

Static Vickers indentation experiments were conducted using a Wilson® Instruments Tukon™ model 2100B, with a 15 second loading duration. An appropriate indentation load range was chosen in order to generate measureable Vickers indents while avoiding excessive cracking and material removal. The indentation load range used on the strengthened glass bars was 0.981 N (100 g) to 19.62 N (2 kg). On the virgin (untreated) glass bars, the load range used was 0.981 N (100 g) to 9.81 N (1 kg). The indents were generated in the center of the glass bars to mitigate any effects of the sample edges. In addition, indentations were conducted on all glass faces to determine representative hardness values at varying indentation loads.

Dynamic Indentation

Dynamic Vickers indentation experiments were conducted using a custom-made dynamic indentation hardness tester (DIHT) shown in Figure 1. The setup consisted of a gas gun, a striker rod and a long (incident) bar. The incident bar included a momentum trap on the end impacted by the striker rod and a standard Vickers indenter tip fixed on the other end which contacted the glass bars. The striker bar was launched from the gas gun towards the incident bar. Upon impact, the incident bar moved toward the sample, generating a Vickers indent. The momentum trap ensures that only one single indentation occurs as the incident bar retracts from the sample. The indentation duration is determined by the length of striker rod used and the indentation load is determined by the velocity of impact of the striker rod. Using a 6 inch striker rod, the loading duration is 60 microseconds. The operating principle of this technique is similar to a modified-split Hopkinson pressure bar and can be found in more detail elsewhere.[6-8] An appropriate indentation load range was chosen in order to generate measureable Vickers indents while avoiding excessive cracking and material removal. The indentation load range was 10.1 N (1.02 kg) to 60.8 N (6.20 kg) on the strengthened glass bars and 10 N (1.03 kg) to 53.3 N (5.43 kg) on the virgin glass bars. The indents were placed in the center region of the indented face in order to diminish any sample edge effects. Furthermore, indentations were performed on all faces in order to determine representative hardness values at varying load levels.

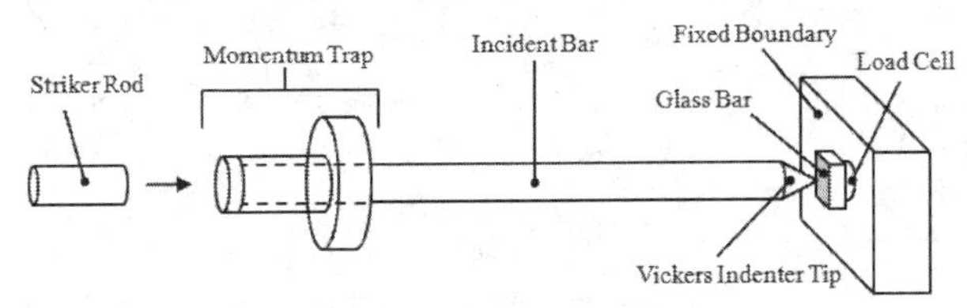

Figure 1. Schematic of the custom-made dynamic indentation hardness tester.

Microscopy

All static and dynamic Vickers indentations were evaluated with an Olympus BX51 optical microscope, under bright-field and differential interference contrast. On average, at least 10 static indents were measured at each load to determine a representative hardness value. Because each dynamic indentation was produced by a slightly different load, dynamic hardness values were not averaged and represented one hardness value at a given indentation load.

RESULTS

For brittle materials, the dependence of Vickers hardness on indentation load (ISE) has been well-documented.[9-12] Figure 2 presents the trends in hardness as a function of indentation load for strengthened and virgin glass surfaces. In all cases the hardness curves showed a decrease in Vickers hardness with increasing load. However, once a certain indentation load was reached, the hardness began to level off. The strengthened glass surfaces showed a 22% increase in static hardness and a 12% increase in dynamic hardness compared to the virgin glass surfaces. The dynamic hardness on the strengthened glass surfaces was 23% greater than the static hardness at comparable loads, and the dynamic hardness on the virgin glass surfaces was 30% greater than the static hardness (only one comparable load). Finally, note that the differences in hardness are lower at higher indentation loads.

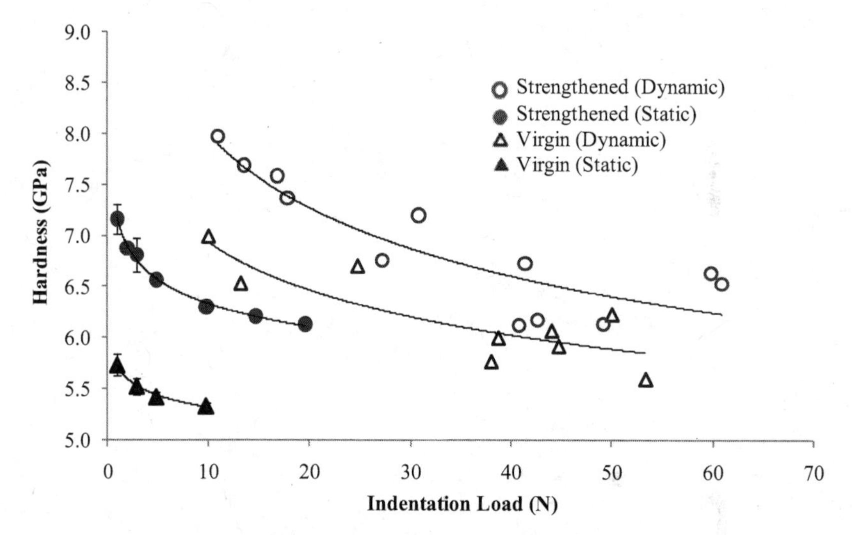

Figure 2. Static and dynamic Vickers hardness data for various indentation loads on strengthened and virgin glass surfaces. The data reveals significant increases in material hardness under dynamic loading for both strengthened and virgin glass samples.

A common method of quantifying the indentation size effect (ISE) is by applying a power law analysis, known as Meyer's Law, to the indentation data.[13-15] The typical Meyer's Law expression is given in Equation 1 as

$$P=Ad^n \tag{1}$$

where P is the indentation load, A is the Meyer's constant, d, is the indent dimension and n is the Meyer's exponent. The indentation size effect is related to the variation of n from a value of 2.[13,14] If n is equal to 2, there is no apparent indentation size effect. Consequently, n is considered the measure of the indentation size effect. The hardness data is shown in Figure 3 as a function of indentation diagonal, d. Under dynamic loads, n is 1.75 and 1.81 for the strengthened and virgin glass surfaces, respectively. Under static loads, n is 1.90 and 1.94 for the strengthened and virgin glass surfaces, respectively. It is clear that the indentation size effect is more prevalent under dynamic loading compared to static loading. In addition, the indentation size effect is more apparent on the strengthened glass compared to the virgin glass. Consequently, it can be expected that Vickers hardness will show increased sensitivity to load when impacted at high strain rates and on strengthened faces.

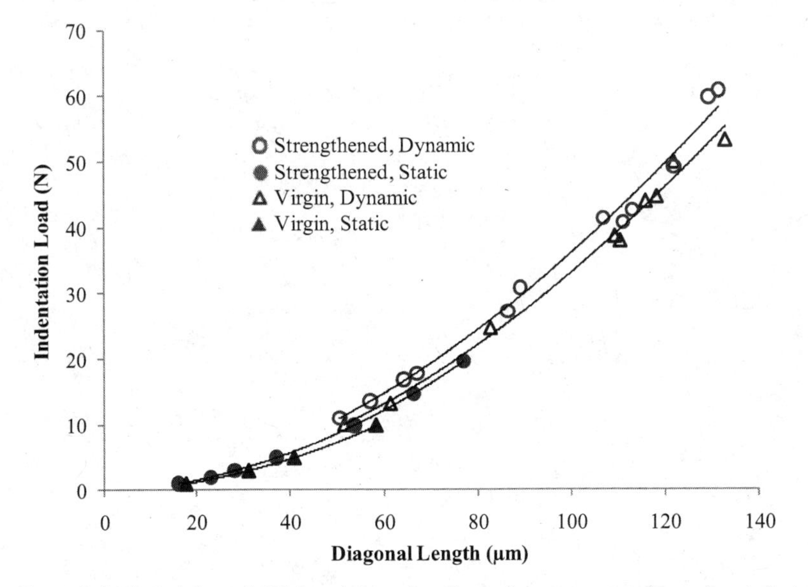

Figure 3. Static and dynamic Vickers indentation diagonal measurements for various indentation loads on strengthened and virgin glass bars. The data reveals increased Meyer's exponent under dynamic loading for both strengthened and virgin glass surfaces compared to static loading.

It is clear from the results that the ion-exchange process significantly enhanced the mechanical properties of the strengthened glass compared to the virgin glass, most notably the hardness. Due to the importance of hardness as a measure of impact and wear resistance, it can be expected that this glass will yield improved effectiveness over traditional glasses in advanced civilian and military applications. Furthermore, as the proposed utility of chemically glasses lies with their improved ability to resist failure during dynamic impacts, such as high-velocity debris, it is important to recognize the noticeable strain-rate hardening present in Ion-Armor™ glass during dynamic indentation.

CONCLUSION

Both static and dynamic indentation studies on strengthened and virgin glass surfaces revealed the presence of an indentation size effect. The strengthened glass surfaces exhibited greater hardness compared to the virgin glass surfaces under static and dynamic loading; however, both surfaces showed increased hardness values under dynamic loading, or a strain rate-sensitivity in hardness. Using power law analysis, the hardness on the strengthened glass bars yielded a higher load dependence compared to the virgin glass bars under static and dynamic loading. It was also observed that the load dependence of Vickers hardness was more prevalent in dynamic indentations.

REFERENCES

[1]A.K. Varshneya and I.M. Spinelli, High-Strength, Large-Case-Depth Chemically Strengthened Lithium Aluminosilcate Glass, *Journal of the American Ceramic Society Bulletin,* **88** [5] 27–33 (2009).

[2]A.K. Varshneya, The physics of chemical strengthening of glass: Room for a new view, *Journal of Non-Crystalline Solids*, **356**, [44-49], 2289-2294 (2010).

[3]P.M. Kelly, Lightweight Transparent Armour Systems for Combat Eyewear, *19th International Symposium of Ballistics*, **7-11**, 969-976 (2001).

[4] J. Gong, J. Wu, Z. Guan, Examination of the indentation size effect in low-load Vickers hardness testing of ceramics, *Journal of the European Ceramic Society*, **19** [15], 2625-2631 (1999).

[5]J.C. LaSalvia, J. Campbell, J.J. Swab, J.W. McCauley, Beyond Hardness: Ceramics and Ceramic-Based Composites for Protection, *Journal of the Minerals, Metals and Materials Society*, **62** [1], 16-23 (2010).

[6] G. Subhash, S. Maiti, P.H. Geubelle, D. Ghosh, Recent Advances in Dynamic Indentation Fracture, Impact Damage and Fragmentation of Ceramics, *Journal of the American Ceramic Society*, **91** [9], 2777-2791 (2008).

[7] R.J. Anton, G. Subhash, Dynamic Vickers indentation of brittle materials, *Wear*, **239** [1], 27-35 (2000).

[8] G. Subhash, Dynamic Indentation Testing''; pp. 519–29 in ASM Handbook on Mechanical Testing and Evaluation, Edited by H. Kuhn and D. Medlin. *ASM International*, Materials Park, OH, **8**, 2000.

[9] J.J. Swab, Recommendations for Determining the Hardness of Armor Ceramics, *International Journal of Applied Ceramic Technology*, **1** [3] 219-25 (2004).

[10]P.M Sargent, Use of the Indentation Size Effect on Microhardness for Materials Characterization, *Microindentation Techniques in Materials Science and Engineering, ASTM STP 889,* eds. P.J. Blau and B.R. Lawn, American Society for Testing and Materials, Philadelphia, PA, pp. 160-174 (1986).

[11]F. Knoop, C. Peters, and W. Emerson, "A Sensitive Pyramidal-Diamond Tool for Indentation Measurements," *Journal of Research of the National Bureau of Standards*, **23**, 39–61 (1939).

[12]C. G. Peters and F. Knoop, Resistance of Glass to Indentation, *Glass Industry*, **20**, 174–176 (1939).

[13]L. Sidjanin, D. Rajnovic, J. Ranogajec, E. Molnar, Measurement of Vickers hardness on ceramic floor tiles, *Journal of the European Ceramic Society*, **27**, 1767-1773 (2007).

[14]N.K. Mukhopadhyay, Analysis of microhardness data using the normalized power law equation and energy balance model, *Journal of Materials Science*, **40**, 241-244 (2005).

[15]H. Li, A. Ghosh, Y.H. Han, R.C. Bradt, The frictional component of the indentation size effect in low load microhardness testing, *Journal of Material Research*, **8** [5], 1028-1032 (1993).

Manufacturing

TRANSPARENT ARMOR FOR THE NEW STANDARD IN BATTLEFIELD PERFORMANCE

Kathie Leighton(1), John Carberry(1), Wiktor Serafin(1), Terrance Avery(2), Douglas Templeton(2)

(1) SCHOTT DiamondView Armor Products, LLC, 1515 Garnet Mine Road, Boothwyn PA 19061
(2) TARDEC 6501 E. 11 Mile Road Warren, MI 48397-5000

ABSTRACT

Armor Transparent Purchase Description (ATPD) 2352 revision P[1] was issued in July 2008 to create a new standard for transparent armor aimed at improving battlefield performance, maintenance costs, equipment survivability, and general durability based on data collected from performance of transparent armor in the battlefield. A transparent armor specifically focused on satisfying all of the ATPD 2352 requirements was invented, developed, and commercialized. A Cooperative Research and Development Agreement with TARDEC resulted in evaluating armor to all the metrics of ATPD 2352.

This paper reports on this initial and subsequent work and;

a) explains the requirements of ATPD 2352 and the challenges they present from a materials properties, armor performance, lifetime testing, transparency, durability, and environmental perspective;
b) presents data, analysis, and preliminary modeling showing the materials and performance properties of a variety of materials to highlight how and why a discontinuously nano-reinforced glass system was able to pass all the requirements;
c) describes the tests and presents test data on the key tests performed for ATPD 2352, including ballistic, environmental, and optical, many never successfully mastered in transparent armor before.

BACKGROUND

The requirements of the new specification for transparent armor, Armor Transparent Purchase Description (ATPD) 2352,[1] were defined over a period of time with an abundance of feedback from the theater in Iraq and other places.[2-6] While the first vehicles put into service in Iraq were frequently unarmored, those armored windows that followed were often found wanting in terms of threat resistance, visibility, and life cycle.

Figure 1: Attack by a small IED shows multiple impacts on the window indicating close proximity and high impact velocity.

The U.S. Army's Tank and Automotive Command (TACOM) conducted a cost benefit study on transparent armor[2] and identified that from a sample of 266 transparent armor damage incidents 62.8% were a result of combat damage. Battlefield reports, for example including news articles and pictures,[3,4] showed that close range rifle and machine gun fire and multiple roof top snipers were an early threat in urban areas where it was learned that in many cases if a window stopped a first round it did not stop subsequent shots. Detonation from improvised explosive devises (IEDs) of various size were ubiquitous and found to impact windows with high velocity spray of fragments,[5,6] for example as shown in Figure 1.

In addition to battle field threats, the harsh environment imposes strong thermo-mechanical challenges to transparent armor degrading the polymer layers resulting in delamination, bubbles, loss of adhesion, clouding and discoloring. Extreme thermal excursions and shocks caused cracking in the glass and also contributed to delamination. Sand abrading against the armor windows produces surface defects and surface defects are known to reduce the strength of glass and lead to cracking.[7,8]

Thermal extremes in Afghanistan have been reported from as low as -46°C (-51°F) and as high as 51 °C (124 °F), and in Iraq extreme highs in the summer can reach 46 °C (115 °F) to 52 °C (125 °F) in the desert areas and have even been reported to 49 °C (120 °F) in the mountain valleys.[9] Thermal extremes of the natural environment combine with thermal shocks and contamination associated with operation and logistics including moving from storage to use, air drops, chemical spray downs, water exposure in fording, and vehicle road dynamics and vibrations.

This severe thermo-mechanical and contamination prone environment is made even worse by the degrading power of the sun. NASA Goddard reports data collected by the Solar Radiation and Climate Experiment[10,11] satellites show that the electromagnetic energy of the sun that hits Earth's atmosphere varies with solar conditions and is about 1368 W/m^2. The insolation, the amount of electromagnetic energy that impinges the surface of the Earth, is less due to cloud cover and surface obliquity and varies with elevation, latitude, time of day and season being greatest at high elevations, tropical latitudes, noon, and in the northern hemisphere summer. The spectrum of insolation also varies with location on the Earth, and due to the fact the direct irradiance from the sun varies more in spectrum than in total energy. NASA's Total Ozone Mapping Spectrometer results indicate the majority of the recent battlefield conflicts take place in regions of the World exposed to the most damaging high energy waves, UV-B in the 290 to 320 nm range.

Replacement and operation needs for transparent armor was running at a cost of $3-$12 million a month during Fiscal Years 2006 - 2008, with a significant percentage related to replacements due to the problems above.

In the same time frame, windows that could offer the necessary higher level of protection were too heavy and too thick. Armor weight strains the mechanical components of a vehicle increasing wear and fuel consumption; in one study 16% of fuel consumption was directly related to road weight of a vehicle.[12] Reducing the weight of a 4.8 liter V8 diesel engine truck can save 0.3-0.9% in fuel costs per 100 lbs of weight savings.[13] Since windows are mounted high up in a vehicle, transparent armor weight was also contributing to the problem of mine resistant vehicles rolling over. Armor weight can slow down transportation to theater and mobility once there.

While life cycle costs are critical and long term budgets require low maintenance costs, the equipment's role in mission effectiveness is the primary and first priority and it is unacceptable for the equipment to fail and compromise a mission. The materials used in many of the first transparent armors delivered to the field absorbed light in the infrared spectrum (IR) making the use of night vision goggles, which function in the near infrared, impossible or impractical requiring such high power levels that glare impaired vision to the point of being useless. The armored window is first a window, so thick windows with distortion, poor visibility, and lost visibility after impact were a problem in many cases.

THE ATPD 2352 REQUIREMENTS

The ATPD 2352 specification addressed all these challenges with well defined requirements related to visible and optical properties in the visible and IR, requiring rifle and fragment penetration resistance at various levels, and providing requirements to maintain necessary visible and optical capabilities after exposure to thermal shock, humidity, solar loads, cycles of high temperature, low temperatures relevant to storage, abrasion on strike face and safe side, and chemical exposures on both sides.

Visible and Optical Requirements

First and foremost a window must offer visibility for situational awareness both in daylight and at night. In this military setting, night vision is critical to maintain situational awareness in the dark and the use of night vision goggles (NVG) is not optional. The ATPD 2352 provides requirements and test protocols for six optical tests.

The first is a visual inspection with defect limitations where, in the most recent version, Paragraph 4.1.1 Allowable Defects in ATPD 2352 Revision R, the inspection is required to be performed looking from the inside through the window, a procedural detail that illustrates that most importantly these are windows to see through and secondly armor.

The next two tests are to measure the transmission of the window in the visible range, luminous transmission, and then in the near IR for NVG compatibility.

ATPD 2352 paragraph 4.4.1 defines how to measure luminous transmittance, "Luminous (photopic) transmittance shall be determined in accordance with the photopic transmission measurement procedure given in MIL-DTL-62420. Transmittance shall be determined before and after the exposure of the Sun Exposure Weathering test, 4.3.5. Spectral transmittance shall be measured at wavelength intervals of 10 nm or less over the 400 to 930 nm band at normal incidence. Luminous visible light transmittance corresponding to daylight vision is determined by integration of individual photopic transmission values in the 400 to 700 nm range, as discussed in MIL-DTL-62420."

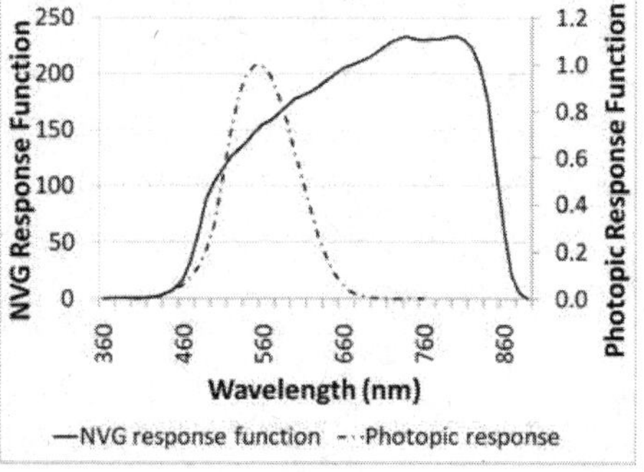

Figure 2: Response functions for photopic and NVG transmittance.

In the ATPD 2352, NVG compatibility was quantified and defined as part of the critical performance of a good window. ATPD 2352 Paragraph 4.4.1.1 states, "The NVG-weighted integrated spectral transmission is determined using the same procedure for determining the luminous transmission, except that the photopic visibility response function is replaced by the NVG-response function and the integration is over the 400-930 nm bandwidths."

Both requirements include calculating integrated transmissions based on response functions, presented in Figure 2. As the response functions show, the photopic transmittance is most heavily weighted by transmission near 555nm, where as visibility through night vision goggles is increasingly weighted approaching the NIR and most heavily from 770nm to 850nm.

The last three required optical properties to measure are haze, deviation, and distortion. Haze is the diffuse transmittance as a percentage of the total transmittance as measured by ASTM D1003[14] (CIE Illuminant A; Method: Procedure B, Diffuse Illumination/Unidirectional Viewing). Deviation is measured by ASTM F801-96[15] or ASTM F2469-05.[16]

The distortion requirement is intended to address the problem that some of the thick windows were distorting the far field images such that telephone poles or edges of buildings would appear curved or wiggly. It is measured in accordance with ASTM F2156[17] where a grid pattern viewed with and without the window is compared and the difference in the straightness of the lines is quantified as a grid line slope (GLS).

Ballistic Requirements

Ballistic protection required stopping kinetic energy rounds in a tight 4-shot "T" pattern, as shown in Figure 3, over temperature range of -43°C up to 63°C.

During 2000/2001 the TARDEC AIL (TARDEC Armor Integration Laboratory) the predecessor of the current TARDEC SABL (Survivability Armor Ballistic Laboratory) performed tests to determine the expected shot spacing of urban attacks on vehicles. This was done by preparing a computer generated generic type SUV vehicle target shown at an angle so that it could contain four identical passengers. The targets were printed full size and presented to groups of shooters of various skill levels. The shooters were NATO troops from several countries and US civilians at our test site.

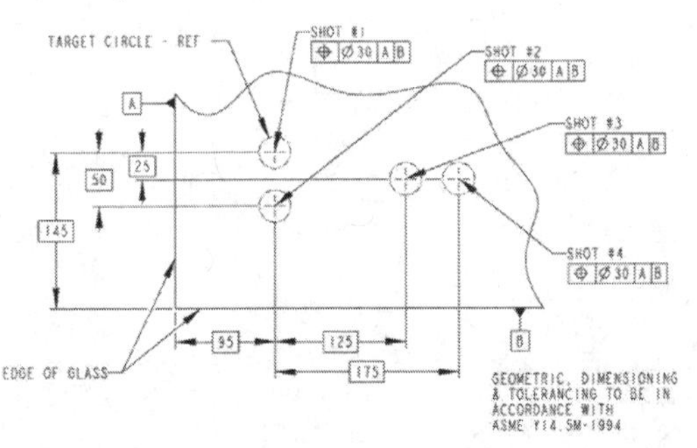

Figure 3: Four-Shot "T" pattern from ATPD 2352 Rev P.

Although the purpose of the test was to determine multi-hit shot spacing the shooters were only told to kill the people in the vehicle during an 18 second time limit at 20 meters. The military shooters used their service weapons while the civilian shooters were allowed to select which military weapons to shoot. The shooters were free to select which method of fire they used, i.e. single shots, burst fire, or fully automatic fire.

The four figures were identical thus allowing the targets to be separated into four "attacks" of 4.5 seconds where the shooters were aiming at the figure seen through the vehicle window. Hundreds of targets were analyzed by measuring the spacing between shots and determining the minimum size triangles that were formed. This data was plotted for three typical threat weapons, the Soviet AK-47, the US M-16, and the German G-3 representing a heavier .30 caliber cartridge. Graphs were made showing minimum shot spacing distance vs. probability of that spacing appearing in this data set. Graphs were also prepared showing the minimum triangle perimeter vs. probability. Thus for example, a shot spacing of 25 mm with the M-16 rifle represents the 10th percentile of that data set. Therefore only 10% or less of the shooters had shot spacing of 25 mm.

This data was used to formulate the NATO AEP-55[18] specification entitled, "Procedures for Evaluating the Protection Level of Logistic and Light Armored Vehicles." The protection level desired, i.e. probability of non-penetration, in this NATO specification is 90% so the shot spacing occurring 10% of the time was selected as the multi-hit requirement. This shot spacing when tested as required by the NATO specification to "exploit the Localized Weak Areas" of the transparent armor (shooting at the edges with a minimum spacing of 25 mm) causes the glass to be made very thick.

When the TACOM ATPD 2352 specification was written the multi-hit shot spacing for rifle/machine gun threats was increased. This was done based on the realization that testing is done at worst case condition for window orientation (zero obliquity) and temperatures. Therefore, the shot spacing selected represents approximately the 40th shooter percentile for the various projectiles used. A four shot "T" shaped pattern was selected with the spaces made of two pairs of shots at the 40th percentile spacing and the spacing between the pairs equaling the long side of the 40th percentile triangle. The requirements of ATPD 2352 were not selected to be equivalent to any particular TACOM vehicle system, but rather it was selected to be the standardized criteria for lot acceptance of transparent armor. It is important to remember that "passing" the ATPD 2352 ballistic requirements do not assure that the product will pass the protection requirements of a particular vehicle system.

Environmental Requirements

Environmental specifications and tests in the ATPD 2352 derive, with some modifications, from the United States Military Standard referred to as MIL-STD-810, "Department of Defense Test Method Standard for Environmental Engineering Considerations and Laboratory Tests" which establish chamber test methods to replicate the effects the environment has on materials and structures rather than on direct simulation of the environment.[19] Two different versions of MIL-STD-810, F and G, are referenced in the ATPD 2352 presumably because both standards were being modified during the same time period. Five different tests are required; Low Temperature, High Temperature, Humidity, Temperature Shock, and Sun exposure weathering. De-icing requirements and tests also test response to thermal stresses. After each test the part is returned to room temperature and ambient conditions and inspected to the six optical, including visual, requirements discussed above, and held to the standards of the original optical requirements.

The low temperature cycle includes a 24 hour hold at -54°C in accordance with MIL-STD-810F Method 502.4 Procedure I. This method was developed to replicate material failures that can occur during low temperature storage of military equipment, Specific failures identified by MIL-STD-810 that are relevant to transparent armor are; hardening and embrittling of polymers leading to cracking and crazing, reduced impact strength, static failure of restrained glass, and condensation and freezing of water. This procedure is intended to test materials in storage conditions and prepare them for additional testing to ensure they meet operating requirements after storage, which in the case of the ATPD 2352 includes the visual inspection and optical tests above.

The high temperature test is in accordance with MIL-STD-810G Method 501.5, Procedure I, A2, Induced. It includes a 24 hour heating and cooling cycle where the chamber varies between 30 and 63°C. The relative humidity is varied from 44% to 5% with the lowest levels at the highest temperatures. Three cycles are required. Failures listed by MIL-STD-810 that can occur under high temperature and relevant to transparent armor include discoloration, cracking or crazing of organic materials, out gassing, and binding due to differential expansion of material with dissimilar coefficients of thermal expansion (CTE). This test is limited to use to evaluate the effects of short term, even distributions of heat without synergistic effects. Procedure I is applicable to storage conditions where the parts are protected from the added heat, +19°C (35 °F),[19,20] and synergistic radiation damage that can be generated by the sun. Its effect on window operation is evaluated by post test visual and optical measurements.

Conformance to optical properties after exposure to warm humid environs is evaluated by exposing windows to five modified cycles of the aggravated humidity profile shown in MIL-STD-810G, Figure 507.5-7. The modified cycle is 48 hours duration at 95% relative humidity and each cycle includes a 30 hr hold at 60°C, and an 8 hr hold at 30°C. After the test, the sample is conditioned at 23°C ± 10°C and 50% maximum relative humidity for 48 hours then inspected to ensure no indication of moisture buildup, bond separation, or any other forms of image degradation per the allowable defects specification. The sample returns to normal ambient conditions and is inspected to the visual and optical specifications.

Temperature shock effects on the transparent armor are evaluated using Method 503.5, Procedure I-C of MIL-STD-810 adapted to include an 18 hour period at -30°C followed by an 18 hour period at +60°C with a transfer time of not more than five minutes. At the conclusion of the thermal shock test the sample is required to conform to the visual and optical requirements. MIL-STD-810 suggests the use of this test when material is likely to experience sudden changes in temperature such as during transfer from climate controlled storage or enclosure to hotter or colder outside temperatures, or when ascending to high altitudes from a high temperature ground environment, or vice a versus such as in an air drop. It is not intended to test for conditions such as water hitting a hot surface or rapid localized heating of a cold surface. Transparent armor exposed to this test may experience shattering of glass, differential contraction or expansion rates or induced strain rates from dissimilar materials, deformation or fracture of components, cracking of surface coatings.

De-icing specifications in the ATPD 2352 require de-icing at -25°C in 60 minutes. A window is cooled to -25°C and held for 12 hours then sprayed with water from a 345kPa pressure gun. The water is allowed to form into ice for 25 minutes before the de-icer is turned on. This ATPD 2352 specified test imposes a combination of thermal stresses and the window is required to be inspected for visual and optical requirements after the test.

Sun exposure weathering tests require the use of Procedure II in MIL-STD-810 Method 505. This procedure was developed to include both the temperature and actinic effects of solar loads. The specified cycle is 20 out of 24 hours at 1120W/m^2 at a constant temperature of 49°C. For four hours each cycle the lights are turned off to induce alternating thermal stressing and allow "dark" processes to occur. The most intense naturally occurring total irradiance on the earth at sea level is represented by the irradiance cycles of Procedure I, which only reach 1120W/m^2 for 2 hours out of each 24 hours. Procedure II accelerates the amount of total irradiance impinging the sample by 2.5 times requiring the 1120W/m2 for 20 of the 24 hours. In addition, Procedure II requires the use of full sun spectrum lamps with 68.3% of the spectrum comprising the high energy UV wavelengths below 400nm so are more active in evaluating actinic material responses which show up in yellowing, discoloration, cracking, or, in extreme cases, mechanical degradation. The acceleration of these actinic processes may be much more than 2.5 times due to the added UV content and require correlation with natural processes and conditions to quantify.

GLASS-CERAMICS TO MEET ATPD 2352

The emerging requirements of APTD 2352 dictate new and critical properties the components and window systems should have:

i) Opaqueness in the ultra-violet (UV), below 370 nanometers to protect the polymer constituents from solar radiation;
ii) Good transparency between 400 and 1200 nanometers for human and night vision visibility;
iii) Very low or no coefficient of thermal expansion (CTE) to promote resistance to thermal shock and cycling;
iv) Unique failure mechanisms to promote multi-shot performance at low weight;
v) Superior ballistic performance to promote low weight against IED fragments.

Glasses used in armor are primarily soda-lime-silicate, which can be improved for infra-red (IR) transmission if made with very low iron and borosilicate glass which has a low density and low CTE. Both offer good transparency in the visible range, but are transparent in the UV (see Table I).

Table I: Properties of soda-lime-silicate and borosilicate glass compared to glass-ceramics.

Material	Density (g/cc)	CTE (ppm/°C)	Transmission at 370 nm	Young's Modulus (GPa)
Soda-Lime	2.49[20]	9.03[21]	>50%[21]	73.1[20]
Borosilicate	2.22[20]	3.25[22]	>85%[23]	63.1[20]
LAS Glass-Ceramic	2.56[25]	0 +/- 0.3[24]	0 [24, 27]	92[25]

Evaluation of commercially available glass-ceramics, focusing on large production of glass-ceramics available in the lithium alumino silicate, Li_2-Al_2O_3-SiO_2 (LAS) family, often used in fire places and cook tops and other appliances, reveals several advantages of this material.[24, 25, 26]

The CTE is 0.3 parts per million per °C but is balanced between the LAS crystal having a negative CTE and the glass a positive CTE.

The glass-ceramic is filled about 65% by volume with nano-crystals of about 70 nanometers and smaller size. This offers the unique advantage that it blocks all wave lengths less than four times this size, but allows the higher wavelength visible light to pass. Properties are listed in Table I.

Its unique microstructure offers advantages in armor as its failure mechanism isolates damage and sets up superior multi-shot performance. For instance Figure 4 compares the damage pattern in a glass-ceramic target to one in a soda lime target, both impacted with an armor piercing round at muzzle velocity. The diameter of the opaque zone which is comprised of a network of radial and circumferential cracks, is 203 mm diameter in the LAS glass-ceramic and 241 mm diameter the soda-lime-silicate glass target, which is 1.41 times larger area with no visibility.

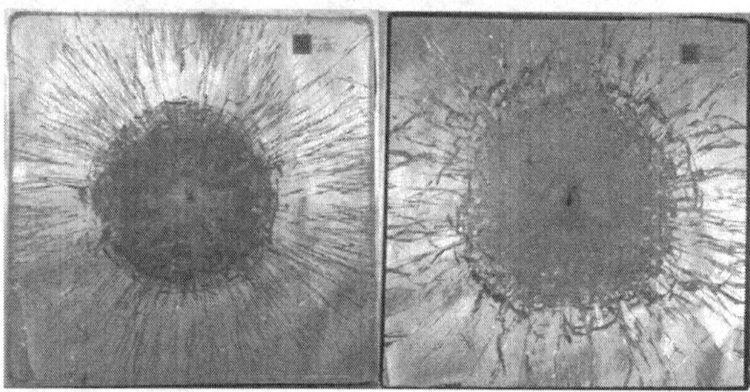

4a: Glass-ceramic target 4b: Soda-lime-silicate target

Figure 4: Damage pattern in comparable glass-ceramic and soda lime targets. Each target is 400 x 400 mm square.

Two disadvantages with production available glass-ceramic were the "orange peel" surface found on both sides due to the nature of the rolling process that pulls it off the melting tank, and transmission in the visible caused by additives and contaminates in the melt.

The glass-ceramic was made clearer by eliminating colorants used in appliance grade LAS glass-ceramic by reducing impurities. The resulting typical transmission as shown in Figure 5 is opaque below 370 nm and transparent from 370 to 1100 nanometers.

The representative orange peel, surface shown in Figure 6, was solved by using an interlayer with an index of refraction suitably but not perfectly matched to the glass-ceramic, taking into account the angles imposed by the geometries of the orange peel. Without the suitable matched interlayer and surface geometries, images viewed through the laminate were fuzzy and blurred. With appropriate interlayer and surface geometry match, images are clear and windows pass all optical requirements of the ATPD 2352.

TESTING LAS GLASS-CERMIC CONTAINING LAMINATED ARMOR TO ATPD 2352

Complex assemblies of plastic and LAS glass-ceramic which offered superior transparency, IR functionality for NVG, and still block UV and survive solar radiation, thermal shock, thermal cycling and all ballistic requirements of ATPD 2352 have been developed. Optical Measurements and Results

Transmission is measured to the requirements of the ATPD 2352 using a UV/VIS/NIR spectrometer with dual beam, dual monochromatic optics having a range from 185 nm to 3300 nm attached to a PC with full spectrum software. Because of the lateral size constraints of this system, specimens are limited to 50 x 50mm up to 100 x 100mm size. Larger samples and full size windows are measured on a custom setup using diffuse illumination and a portable fiber optic spectrometer that is calibrated by correlating with the calibrated spectrometer.

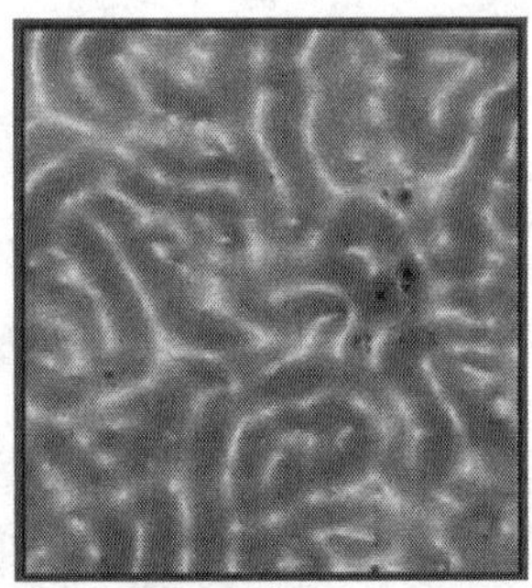

Figure 6: Characteristic orange peel on the surface of rolled glass-ceramic. The image, 15 mm wide, was created with a 3D optical surface profiler.

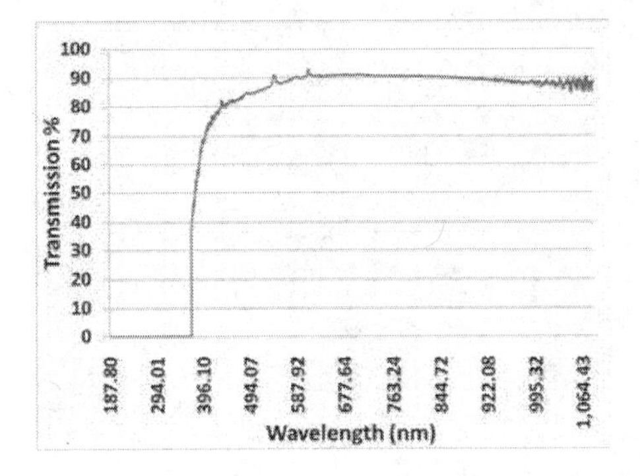

Figure 5: Transmission of LAS Glass-ceramic, 8 mm thick.

The transmission characteristics of four different sample constructions are presented in Table II. The cut-on wavelength is the value at which the sample begins to show positive transmittance. Below this wavelength the samples block 100% of the electromagnetic radiation. All the lay-ups meet the optical requirements, exceed the NVG transmittance, and provide UV blocking.

Table II: Transmittance for various laminates using glass-ceramic layers.

Sample ID	Thickness (mm)	Areal Density (kg/m^2)	Wavelength Cut-on (nm)	Photopic Transmittance (%)	NVG Weighted Transmittance (%)
daptms001	50	103	395	71	80
dap1772	90	161	400	73	74
daptms002	90	186	405	55	68
dap200081	109	215	399	62	77

Production grade processes controlling dust and debris in lay-up rooms and consistency with glass production increase the photopic transmission from an initial value ranging from 58.9% to 60.8% to an average production value of 66.5% with a minimum of 61%.

Environmental Tests and Results

Sun weathering tests are performed in a Xenon test chamber using B/B filters and Xenon Arc lamps to produce the required spectrum and energy. It is capable of controlling the ambient temperature to the required 49°C and has on the fly calibration performed with dedicated radiometer and thermometer traceable to NIST standards. The first samples are transparent armor, comprised of glass-ceramic and polycarbonate laminated into a 171 kg/m^2 system, approximately 81mm thick. The sample is 300mm x 300mm, wrapped in silicone foam edging and gasket material, and framed in an aluminum casing. Temperature probes are attached to the front and back surface. The sample is placed with the "safe side" laying on an aluminum tray and the glass "strike face" facing the lights. It is exposed to 56 cycles; each cycle is 24 hours long comprised of 20 hours with the light on and 4 with the light off. The air temperature control was set at 49°C. The surface temperature of the glass face reached a maximum of 73°C during the light on phase and dropped back to the chamber air temperature during the lights off phase of each cycle. The surface of the safe side was only 6 mm away from the aluminum tray and the maximum surface temperature on the safe side was highest, 82°C, when the sample was rotated to a position where the thermocouple was located with less than 30 mm of air space between the tray and bottom of the chamber. MIL-STD-810G part one c-2 states, with respect to hottest climates on earth, "except for a few specific places, outdoor ambient air temperatures will seldom be above 49°C (120°F). ... The thermal effects of solar loading can be significant for material exposed to direct sunlight, but will vary significantly with the exposure situation. The ground surface can attain temperatures of 17 to 33°C (30 to 60°F) higher than that of the free air, depending on the type/color of the ground surface," So the surface temperatures measured for the sample are on the upper limit of what might be seen in the field. Post test, the sample passes the allowable defect specification of the ATPD 2352 having grown only one small bubble, <1.6mm diameter, which is less than half the minimum size allowed.

A second series of experiments compare glass-polycarbonate laminates with various amounts of glass-ceramic, from none (dap004446), one layer just under the strike face (dap004435, and all except the strike face (dap004440). All specimens are 100 x 100 mm and have edges finished with a two part polyurethane sealant. Samples are inspected every week. Post test, all samples show delamination lines related to contamination by the sealant. Results summarized in Table III, show that the addition of one layer of glass-ceramic nearly doubles the life time of resistance to delaminating even in an environment of contamination.

Table III: Effect of glass-ceramic layers on solar loading test results.

Sample ID	Sample	Thick (mm)	Delam Day	Pre-Photopic (%)	Post-Photopic (%)	Pre-NVG (%)	Post-NVG (%)
dap004446	Soda-lime	125	21	74	57	54	42
dap004435	One layer glass-ceramic	128	56	49	49	35	34
dap004440	Multiple glass-ceramic	101	56	56	55	71	69

Evaluation against the remaining environment tests, high temperature, low temperature, thermal shock, and humidity is performed on two types of full size window samples; one (dap-GC) made of mostly glass-ceramic weighs 201 kg/m^2 (41.1 psf), the other (dap-SL) uses one layer of glass-ceramic just under the strike face and weighs 244 kg/m^2 (49.8 psf). Both systems, post test, meet the allowable defect specification, and retain the required levels of photopic and NVG transmittance (see Table IV).

Table IV: Optical properties of two different transparent armor systems after exposure to environmental tests.

TA Type	Sample ID#	Post Test Luminous Transmittance (%) Min: 55%	Post Test NVG (%) Min: 30%	Post Test Allowable Defects	Environmental Test
dap-GC	4575	56.3	70.0	pass	High Temp
dap-GC	4578	56.7	71.1	pass	High Temp
dap-GC	4563	56.3	70.1	pass	Humidity
dap-GC	4566	57.6	69.3	pass	Humidity
dap-GC	4576	56.1	70.2	pass	Low Temp
dap-GC	4562	56.1	70.0	pass	Low Temp
dap-GC	4569	57.0	70.0	pass	Temp Shock
dap-SL	4611	67.5	54.9	pass	Low Temp
dap-SL	4615	67.0	54.4	pass	Low Temp
dap-SL	4616	65.8	56.3	pass	Temp Shock
dap-SL	4617	67.2	55.0	pass	Temp Shock
dap-SL	4618	67.9	55.8	pass	Humidity
dap-SL	4622	66.8	55.8	pass	Humidity
dap-SL	4628	66.7	54.4	pass	High Temp
dap-SL	4625	68.7	56.6	pass	High Temp

BALLISTIC PERFORMANCE

Ballistic weight efficiencies of the developed LAS glass-ceramic containing transparent armor recipes are typically 20% - 40% lighter than incumbent soda-lime based transparent armor depending on the specific threats of interest. Examples are listed in Table V.

Table V. Ballistic performance of various LAS glass-ceramic based armor recipes*

Sample #	Areal Density (kg/m² (psf))	Thickness (mm)	Projectile	Impact Velocity Range (m/s)		Multi-hit	Test Temp
1829	255 (52)	168	7.62 AP	968	981	4-shot T	65 °C
8092B	255 (52)	114	7.62 AP	966	977	4-shot T	-43 °C
8068L	254 (52)	114	7.62 AP	962	973	4-shot T	Ambient
ddm1226	231 (47)	107	20mm FSP	1509	1522	3-shot 160 mm Triangle	Ambient
ddm1827	202 (41)	142	7.62 AP	877	882	4-ahot T	65 °C
ddm1031	188 (38)	89	7.62 AP	875	889	4-shot T	Ambient
ddm0983	169 (35)	82	20mm FSP	1054	1080	3 shots 150 mm triangle	Ambient
ddm0971	168 (34)	79	7.62 AP	884	895	3 shots 120mm triangle	Ambient
ddm0947	173 (35)	83	7.62 Ball	871	878	4-shot T	Ambient
ddm0923	103 (21)	51	7.62 Ball	831	853	5-shot NIJ 0108.01 III	Ambient
ddm0944	103 (21)	48	7.62 Ball	724	729	4-shot T	Ambient
ddm0925	103 (21)	51	0.50 Cal FSP	1226		1-shot	Ambient
ddm0926	103 (21)	55	0.30 Cal AP-M2	844		1-shot	Ambient
ddm0927	103 (21)	48	7.62 AP	836		1-shot	Ambient
ddm1012	95 (19)	43	7.62 AP	872		1-shot	Ambient
ddm1470	84 (17)	42	7.62 AP	781	794	2-shot in 12"	Ambient
ddm1472	66 (14)	35	7.62 Ball	826	837	3-shot 120mm triangle	Ambient
p39	60 (12)	34	0.50 FSP	V50 = 1089			Ambient
ddm696	60 (12)	33	7.62 Ball	888		1-shot, UL 752 level 5	Ambient
ddm693	60 (12)	33	7.62 Ball	859		1-shot, UL 752 level 5	Ambient
ddm752	60 (12)	33	5.56 x 45 M855	908	919	3 shots in 8" dia. circle, SD-STD-01.01	Ambient
hat-4D	54 (11)	27	12.7 AP @ 60°	496		1-shot	Ambient
hat-5c	39 (8)	21	7.62 AP @ 60°	773		1-shot	Ambient

*Note: More detailed ballistic test results are available under limited distribution.

CONCLUSIONS

An LAS glass-ceramic based transparent armor was developed which is the lightest weight transparent armor recipe to date that is ballistically qualified to the 3a all temperature level of the ATDP 2352. In addition, it is capable of passing all other requirements of ATPD 2352 Rev R weighing 201 kg/m² (41.1 psf).

ACKNOWLEDGEMENTS

The authors appreciate the support of TARDEC under CRADA #08-18, the funding from the Technical Support Working Group through EMRTC that supported the environmental test on the full size windows, and the extensive funding and contributions of materials and the transparent armor inventions from DiamondView Armor Products (DAP) and then SCHOTT DiamondView Armor Products which acquired DAP. We acknowledge and appreciate the numerous discussions with TARDEC staff working to develop the ATPD 2352 purchase description including David Hanson, David Sass, Steve Hoffman and Robert Goedert who also provided luminous and NVG transmittance calculation templates. The authors recognize and appreciate that the optical measuring equipment and techniques were developed and exercised by optical experts at SCHOTT's R&D facility in Duryea PA, Carsten Weinhold, David Badack, Joe Granko, and Beth Gober-Mangan. The authors are indebted to Edgar Aleshire who made numerous contributions to the designs and fabrication of the many test coupons and to developing production grade processes for improved quality. The authors are grateful to Rebecca Neill for overseeing tests performed at Dayton T. Brown and for helpful suggestions in editing the manuscript. Authors acknowledge Zygo Corporation for making the surface profile measurements shown in Figure 6 free of charge and Tim Talladay for photographs in Figure 4.

REFERENCES

1. "Purchase Specification Transparent Armor ATPD 2352", TACOM-LCMC, Department of the Army, dami_standardization@conus.army.mil. Rev P 7 July 2008, Rev R 26 April 2010.
2. David Holm, Raymond Kleinberg, Lisa Prokurat Franks, "Transparent Armor Cost Benefit Study Fleet Update", TACOM Cost & Systems Analysis Directorate and RDECOM, TARDEC, August 2007.
3. "Assassination Rocks Mideast U.S. Strategy", Combined Wire Services, The Hartford Courant, 22 Nov. 2006.
4. David Swanson, Picture of multi-hit impacts on front windshield of HMMWV, Knight Ridder-Tribune, front page Detroit Free Press, 7 April 2004.
5. Joshua Partlow, "One Month Two Brushes With Death - In Iraq Lucky is Difficult to Define", Washington Post Foreign Service, 23 July 2007.
6. Meg Jones, "Saved by 4" Thick Glass", Journal Sentinel Online, 13 June 2005.
7. A.M. Muller and D.J. Green, "Elastic Indentation Response of Float Glass Surfaces", J. Am. Cer. Soc. 93, [1] 209-216, (2010).
8. A. A. Wereszczak, T. P. Kirkland, M. E. Ragan, K. T. Strong, Jr., H-T Lin, P. Patel "Size Scaling of Tensile Failure Stress in a Float Soda–Lime–Silicate Glass", International Journal of Applied Glass Science 1 [2] 143–150 (2010).
9. "Climatology for Southwest Asia", http://www.ncdc.noaa.gov/oa/climate/afghan/#intro, joint web site by National Climatic Data Center, Air Force Combat Climatology Center, and the Navy hosted by National Oceanic and Atmospheric Administration, Ashville NC 28801.
10. "Solar Radiation and Climate Experiment", http://earthobservatory.nasa.gov/Features/SORCE/, Earth Observatory, NASA Goddard, updated 18 Dec 2010.
11. R. F. Cahalan, G. Wen, J. W. Harder, and P. Pilewskie, "Temperature responses to spectral solar variability on decadal time scales", Geophysical Research Letters, Vol. 37 (2010).
12. N. Lutsey, "Review of technical literature and trends related to automobile mass-reduction technology", Institute of Transportation Studies, University of California - Davis. UCD-ITS-RR-10-10, http://pubs.its.ucdavis.edu/publication_detail.php?id=1390, (2010).
13. A. Casadei and R. Broda, "Impact of Vehicle Weight Reduction on Fuel Economy for Various Vehicle Architectures", Project FB769, ©Riccardo Inc. 20 Dec 2007.

14. "Standard Test Method for Haze and Luminous Transmittance of Transparent Plastics," ASTM D1003.
15. "Standard Test Method for Measuring Optical Angular Deviation of Transparent Parts," ASTM F801-96.
16. "Standard Test Method for Measuring Optical Angular Deviation of Transparent Parts Using the Double-Exposure Method," ASTM F2469-05.
17. "Standard Test Method for Measuring Optical Distortion in Transparent Parts Using Grid Line Slope," ASTM F2156.
18. "Procedures for Evaluating the Protection Level of Logistic and Light Armored Vehicles", AEP-55 Vol. 1 & 2, NATO/PfP unclassified publication, Edition 1, February 2005.
19. "Environmental Engineering Considerations and Laboratory Tests", Mil-STD-810F (1/1/2000) and G (10/31/2008).
20. A. A. Wereszczak, K. E. Johanns, T. P. Kirkland, C. E. Anderson, Jr., T. Behner, P. Patel, D. W. Templeton, " Strength and Contact Damage Responses in a Soda-Lime-Silicate and a Borosilicate Glass", 25th Army Science Conference, Orlando FL (2006).
21. PPG Starphire® mechanical properties listed on Precision Glass and Optics web site: http://www.pgo.com/pdf/ppg_starphire.pdf.
22. Borofloat® 33, thermal properties listed on SCHOTT web site: http://www.us.schott.com/borofloat/english/attribute/thermic/index.html.
23. Borofloat® 33, UV transmission, listed on SCHOTT web site: http://www.us.schott.com/borofloat/english/attribute/optical/index.html.
24. H. Schiedler, E. Rodek, "Li_2O-Al_2O_3-SiO_2 Glass-ceramics", Ceramic Bulletin, 68 [11] 1926-1930 (1989).
25. M. Hiltl, H. Nahme, "Dynamic Behavior of a Shock-Loaded Glass-ceramic based on the Li_2O-Al_2O_3-SiO_2 System," J. Phys. IV France 7, 587-592 (1997).
26. Robax® glass-ceramic properties, listed on MatWeb web site: http://www.matweb.com.
27. W.E. Pannhorst, "Low expansion glass-ceramics: review of the glass-ceramics Ceran® and Zerodur® and their applications." In Ceramic Transactions Nucleation and Crystallization in Liquids and Glasses, edited by M.C. Weinberg, The American Ceramic Society, Ohio Vol. 30, 267 – 276 (1993).

CHARACTERIZATION OF RESIDUAL STRESSES IN SIC BASED CERAMIC TILES

Brian Munn and Keyu Li
Department of Mechanical Engineering
Oakland University, Rochester, Michigan 48309, USA

James Zheng and Karl Masters
US Army PM Soldier Equipment
15395 John Marshall Highway
Haymarket, VA 20169, USA

ABSTRACT

Recent studies have documented the existence of residual stresses in a variety of engineered ceramic components. Most ceramic components are brittle by nature and any positive (tensile) residual stresses could be detrimental to their performance in service. Due to these recent findings, there is a need to develop practical method(s) of characterizing residual stresses in engineered ceramic components. This study investigates the viability and practicality of the hole drilling method in characterizing residual stresses in monolithic silicon-carbide (SiC) tiles of varying thicknesses. The final result of this study not only confirmed the viability of the hole drilling method, but the existence of positive residual stresses in all four tile configurations tested. The residual stresses were also found to vary in magnitude based on tile thickness. A mathematical relationship has been derived to describe the relationship. The impact positive (tensile) residual stresses could have on component performance has also been discussed in terms of a safety factor.

INTRODUCTION

Today it is well known that common manufacturing processes can induce residual stresses in engineered components. Residual stresses are pre-existing; free of external load and self-balanced.[1] In other words, residual stresses do not induce a net external force on the engineered component. Therefore, any regions of compressive residual stress are counter balanced by other regions of tensile residual stress.[2] These stresses are also considered to be 'static' in nature and remain in place throughout the service life of the engineered component. Furthermore, severe residual stresses are located at or near the component surface and can be either positive (tensile) or negative (compressive) in nature.[3]

It is also common knowledge that compressive stresses at the surface of a component have a positive influence on mechanical performance. In fact, secondary operations such as shot peening have been introduced to intentionally create a negative residual stress on the surface of numerous metal components to improve mechanical performance.[4] Unfortunately, positive or tensile stresses at a material surface are known to have a detrimental effect on mechanical performance. A tensile residual stress can enhance or promote the initiation and growth of cracks at the surface which can ultimately lead to component failure in service. Hard grinding is an example of a secondary process that induces tensile residual stresses that negatively impact component performance.[5]

Until recently, residual stresses were not considered significant in components made from monolithic ceramic materials. This conclusion was based on the following two assumptions:

1. The lack of secondary manufacturing processes for ceramic components such as shot peening and grinding.
2. Most ceramics have a very high elastic modulus rendering any residual stresses insignificant.

However, recent studies have shown that residual stresses can be induced through mechanical means such as grinding and lapping in ceramics. These machined induced residual stresses were found to be tensile in nature with magnitudes large enough to impact component performance in service.[6]

Another means of inducing residual stresses in engineered ceramics is through the development of thermal strains. Thermal strains arise as a result of thermal expansion anisotropy and crystallographic misorientation across grain boundaries during the cooling stage of the sintering operation.[7,8] As a result of these recent findings, it has become increasing important to determine the existence and to characterize the type of residual stresses present in engineered components made of high grade ceramic materials.

There are several analytical techniques for measuring residual stresses in engineered components. For metals, there is an extensive amount of published data on each analytical technique. Unfortunately, for engineered ceramic components the published data is somewhat limited. Pfeiffer and Hollstein[9,10] were the first to publish residual stress data on surfaced machined silicon-nitride (Si_3N_4) using X-ray diffraction (XRD) techniques. Their results showed that the hard contact created by grinding and lapping created positive residual stresses at the surface of Si_3N_4 components. Pfeiffer followed up his initial findings by applying XRD to measure surface residual stresses in a Si_3N_4 substrate that had been subjected to a shot peening process.[11]

There is a lack of published research on the characterization of residual stresses in monolithic silicon-carbide (SiC) components. This is primarily due to the limited number of current applications for components made from monolithic SiC.[12] However, due to improvements in near-net-shape technology there has been a renewed interest in the use of monolithic SiC components. As a result, there is a need to determine practical methods for characterizing residual stresses in SiC components. The following study investigates the viability and practicality of applying resistance strain gages and the hole drilling method to characterize residual stresses in monolithic SiC plates (tiles).

THEORY OF THE EXPERIMENTAL APPROACH

Residual stresses are self-equilibrated and can exist free of any applied service load. To measure residual stresses with a strain gage, it becomes necessary to relieve or unload the material to release residual stresses after the strain gage has been applied to the specimen surface. The simplest and most common technique to relieve stresses is via the hole drilling method. For this particular application, a rectangular strain gage rosette (model #CEA-XX-062UM-120) was used. This type of gage has the configuration shown in Figure 1, with a total of three grids having the second and third grids angularly displaced from the first grid by 45° and 90° respectively. The grids are fully encapsulated with large copper-coated soldering tabs and special trim alignment marks also shown in Figure 1. The trim line is spaced 0.068 in (1.73 mm) from where the hole is to be drilled through the gage.

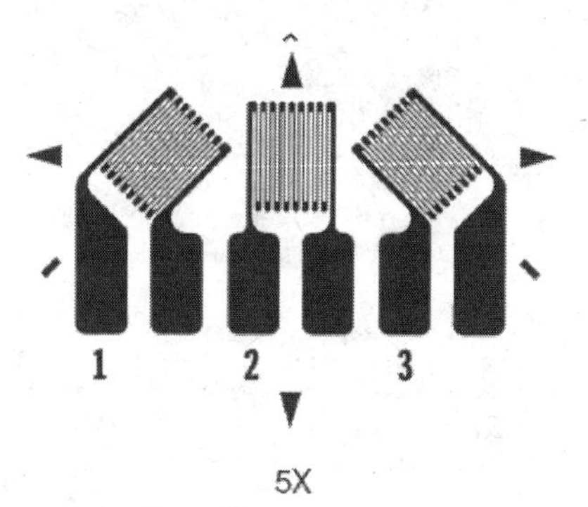

Figure 1. Basic configuration of a 45° rectangular strain gage rosette

A hole drilled of radius, r in a stressed material will release the residual stresses and produce a localized change in strain around the hole. If these strains are uniform with depth, then the strain magnitude can be determined by the following equation[13]:

$$\varepsilon = \frac{(1+\upsilon)\bar{a}}{2E}(\sigma_{max} + \sigma_{min}) + \frac{\bar{b}}{2E}(\sigma_{max} - \sigma_{min})\cos 2\beta \tag{1}$$

where ε is the measured strain, σ_{max} and σ_{min} are the principle stresses, β is the angle to maximum principle stress direction, E is Young's modulus, υ is Poisson's ratio, and $\bar{a}$, $\bar{b}$ are calibration constants.

However, in most practical applications, the released stresses are not uniform with depth. In this case, the measurement procedure involves drilling a series of small-depth increments and measuring strains at each depth increment. This incremental change in strain can be determined by a modification to equation (1):

$$\{\varepsilon\} = \frac{(1+\upsilon)}{2E}[\bar{a}]\{(\sigma_{max} + \sigma_{min})\} + \frac{1}{2E}[\bar{b}]\{(\sigma_{max} - \sigma_{min})\cos 2\beta\} \tag{2}$$

where the braces represent vector quantities and the square brackets matrices. The vector, $\{\varepsilon\}$ is the set of strains measured at each incremental hole depth with $\{\sigma_{max}, \sigma_{min}\}$ the principle stresses at each depth increment. The matrices $[\bar{a}]$ and $[\bar{b}]$ are by definition lower triangular. In order to calculate the stresses at each depth increment, the matrices in equation (3) must be solved for each of the three strain gages in the rosette. The solution to the matrices will be discussed in detail in a later section of this paper. By measuring the strain change due to the removal of material a practical and efficient method to back calculate residual stresses has been provided.

From a theoretical perspective, the hole drilling method has one main drawback that can impact accuracy of strains measured and subsequent calculations of residual stresses. As discussed in the previous section, the material must be linearly elastic in the range that the residual stresses are measured. All applied equations are based on elastic theory with only elastic parameters inputted into the equations. As a result, the magnitude of the residual stresses must be less than one-half of the material yield strength to ensure no localized plasticity has occurred.

From an applications perspective there is a limitation that involves residual stress uniformity through the material thickness. Variations in residual stresses through the thickness have often been observed in practical applications.[14] There are also some physical limitations to this method such as measurements must be taken at or near room temperature, a smooth surface is required for strain gage application and the technique is limited to near surface measurements.

EXPERIMENTAL PROCEDURE

The initial work involved developing and optimizing the equipment and techniques essential to the operation of a practical measurement system. In other words, a methodology had to be developed to obtain consistent results from measurement to measurement. In particular, drilling a clean hole with no cracks or chipping needed to be achieved in order to obtain accurate strain measurements. The details of this effort will be outlined next.

Four squared/rectangular plate specimens with different thickness were received for this study. Dimensional characteristics are provided in Figure 2 for the as-received plates.

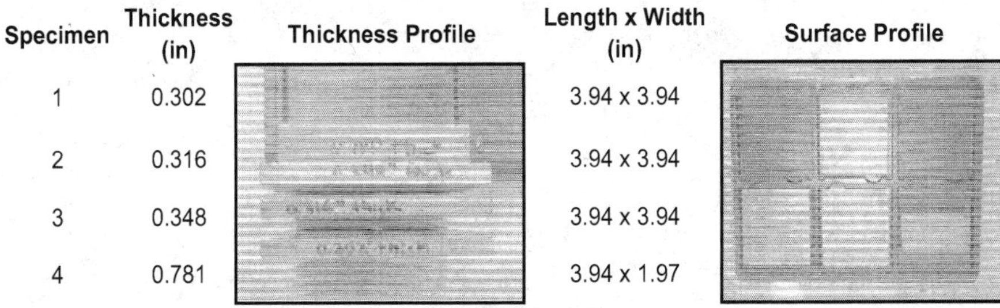

Figure 2. As-received tiles.

The samples were reported as alpha silicon carbide, having been produced via a pressureless sintering process with submicron SiC powder. The result is a self-bonded, fine grain (less than 10 microns) microstructure with a theoretical density greater than 98%. In addition, no secondary operations such as grinding or polishing had been performed on these specimens prior to receipt. The provided physical property data is listed in Table 1 and applies to all four as-received specimens.

Table 1: Physical/Mechanical Properties

Property	Units	Typical Value
Composition	-	SiC
Poisson Ratio	-	0.14
Flexural Strength (3pt) Flexural Strength (4 pt)	MPa	550 380
Compression	MPa	3900
Modulus	GPa	410

To confirm the chemical composition, a Scanning Electron Microscope (SEM) with elemental analysis capability was used to examine the substrate surface. Energy dispersive X-ray spectroscopy (EDS) is an analytical technique used for the elemental analysis of any given sample. As a type of spectroscopy, this technique relies on the interactions between electromagnetic radiation and the sample of interest, analyzing x-rays emitted by the specimen in response to being hit with charged particles. The charged particles in this case were a high beam of electrons aimed at the SiC specimen. The number and energy of the X-rays emitted from the SiC specimen were measured by an energy dispersive spectrometer. Each element emits its own characteristic X-ray signature when bombarded by charged particles. The spectrometer measures the individual counts and wavelength (KeV) of the X-rays being emitted from the SiC specimen. The EDS analysis result shows only two strong (visible peaks in the elemental spectrum which are Silicon, Si and Carbon, C respectively.

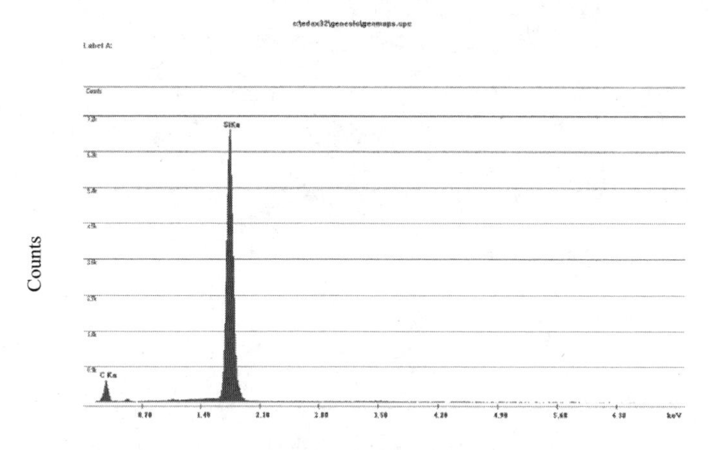

Wavelength (KeV)
Figure 3. Elemental analysis of SiC substrate.

The SEM was also used to examine microstructure characteristics as shown in Figures 4 and 5 respectively. At a magnification of x700 a fine grained, self-bonded SiC was observed with intermittent porosity (dark spots) throughout the microstructure.

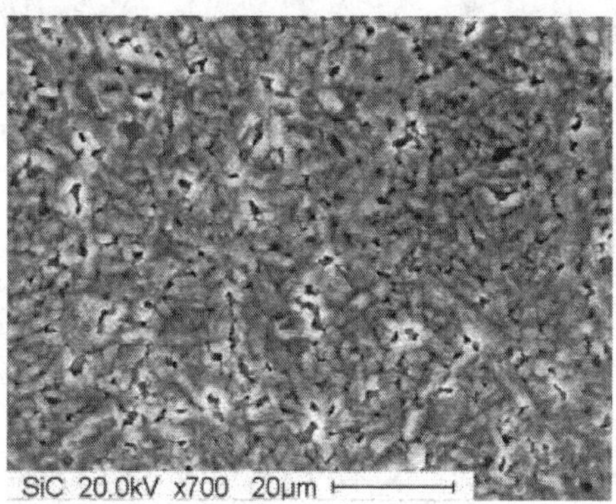

Figure 4. SEM image (x700) showing fine particles and porosity on the SiC tile surface.

The SiC particulate size at x3000 appears to be on the order of several microns (as specified) with porosity size at the same order of magnitude. The particulates are also arranged randomly with no observable directional orientation. In general, the microstructural characteristics were typical of those produced by pressureless sintering of a submicron sized powder.

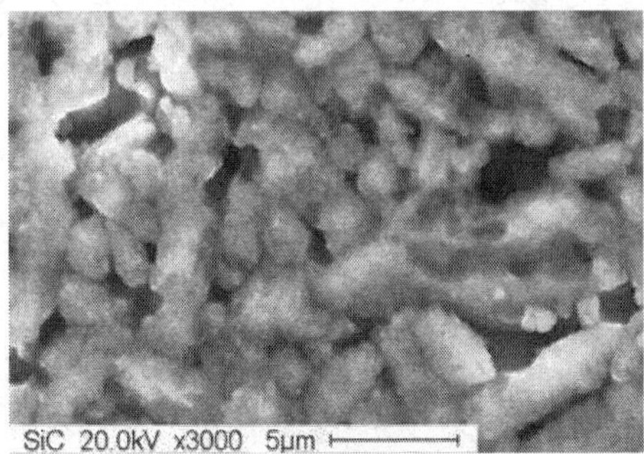

Figure 5. SEM image (x3500) of the surface on the SiC tile.

The four specimens were marked on a side with thickness values, e.g., 0.302 inches, 0.316 inches, 0.348 inches and 0.781 inches respectively. All samples were square except for one which was rectangular. The measurement location was selected at the center of each sample. In specimens this small, the most likely location for residual stresses would be at the center of each plate.

An epoxy-backed, etched-foil strain gage (model #CEA-XX-062UM-120) was used. A good gage-to-ceramic bond was also required since the bonding material shear strength must be able to withstand the high stress gradients associated with the hole drilling process. Type M-200 cement was applied uniformly across the entire gage area to ensure no foil peeling or the formation of air bubbles underneath the foil. The gage assemblies were then held for several minutes, at room temperature, under a light thumb-pressure of at least 5 psi.

Hole properties such as location and alignment were controlled through the development of hole-milling fixturing. The hole milling set-up with fixturing is shown in Figure 6.

Figure 6. Hole drilling set-up.

The set-up in Figure 6 provided the capability to produce a straight untapered hole that was perpendicular to the ceramic surface. A hole as small as 0.062 inches in diameter was to be milled into the center of the strain gage rosette with the allowed tolerance of the hole being +0.0005 and -0.0000 inches. The hole cutting was to remove material from the bottom of the hole at increments of 0.005 inches without disturbing the hole wall. The drill bit was to produce a wall and bottom surface that were smooth, free of blemishes, cracks or tears.

The milling unit, included the milling cutter (drill bit), flexible coupling that provided a positive drill force, and a micrometer for incremental measurements. A hand drill was centered on the specimen by the milling fixture (see Figure 6). Clearance was provided beneath the milling fixture for strain gage leads that were hooked into strain indicators. With the milling device aligned over the center of the strain gage, the drill bit was carefully bored through the gage foil before a strain measurement was taken. An increment of hole was then bored into the ceramic by hand with a high speed air-turbine drill. The drill bit was removed from the hole and the strain indicators allowed to stabilize before a strain reading was taken. This sequence continued until the target depth for the hole was reached and, at this point, all milling stopped. The wires connecting the gages were cut after each test was completed as shown in Figure 7. Six sets of strain readings were taken on six steps of drilling with each strain gage rosette. The error range of the strain data is "±1" micro-strains as measured by the strain indicator.

Figure 7. Strain gage rosette applied at center of each plate.

Calculating Residual Stresses from Strain Measurements

The stains measured from the hole drilling method must now be used to back calculate stresses. This was accomplished through the use of a Matlab program specifically developed for this purpose. The Matlab program takes the released strains measured by the hole drilling method and uses them to calculate stresses by applying the following equation.[15]

$$\begin{bmatrix} \sigma_1 \\ \tau_{13} \\ \sigma_3 \end{bmatrix} = \begin{bmatrix} A+B, 0, A-B \\ A, -2B, A \\ A-B, 0, A+B \end{bmatrix}^{-1} \begin{bmatrix} \varepsilon_1 \\ \varepsilon_2 \\ \varepsilon_3 \end{bmatrix} \tag{3}$$

where the relieved strains ε_1, ε_2, and ε_3 are measured by the three strain gages 1, 2 and 3, the residual stress components σ_1 and σ_3 are normal stresses in the direction of gages 1 and 3 respectively. The shear stress component, τ_{13} is in the direction of the 1 and 3 coordinate system. The residual stresses in the multiple layers were calculated by using the measured strains from each layer and implementing equation (3) to calculate the stress at each incremental depth.[16,17,18]

As shown in Equation (3) the calculation from strain to stress also required a coefficient matrix (CM). The coefficient matrix is made up of two calibration constants A and B. The constants A and B include Young's modulus, E and Poisson's ratio, υ respectively. These coefficients can be calculated using a calibration test, elasticity theory, or finite element analysis, and are recorded in numerous published resources, e.g., references.[19,20] The directional residual stresses σ_1, σ_3 and τ_{13} were solved from equation (3). The maximum and minimum principal stresses, σ_{max} and σ_{min} respectively can then be calculated from the residual stress results σ_1, σ_3 and τ_{13} by using the Mohr Circle method.

TEST RESULTS

The mechanical properties used in the stress analysis were picked out of the mechanical property data listed in Table 1 (E = 410 GPa and υ=0.14) Additional information required was the mean diameter of the strain gage, 0.20 inches (5.13 mm) and the diameter of the hole drilled i.e. 0.063 inches (1.6 mm). These parameters were all inputted into the Matlab program to obtain residual stress results from the measured strains collected from the hole drilling test.

The thinnest sample at a thickness of 0.302 inches was tested first, followed by the sample at a thickness of 0.306 inches, in order of thickness with the 0.718 inch thick plate tested last. Because the SiC material is extremely hard, a normal carbide drilling bit failed to cut into the substrate cleanly. As a result of this set back, a specially designed diamond coated drill bit was used to perform each hole drilling operation. Even with these special bits, one diamond drilling bit per hole was needed. Special care was also given to monitoring the drilling speed to avoid generating excessive heat that could induce residual stress fluctuations. The boring into the SiC substrate itself yielded a high quality hole. The hole walls and bottom had no evidence of cracks or chipping and no other types of imperfections were observed. The result of applying a diamond coated drill bit with a high performance air-turbine drill produced the desired results as shown in Figure 8.

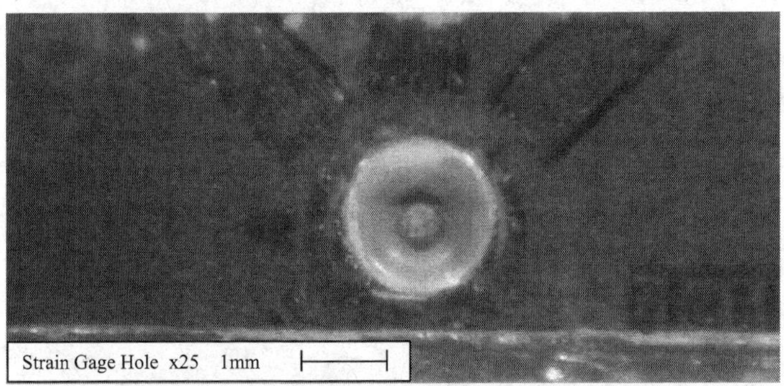

Figure 8. Example of a high quality hole bored into the substrate.

Measured Strain Results

Table 2 is a comparison of the measured strains in all four plate thicknesses of 0.302 inch, 0.316 inch, 0.348 inch and 0.781 inch. The unit of strain measurement is micro-strains i.e. μm.

Table 2: Strain data comparison from all four plates

Thickness (inch)	0.302			0.316			0.348			0.781		
Stain (με) / Depth(mm)	ε_1	ε_2	ε_3	ε_1	ε_2	ε_3	ε_1	ε_2	ε_3	ε_1	ε_2	ε_3
0.127	-5	-5	-4	-3	-4	-4	-4	-5	-3	-2	-1	0
0.191	-4	-3	-3	-3	-5	-5	-4	-4	-2	-2	-1	-1
0.254	-5	-4	-3	-4	-6	-6	-4	-4	-2	-2	-1	-1
0.381	-8	-6	-6	-6	-7	-6	-5	-6	-3	-4	-4	-2
0.508	-10	-9	-8	-7	-9	-7	-6	-7	-4	-3	-2	-1
0.762	-9	-7	-6	-8	-11	-9	-9	-10	-7	-6	-4	-3

The measured strains in each plate, at all six increments of drilling depth, are also compared in Figure 9. The dark diamond points are for the 0.302 inch thick plate, the squared points are for the 0.316 inch thick plate, the triangular points are for the 0.348 inch thick plate and the cross points are for the 0.781 inch thick plate. On the plot, the solid lines fit the data points of the strain ε_1 measured by gage 1, the "dot-dash" lines fit the data points of the strain ε_2 measured by gage 2, and the "dashed" lines fit the data points of the strain ε_3 measured by gage 3.

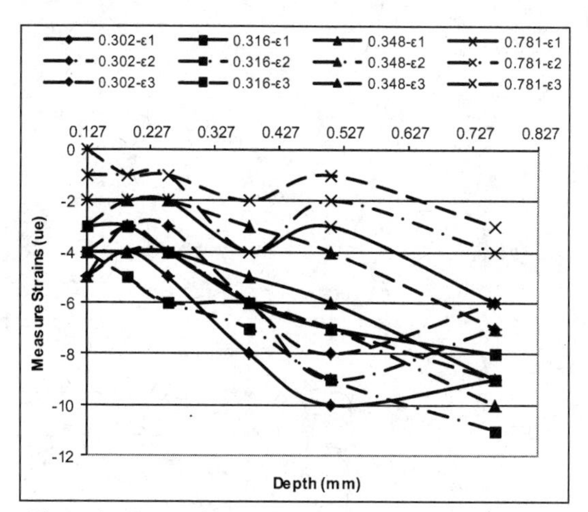

Figure 9. Measured strain for all four plate configurations.

From the strain data collected, the following observations can be made; all strains measured were negative, the smallest negative strains were at the surface and accumulated to larger negative strains as the depth increased from the surface. This indicates that tensile stresses pre-existed in the material which was removed to cause compressive strains around the hole drilled.

Calculated Residual Stresses

The measured strains from each plate were then inputted into the Matlab program to determine the directional and principle stresses at each incremental drill depth. The calculated principal stresses (σ_{max} and σ_{min}) for all four plates are compared in Table 3 and Figure 10 versus depth.

Table 3. Stress data comparisons for all four plates

Stresses	σ_{max} (MPa)				σ_{min} (MPa)			
Thickness (inch) / Depth (mm)	0.302	0.316	0.348	0.781	0.302	0.316	0.348	0.781
0.127	148.7998	118.5959	134.5187	48.4217	123.0361	92.8321	76.9092	11.9863
0.191	88.5256	109.5434	87.0408	43.5204	68.9926	70.4773	47.9747	23.9874
0.254	58.2031	74.0163	48.7588	24.3794	42.8271	52.2713	27.0138	13.5069
0.381	61.5492	51.7659	41.8376	30.1511	48.3443	42.4285	20.9587	16.9461
0.508	54.9913	46.8449	36.1522	14.8128	48.3249	33.5121	21.2457	8.1464
0.762	33.5261	39.6651	36.9718	21.509	26.5594	28.4317	27.1194	14.5423

As before, the dark diamond points are for the 0.302 inch plate, the square points are for the 0.316 inch plate, the triangular points are for the 0.348 inch plate and the cross points are for the 0.781 inch thick plate. The solid line fitting is for the maximum principal stress and the "dotted" line fitting is for the minimum principal stress.

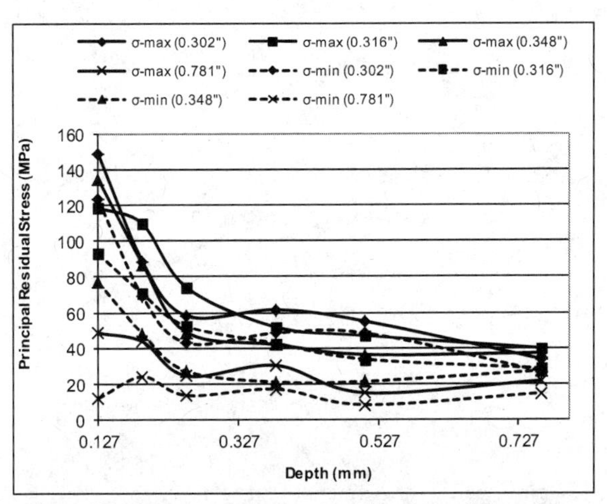

Figure 10. Stress comparisons among data from all four plates.

The calculated stresses in all four plates (Table 3 and Figure 10) had positive principal stresses. This is indicative that tensile stresses existed in the sub-surface of each specimen. These tensile stresses were at or near the surface of each specimen and decreased with depth. The largest principal stress of 149 MPa was at the surface of the 0.302 inch thick specimen. While the smallest principal stress of about 10 MPa was at the surface of the 0.781 inch thick specimen. Comparing the maximum and minimum principal stresses in each of the three square plates showed values relatively close in magnitude. This is an indication that an equal biaxial stress state exists in each one of the square plates. The rectangular plate did not exhibit this characteristic. The maximum principal stresses in the four specimens are further compared in Figure 11. This was done to better understand any trends in the maximum principal stresses. A thicker solid trend line was plotted for each data set using a "power function" in Microsoft Excel.

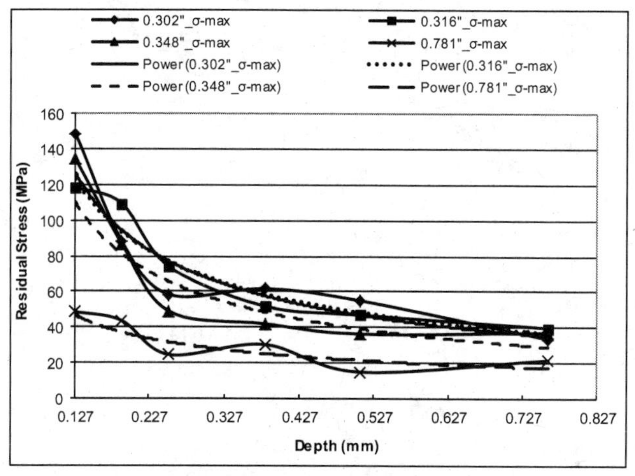

Figure 11. Maximum principal stress comparisons.

Figure 12 shows the comparison of the minimum principal stresses (dotted line configurations from Figure 11). The trend lines are plotted again using the "power" function to fit to the data. Comparing the plotted trend lines showed principal stresses went from a high at or near the surface to a low at the farthest point measured from the surface. The stress values fluctuate around the "trend" lines for all four specimens which is indicative of a stress distribution that is non-uniform in nature. The stress level was at its highest in the 0.302 inch thick plate, followed in order by the 0.316 inch plate, 0.348 inch plate and the 0.718 inch plate respectively. In addition, there was some overlap of the maximum principal stresses between the 0.302 inch and 0.316 inch thick plates. This would be expected since these two plates are closest in thickness and the accuracy in stress measurements has since been determined to be on the order of ±20 MPa.

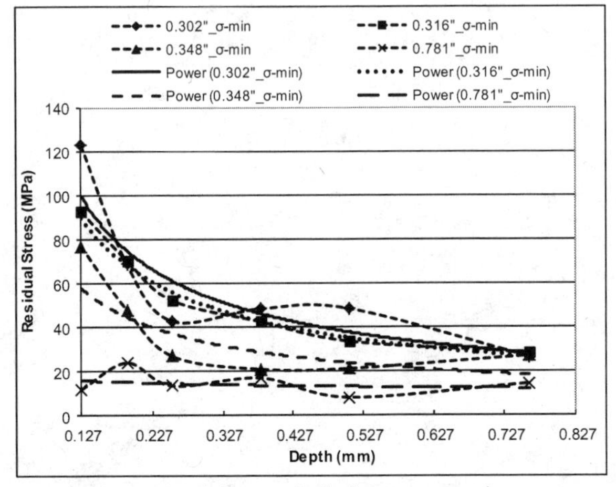

Figure 12. Minimum principal stress comparisons.

A total of eight fitted lines to the power function were made and plotted in Figures 11 and 12. The data for the eight power functions are shown in Table 4. The eight pairs of parameters for the power coefficient, K and power exponent, n were also determined for the maximum and minimum principal stresses and are shown in Table 4. By applying the equation for the power function, the stress value at any depth from the surface can then be estimated using equation (4).

$$\sigma = K \times (Depth)^n \tag{4}$$

The power function parameters and estimated stresses at a drill depth of 0.005 inches (0.127 mm) are listed in Table 4.

Table 4. Power function fitting of the maximum and minimum stresses

Power Fitting					
$\sigma = K \times (Depth)^n$					
Thickness (inch)		0.302	0.316	0.348	0.781
σ_{max}	**K(MPa/mm^n)**	28.8	30.5	23.6	14.6
	n	-0.718	-0.678	-0.747	-0.559
	σ_{max} (MPa) Depth=0.127mm	126.7	123.5	109.9	46.4
σ_{min}	**K(MPa/mm^n)**	23.4	22.2	15.5	11.9
	n	-0.703	-0.677	-0.639	-0.143
	σ_{min} (MPa) Depth=0.127mm	99.8	89.8	57.8	16.0

The stress values calculated in Table 4 at a depth of 0.005 inches (0.127 mm) are plotted as a function of thickness in Figure 13. Power function fitting is used in Figure 13 to fit a set of four data points from each of the four plates. The open-solid black circles are for stress values of the maximum principal stresses calculated in Table 4. The black solid line is the power function fitting line for the maximum stresses. The open dashed circles are the stress values for the minimum stresses calculated in Table 4. The black dotted line is the power function fitting for the minimum stresses.

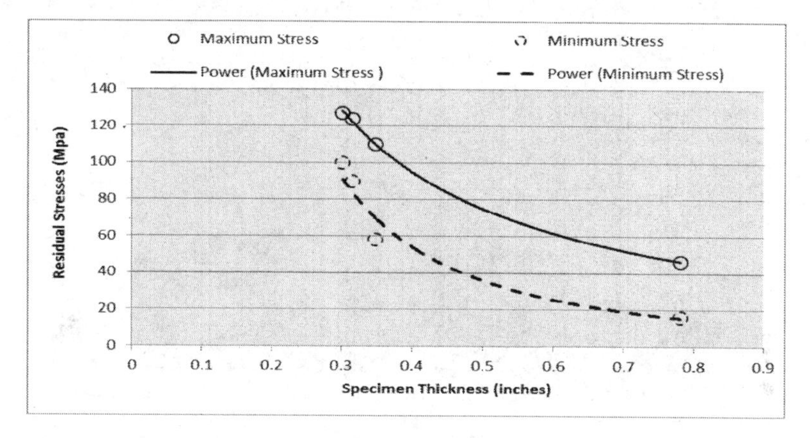

Figure 13. Maximum and minimum residual stress versus specimen thickness.

From Figure 13, both maximum and minimum stress values decreased as the thickness of the plate increased. In addition, the power function parameters shown in Table 5 were successfully used to plot the data shown in Figure 13. Using the power function in Table 5 approximate stress values at a depth of 0.005 inches (0.127 mm) for any given thickness value can be estimated.

Table 5. Power fitting functions for the stresses vs. thicknesses

Power Fitting $c = K \times thickness^n$		
	σ_{max}	σ_{min}
K (MPa/in^n)	35.6	9.9
N	-1.068	-1.854

The power function trend suggests that the stress values decrease rapidly only a short distance from the surface. This trend can be seen by the steepness of the tangent slope on each curve. After the thickness reaches about 0.02 inches (0.5 mm), the maximum stress value fall below 80 MPa and the change in the stress values tend to moderate.

DISCUSSION

Positive residual stresses were found at the center in all four plates. Positive stresses are tensile stresses which tend to pull or stretch the material. This type of action enhances the initiation and formation of surface cracks especially in brittle material. The maximum tensile stresses measured were near or at the surface of each plate. As a result, these positive stresses could negatively impact the in service performance of the plates. The maximum tensile stress found near the surface of the plates was approximately +149 MPa.

The severest residual stresses and stress fluctuations were in the 0.302 inch thick specimen. The least severe residual stresses and fluctuations were in the 0.781 inch thick specimen. In general as the thickness increased, the severity of fluctuations in the residual stress decreased. There also appears to be a decrease in the magnitude of residual stresses when the thickness of the plate increases as shown in Figures 11 and 12. In other words, there appears to be an inverse relationship between the magnitude of residual stress and specimen thickness, e.g. a power function as shown in Table 5. To examine this trend in further detail, a power fitting method was applied to study how the stress value changes as a function of plate thickness as shown in Figure 13.

As stated earlier, the highest residual stress was at or near the surface of each plate. To quantify these values further, Table 6 below compares the highest residual stress values to the 3 point and 4 point flexural strengths at room temperature. For the monolithic SiC material, Table 1 has the 3 and 4 point flexural strengths at 550 MPa and 380 MPa respectively.

Table 6. Comparison of the highest residual stress in each plate with the flexural strength

Plate thickness (inches)	Maximum Stress at Surface (MPa)	Percentage in 3 pt Flexural Strength (550MPa)	Percentage in 4 pt Flexural Strength (380Mpa)
0.302	149	27%	39%
0.316	119	22%	31%
0.348	135	24%	36%
0.781	48	8.7%	13%

The worst case comparison is between the 4 pt flexural strength and the maximum residual stress measured in the 0.302 inch thick plate. In this case the residual stress value of 149 MPa is about 39% of the flexural strength. The best case comparison can be found in the 0.781 inch plate, where the percentage drops significantly to approximately 13% of the 4-point flexural strength. Overall the SiC plates had some significantly large tensile residual stresses up to 39% of the 4-point flexural strength of the material.

The flexural strength is the maximum stress that the ceramic plate can sustain under a bending situation which will be discussed in detail. Any combination, of inherent and service tensile stresses, greater than the flexural strength, will cause failure of the plates by a crack initiation-crack propagation process. Normally, the operational stress should be significantly less than the flexural stress. In other words, a built in safety factor (SF) is required to avoid any type of fracture or failure.

The safety factor is used to provide a design margin over the theoretical design capacity to allow for uncertainty in the design process. The value of the safety factor is proportional to the lack in confidence of the design and manufacturing processes. The simplest interpretation of the safety factor is as follows:

$$s_w = S_m / f_s \tag{5}$$

where s_w is the allowable working stress, S_m is the material strength and f_s is the factor of safety. The selection of the appropriate factor of safety to be used in design of components is a compromise between the additional cost and weight for a benefit of increased safety and/or reliability.

For example, taking $f_s = 2$ (a typical engineering design), the maximum design stress should be less than 380 MPa divided by a factor 2 which equals a maximum service load of 190 MPa. Considering the highest measured residual stress in one plate (0.302 inch) was 149 MPa, the maximum bending stress that the plate can sustain in service is 190 MPA – 149 MPA = 51 MPA. In other words, the large tensile residual stress significantly reduces the in service load a component can withstand.

CONCLUSIONS

The residual stresses found in the sub-surfaces of the four specimens were all tensile in nature. A tensile stress at the surface of a ceramic could lead to a failure mode involving the formation of cracks and the rapid growth of these cracks under any type of service load. The highest residual stresses were found at or very near the surface of all measured specimens. Therefore, any cracks that form will initiate at the surface. Depending on the size of the crack, visual inspection may be able to detect them.

The highest residual stress was 149 MPa in the thinnest specimen, the 0.302 inch thick plate. The magnitude of this stress relative to the 4-point flexural strength was determined to be 39%. This large a residual, tensile stress will significantly lower the allowable working stress of the SiC tile in-service.

There was also a trend for the residual tensile stresses to decrease with a thickness increase in the specimens. The lowest magnitude of a surface residual stress was in the thickest 0.781 inch plate. This was only 13% of the 4-point flexural strength. The relationship between the stress and thickness is studied and an equation is derived based on data fitting methods. The equation trend suggests that the stress values decrease rapidly at initial thickness deduction when the tangent slope of the curve is steep. After the thickness reaches 0.5 inches, the maximum residual stress values fall below 80 MPa. At this point the residual stress values tend to moderate. Unfortunately, an increase of thickness adds weight to the plate which may not be desirable for the current product design. Therefore, an optimization of both stress and weight must be considered in order to achieve best performance and product quality.

ACKNOWLEDGEMENT

The authors wish to acknowledge the assistance of R. Petrach with experimental work and R. She with experimental data analysis. The authors also wish to thank the US Army (PM Soldier Equipment) for their funding and support in developing this paper.

REFERENCES

[1]Dally and Riley, *Experimental stress analysis*, College House Enterprises, Fourth edition, 2005.
[2]Totten, G., Howes, M. and Inoue, T., Editors (2002) *Handbook of Residual Stress and Deformation of Steel.* ASM International, Materials Park, Ohio.
[3]McClung, R., C., 2007, *A literature survey on the stability and significance of residual stresses during fatigue*, Fatigue Fract Engng Mater Struct, vol. 30, 173–205 p.
[4]Herzog, R., *Auswirkungen bearbeitungsbedingter Randschichteigenschaften auf das Schwingungrißkorrosionsverhalten von* CK45 und X35CrMo 17, Shaker Verlag, Aachen, 1998.
[5]Sollich, A., *Verbesserung des Dauerschwingverhaltens hochfester Stahle durch gezielte Eigenspannungserzeugung*, Fortschrittsberichte VDI, Reihe 5, Nr. 376, VDI-Verlag, Dusseldorf, 1994.
[6]Jahanmir, S., Ramulu, M., Koshy, P., 1999, Machining of Ceramics and Composites, CRC Publishing, 704 p.
[7]www.osti.gov/bridge/servlets/purl/7262-MHB5hG/.../7262.pdf.
[8]Pan M., Green, D. J., and Hellmann, J., R., 1997, *Influence of Crystal Anisotropy on Residual Stresses in Ceramic Composites*, Scripta Mateaialia, Vol. 36, No. 10, 1095-1 100 p.
[9]Pfeiffer, W., Hollstein, T., 1996, *Characterization and Assessment of Machined Ceramic Surfaces*, 2nd International Conference on Machining Advanced Materials (MAM), VDI-Verlag, Dusseldorf, Germany, VDI Bulletin #1276, 587-602 p.
[10]Pfeiffer, W., Hollstein, T., and Sommer, E., 1995, Strength properties of surface-machined components of structural ceramics, Fracture Mechanics, Vol. 25, 19-30 p.
[11]Pfeiffer, W. and Rombach, M., 1997, *Residual stresses and damage in ceramics due to contact loading*, Proceedings of the ICRS5, Linkopping, Sweden.

[12]Sanjay, A., and Venkateswara R., 2008, *Experimental investigation of surface/subsurface damage formation and material removal mechanisms in SiC grinding*, International Journal of Machine Tools & Manufacture, Vol. 48, 698–710 p.
[13]Schajer, G. S. and Tootoonian, M., 1997, *A New Rosette Design for More Reliable Hole-drilling Residual Stress Measurements*, Experimental Mechanics, Vol. 37 (3), 299-306 p.
[14]Beghini, M., and Bertini, L., 1998, *Recent Advances in the Hole Drilling Method for Residual Stress Measurement*, Journal of Materials Engineering and Performance, Vol. 7 (2), 163-172 p.
[15]Schajer, G. S., 1988, *Measurement of Non-Uniform Residual Stresses Using the Hole Drilling Method*, Part I—Stress Calculation Procedures. ASME Journal of Engineering Materials and Technology, 110, No. 4, 318–342 p.
[16]Schajer, G. S., 1988, *Measurement of Non-Uniform Residual Stresses Using the Hole Drilling Method*, Part II—Practical Application of the Integral Method. ASME Journal of Engineering Materials and Technology, 110, No. 4, 344–349 p.
[17]Wern, H., 1995, *Measurement of the non-uniform residual stresses using the hole drilling method, a new integral formalism*, Strain, Vol.31-2, 63-68 p.
[18]Stefanescu, D. Truman, C.E. Smith, D.J. etc., 2006, *Improvements in Residual Stress Measurement by the Incremental Centre Hole Drilling Technique*, Experimental Mechnics, Vol. 46, page 417-427.
[19]Aoh, Jong-Ning. and Wei, Chung-Sheng, 2002, *On the Improvement of Calibration Coefficients for Hole-Drilling Integral Method: Part 1—Analysis of Calibration Coefficients Obtained by a 3-D FEM Model*, ASME Journal of Engineering Materials and Technology, Vol. 124.
[20]Aoh, Jong-Ning. and Wei, Chung-Sheng, 2003, *On the Improvement of Calibration Coefficients for Hole-Drilling Integral Method: Part 2— Experimental Validation of Calibration Coefficients*, ASME Journal of Engineering Materials and Technology, Vol. 125.

Microstructural Design for Enhanced Armor Ceramics

MICROSTRUCTURAL DESIGN FOR Si-B_4C-DIAMOND SYSTEM

P. G. Karandikar and S. Wong
M Cubed Technologies, Inc.
Newark, DE, USA

ABSTRACT

Reaction bonded SiC (RBSC) and reaction bonded B_4C (RBBC) materials have been used successfully for armor applications over the last decade. In reaction bonded B_4C, typically three phases exist viz. B_4C, reaction formed SiC, and residual Si. Our previous work has shown that ballistic performance of RBBC increases as residual Si decreases. Finally, the properties of RBBC materials can be further tailored by using diamond reinforcement. Diamond is an extremely effective reinforcement due to its very high elastic modulus (1050 GPa), and hardness (12,000 kg/mm^2). Diamond, however, is heavier (density = 3.52 g/cc) and more expensive, and thus, its content needs to be minimized to keep the armor light and affordable. Also, similar to B_4C, diamond can react with molten Si, forming SiC. Thus, a key question is whether diamond is important as residual reinforcement or as a very efficient SiC former (which reduces residual Si). Due to the four-phase (B_4C, SiC, residual Si, and residual diamond), complex microstructure, and a significant number of potential process variables, a design of experiments (DoE) approach was used to fabricate RBBC-diamond test samples. These were subjected to microstructural (optical, SEM, EDAX, and X-ray diffraction), physical, chemical, and elastic properties characterization. Theoretical modeling was carried out to predict composite density as a function of residual Si (and diamond). Experimental data were compared with predictions to shed further light on the microstructural design.

BACKGROUND

Sintered and hot pressed Al_2O_3, B_4C and SiC have been traditionally used in personnel armor systems. Reaction bonded SiC (RBSC) and reaction bonded B_4C (RBBC) have been reported in the literature as far back as 1940s[1-3]. However, their use in armor systems started in the late 1990s[4-11]. B_4C is the lowest density and highest hardness ceramic typically used for personnel armor. While RBBC offers good performance for currently fielded body armor, there is a continuous desire to obtain the same performance with lighter weight systems.

In the reaction bonding process (silicon-based matrices), good wetting and a highly exothermic reaction between liquid silicon and carbon are utilized to achieve pressure-less infiltration of a powder or fiber preform. This process is given many names such as reaction-bonding, reaction-sintering, self-bonding, and melt infiltration. The reaction bonding process was described in detail by the authors previously[10, 11]. The steps in the process can be briefly described as follows: (1) mixing of B_4C (or SiC) powder and a binder to make a slurry, (2) shaping the slurry by various techniques such as casting, injection molding, pressing etc., (3) drying and carbonizing of binder, (4) green machining, (5) infiltration (reaction bonding) with molten Si (or alloy) above 1410°C in an inert/vacuum atmosphere, and (6) solidification and cooling. During the infiltration step, the carbon in the preform reacts with molten Si forming SiC around the original ceramic particles, bonding them together – hence the term reaction bonding. A typical microstructure of RBBC consists of nominally B_4C particles, reaction formed SiC, and some residual silicon. In reaction bonded SiC, the starting SiC grains stay unreacted through the reaction bonding process. However, as shown previously[10, 11], B_4C grains react with molten silicon (and alloying elements in it) and convert to B-Si-C phases or B-Al-C phases. The original B_4C grain shapes and volumes are mostly maintained. It was also shown by the authors previously[10, 11] that the ballistic performance of RBBC increases as the content of residual silicon decreases.

Thus, a key objective in processing RBBC for ballistic applications is to reduce the residual silicon. Further, additions that can increase elastic modulus and hardness are desirable. Diamond is an extremely effective reinforcement due to its very high elastic modulus (1050 GPa) and hardness (12,000 kg/mm^2). Diamond, however, is heavier (density = 3.52 g/cc) and more expensive then B_4C, and thus, its content needs to be minimized to keep the armor light and affordable. Also, similar to B_4C, diamond can react with molten Si forming SiC. Thus, in processing of the Si-B_4C-diamond system, it is necessary to determine if performance is optimized due to the presence of diamond (prevention of reaction of diamond) or consumption of diamond to minimize residual silicon. This study focuses on development of process-microstructure-property relation in the Si-B_4C-diamond system. The resultant composite has an extremely complex microstructure which includes four phases viz. B_4C, reaction-formed SiC, residual diamond, and residual Si. In addition, as shown previously[10, 11], the B_4C particles can react with molten silicon generating B-Si-C phases. Due to this complexity, a design-of-experiments (DoE) approach[12] was used to identify process and composition parameters to assess their effects on residual diamond and residual Si content and in turn, maximize the ballistic performance. Based on the results of the DoE, a desirable Si-B_4C-diamond system was selected for more detailed processing experiments and characterization. In addition, theoretical calculations were carried out for predicting composite densities and residual silicon by varying the extent of diamond conversion (0 to 100%) to SiC. Finally, experimental data were compared with predictions.

EXPERIMENTAL PROCEDURE

RBBC plates (100 x 100 x 10 mm) were made with different process and composition parameters per a design-of-experiments (DoE) approach. Based on the results of the DoE, a particular material composition was also selected for more detailed processing experiments at various infiltration times. Microstructure and phases in the samples were characterized by optical microscopy, SEM, EDAX, and X-ray diffraction (XRD). Samples containing finer diamond (3 m) could be polished by conventional grinding and polishing techniques. However, samples containing coarser diamond could not be polished due to their extreme hardness. SEM analysis was used to observe fracture surfaces of the samples containing coarse diamond. For XRD, each comminuted sample (WC ball mill) was placed into a standard sample cup and put into a Panalytical X'Pert MPD Pro diffractometer using Cu radiation at 45KV/40mA over the range of 20° - 80° with a step size of 0.0156° and a count time of 500 seconds per step. The phases were then identified by using the Powder Diffraction File published by the International Centre for Diffraction Data. The detected phases were quantified with the aid of Rietveld analysis. Densities (Archimedes principle, ASTM B311), and elastic moduli (ultrasonic pulse-echo, ASTM E494-05) of the resultant materials were also characterized. Residual Si in the RBBC composites was measured by chemical (mineral acid) digestion followed by inductively coupled plasma atomic emission spectroscopy (ICP-AES).

DESIGN OF EXPERIMENTS (DoE)

A design of experiments (DoE) approach[12] was selected to assess the effects of various process parameters and the starting constituents on density, elastic modulus, residual diamond content, and residual Si content. A 2^{4-1} fraction factorial design was selected as shown in Table I. Four variables (process temperature, process time, starting diamond content %, and starting diamond size, were controlled at two levels (high and low). Thus, eight (2^{4-1}) processing experiments were carried out. For these experiments, preforms were made in an identical manner.

Measured properties of the tiles made per the DoE matrix are summarized in Table II. The data in Table II were used to compute the effect of increasing each process parameter on four properties (density, elastic modulus, residual Si content, and residual diamond content). For a given variable, the effect (or change) on a property is calculated by subtracting the average value of the property when the variable was at low setting from the average value of the property when the variable was at high setting. Results of these calculations are shown in Table III. The effect of combined variable A (temperature) and B (time) was also computed.

Table I. Experimental design for RBBC-diamond composites

Levels	2	High (+)	Low (-)		
Variables	4	A Temperature (°C)	B Time (minute)	C Diamond %	D Diamond Size (m)
Level	High (+)	1470	210	10	22
	Low (-)	1400	60	5	3

Table II. Summary of processing parameters and properties of experimental tiles

Expt. No.	Temp. (°C)	Time (min)	Diamond (wt. %)	Diamond Size (m)	Density (g/cc)	Elastic Modulus (GPa)	Residual Si (wt. %)	Residual Diamond (wt. %)
1	1400 (-)	60 (-)	5 (-)	3 (-)	2.588	401	13.84	1.3
2	1470 (+)	60 (-)	5 (-)	22 (+)	2.596	396	14.27	2.2
3	1400 (-)	210 (+)	5 (-)	22 (+)	2.603	397	14.43	2.0
4	1470 (+)	210 (+)	5 (-)	3 (-)	2.618	408	11.43	0.8
5	1400 (-)	60 (-)	10 (+)	22 (+)	2.620	415	10.64	5.1
6	1470 (+)	60 (-)	10 (+)	3 (-)	2.627	425	8.29	2.3
7	1400 (-)	210 (+)	10 (+)	3 (-)	2.628	420	8.61	1.9
8	1470 (+)	210 (+)	10 (+)	22 (+)	2.635	417	12.94	5.4

Table III. Effects of increasing of process variables on properties

Variable Designation	Variable	Effect on Density (%)	Effect on Modulus (%)	Effect on Residual Si Content (%)	Effect on Residual Diamond (%)
A	Temperature	0.8	0.4	-1.2	3.9
B	Time	0.3	0.5	0.8	-7.3
C	Diamond %	4.7	1.0	-25.0	133.3
D	Diamond Size	-1.8	-0.1	24.0	133.3
AB	Temperature & Time	0.9	1.0	-0.4	-3.1

As can be seen from the data in Table III, higher temperature increases density by 0.8%, elastic modulus by 0.4%, reduces residual Si content by 1.2%, and increases residual diamond by 3.9%. As process temperature increases, it is expected that more diamond will react, making more SiC, reducing residual Si. The result on diamond content is some what unexpected. However, it is important to remember that these are small percent effects and are subject to experimental errors. The results show that the biggest effect is produced by two variables viz. diamond content and diamond size. As more diamond is added, Si content is reduced by 25%. This is reasonable as more diamond is available to form SiC and take up space that would have been otherwise occupied by residual silicon. Residual diamond content is also increased. Correspondingly, density and modulus also increased. As diamond size is increased, residual silicon content is significantly increased (24%) and residual diamond is significantly increased. In other words, it is desirable to reduce diamond size to reduce residual silicon content. Thus, the DoE results suggest that higher diamond content and finer diamond are desirable.

The effects of process temperature and time are more subtle. Based on these results, an RBBC + 10% 3 m diamond system was selected for more detailed experiments with respect to process temperature and time.

MICROSTRUCTURE AND COMPOSITION

Optical micrographs of four composites (RBBC + 10% 3 m diamond) made at different processing conditions are shown in Figure 1. In all the microstructures, B_4C, diamond, and residual Si can be easily seen. Reaction formed SiC is slightly harder to distinguish as it appears with a gray shading in between that of B_4C and diamond and has much finer grain size. Figure 3 shows back scattered scanning electron micrographs of a sample from Experiment 6 (RBBC + 10% 3 m diamond, 1470°C, 60 minutes). In the higher magnification micrograph on the right, B_4C grains (dark), diamond particles (dark gray), SiC surrounding diamond particles (light gray), and residual Si (light) can be seen. In the back scattered electron images contrast is derived from the difference in atomic weights.

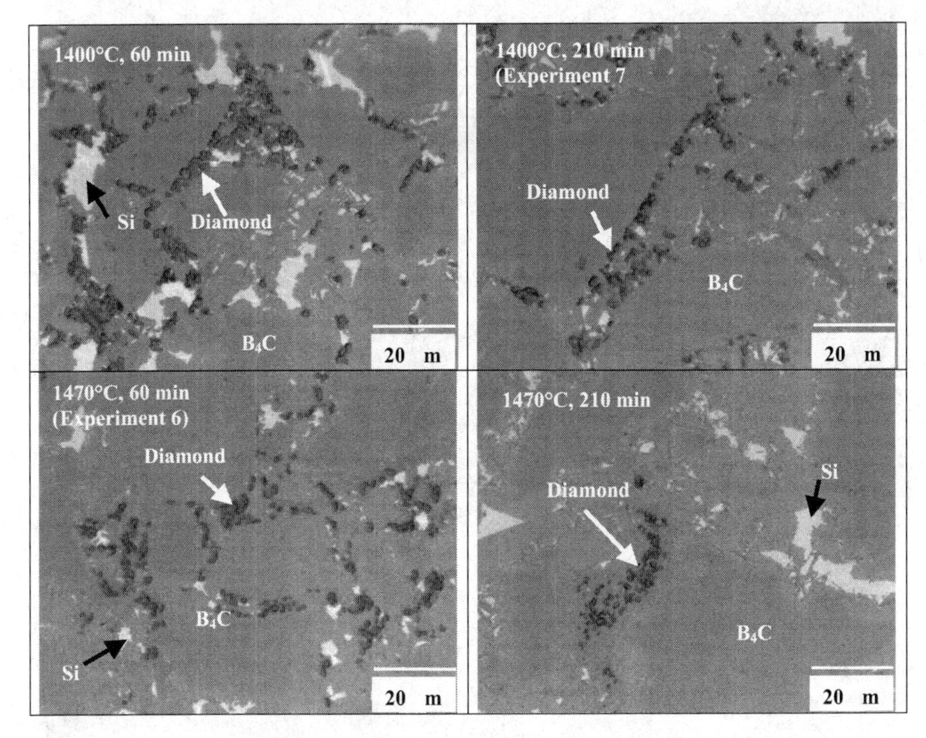

Figure 1. Optical micrographs of RBBC + 10% 3 m diamond composites made at different processing conditions.

Chemical compositional analysis was carried out using energy dispersive analysis of X-rays (EDAX) on several samples. Figure 3 shows a back scattered electron image of a sample from

Experiment 8 (RBBC + 10% 22 m diamond, 1470°C, 210 minutes). Three areas (A, B, C) with different gray shadings in this microstructure were selected for EDAX analysis. The corresponding EDAX patterns and atomic compositions are also shown in Figure 3.

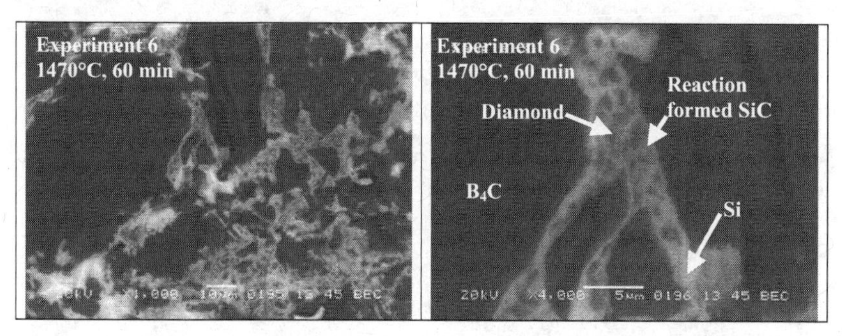

Figure 2. Low and high magnification back scattered electron (elemental contrast) micrographs of a sample from Experiment 6 (RBBC + 10% 3 m diamond, 1470°C, 60 minutes). Fine diamond particles between the B_4C grains are observable. The diamond particles are surrounded by light gray reaction-formed SiC.

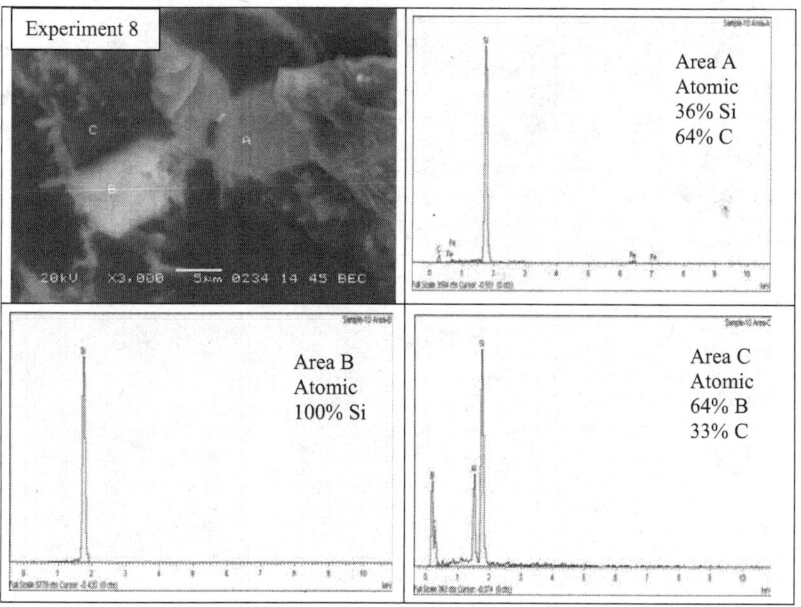

Figure 3. SEM of a sample from Experiment 8 (RBBC + 10% 22 m diamond, 1470°C, 210 minutes) and EDAX compositional analysis.

It is critical to note that X-rays from light elements such as B and C are hard to detect and also have difficulty in reaching the detector as they are absorbed by surrounding heavier elements (in this case, Si). Thus, the peaks of B and C appear to be very small. A correction has to be applied to estimate the atomic content of the light elements to compensate for absorption of the X-rays by surrounding heavier elements. The compositions thus obtained are also listed with each EDAX pattern in Figure 3. Based on the atomic composition, Area A is closer to a SiC composition, Area B is residual Si, and Area C is B_4C.

Analysis on a sample from Experiment 2 (RBBC + 5% 22 m diamond, 1470°C, 60 minutes) presented a region that appeared to be a partially converted diamond particle. This is shown in Figure 4. To assess if that was indeed the case, EDAX analysis was conducted on the darker center and the lighter surrounding area. The EDAX patterns obtained are also shown in Figure 4, along with the atomic compositions. The elemental compositions confirm that the dark center is predominantly carbon (diamond) and the surrounding lighter area has a composition corresponding to SiC.

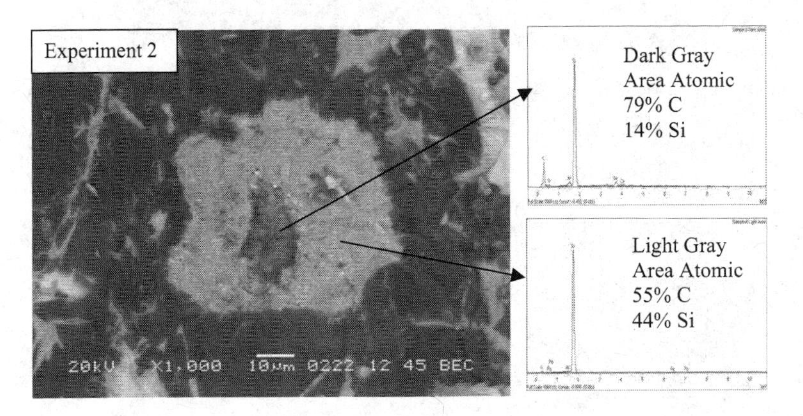

Figure 4. SEM of a sample from Experiment 2 (RBBC + 5% 22 m diamond, 1470°C, 60 minutes) and EDAX compositional analysis. This analysis suggests that this diamond particle was partially converted to SiC.

DETAILED STUDY OF A SELECTED COMPOSITION

Samples of RBBC + 10% 3 m diamond composites were made by varying the process time (60, 120, 240, 480 minutes) at 1400°C. Density, elastic modulus, residual Si and residual diamond were measured for these specimens. Variations of composite densities and elastic moduli as a function of process time are shown in Figure 5. While both composite density and modulus increase as the process time increases; modulus shows a slight decrease at the highest process time. Variation of residual silicon and residual diamond are shown in Figure 6. Both residual diamond and residual silicon decreased as process time increased. Thus, at a given composition (diamond size and starting content), the residual Si could be reduced appreciably by increasing the process time. This is expected as more and more diamond reacts to form higher-volume SiC with increasing process time, which results in lower residual Si.

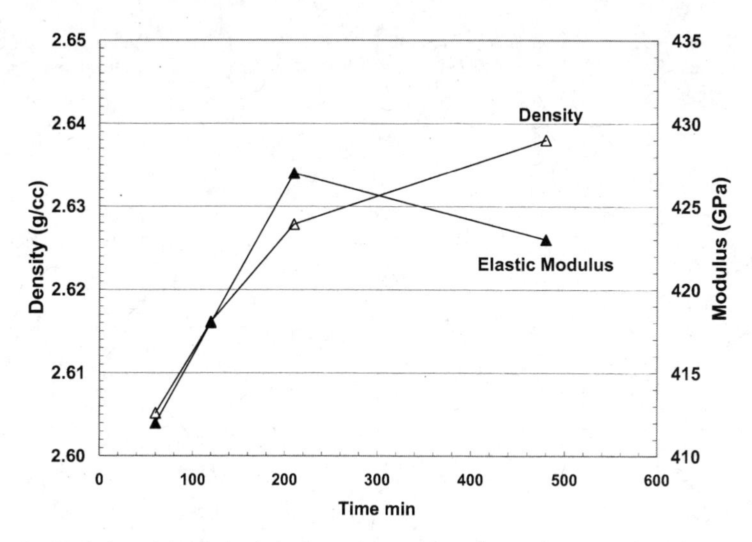

Figure 5. Variation of density and elastic modulus with process time (at 1400°C) for RBBC + 10% 3 m diamond composite.

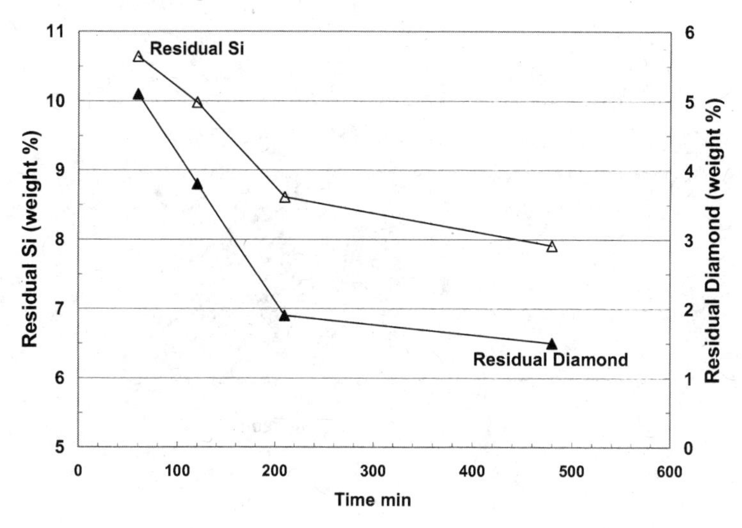

Figure 6. Variation of residual Si and diamond with process time at 1400°C for RBBC + 10% 3 m diamond.

THEORETICAL ANALYSIS AND COMPARISON WITH EXPERIMENTAL DATA

In the reaction bonding process, carbon reacts with molten silicon to form SiC per the following reaction:

$$C + Si \rightarrow SiC \quad (1)$$

Thus, 1 mol of carbon (12.0107g) reacts with 1 mol of silicon (28.0855g) to form 1 mol of SiC (40.0962g). Thus, 12.0107g of carbon forms 40.0962g of SiC. Carbon can be present in many forms such as amorphous (1.6 g/cc), graphitic (2.0 g/cc), and diamond (3.52 g/cc). Using the respective densities, the volume change in reaction 1 can be calculated as shown in Table IV.

Table IV. Volume change in the C → SiC reaction

Form of Carbon	Density (g/cc)	Volume of 1 mol of Carbon (cc)	Volume of 1 mol of SiC (cc)	Volume Magnification
Amorphous	1.60	7.51	12.49	1.66
Graphitic	2.00	6.01	12.49	2.08
Diamond	3.52	3.41	12.49	3.66

Thus, for the objective of minimizing the residual silicon, the denser form of carbon is preferred as it leads to higher volume magnification. Hence, it is desirable to add denser carbon to the preform to convert to higher amount of SiC during the reaction bonding process. For the theoretical calculations of contents of various phases, densities, residual Si and residual diamond present, the unit cell shown in Figure 7 was used.

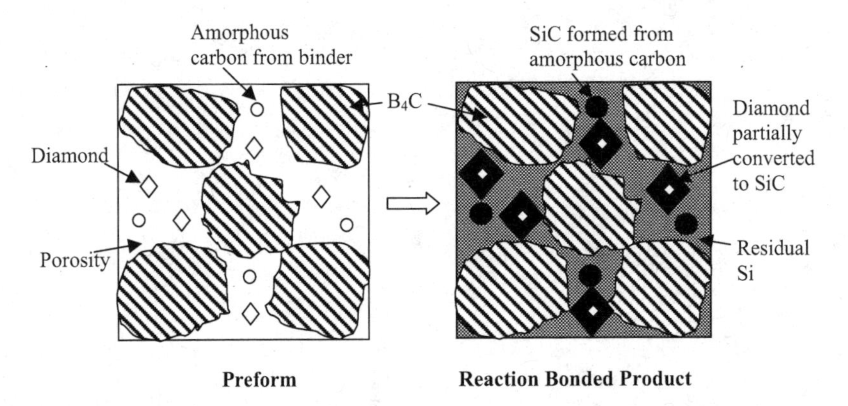

Figure 7. A unit cell for the theoretical calculations for the Si-B_4C-diamond system.

For the theoretical calculations, the initial B_4C packing volume percent in the unit cell was varied from 65% to 85%. The initial diamond weight was 10% of the B_4C weight. Amorphous carbon from the binder was assumed to be fully converted to SiC. Diamond conversion to SiC was varied from 0% to 100%. The B_4C phase was assumed to maintain its original volume fraction. Thus, the final composite has four phases, starting B_4C, reaction-formed SiC, residual Si, and residual diamond. Using this approach, the amounts of reaction-formed SiC, residual diamond, and residual Si were calculated by creating an Excel spreadsheet with appropriate formulas. Using the calculated composition and the individual phase densities, the density of the resultant composite was calculated by the rule-of-mixtures. A plot of calculated density as a function of residual silicon at various initial B_4C packing volume percents is shown in Figure 8. For each predicted line, the low end of density represents 0% diamond conversion, and the high end represents 100% diamond conversion. Experimentally measured data (obtained at various processing conditions but constant initial diamond content) are also shown on this plot as circles.

The theoretical predictions and comparison with experimental data indicate the following:

(1) As expected, as the initial B_4C packing increases, residual silicon decreases.
(2) As the extent of diamond conversion increases, density increases and residual silicon decreases.
(3) Comparison of experimental data with predictions suggests that initial B_4C packing is between 75 and 80%. This is slightly higher than expected and measured (70-75%) using mercury porosimetry. One explanation of this small difference is the fact that the B_4C phase is assumed to be non-reactive in the theoretical calculations. However, it has been shown by the authors[10-11] that some Si-B_4C grain reaction does occur. Also, any residual carbon in B_4C is also converted to SiC (a higher density phase).

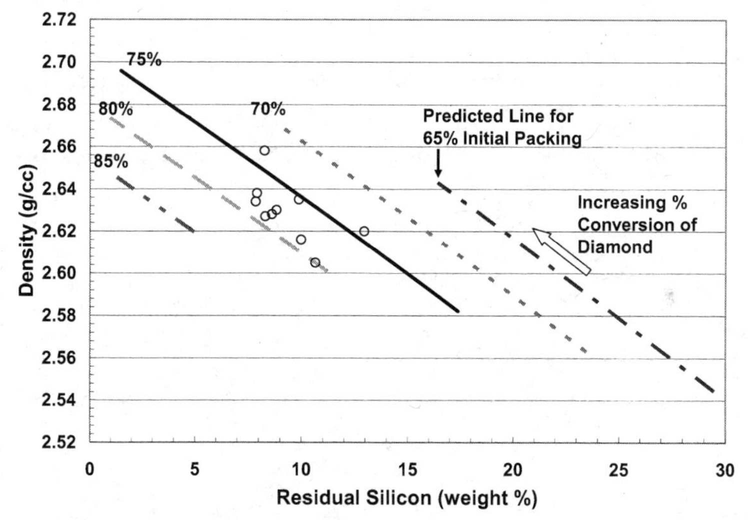

Figure 8. Theoretical prediction of density as a function of residual silicon at various initial B_4C packings and comparison with experimental data (o).

SUMMARY AND CONCLUSIONS

A design-of-experiments approach was used to assess the effects of composition and processing parameters on the properties of RBBC with diamond. Results of the DoE suggested that diamond amount and size had the most effect on the resultant properties, and particularly on the residual silicon. Higher diamond amount and finer diamond size were desirable for minimizing residual silicon in the composite. Based on these results, an RBBC + 10% 3 m diamond system was identified. For this system, processing parameters were developed to achieve residual Si content of less than 8%. Theoretical analysis was used to successfully predict a relationship between composite density and residual silicon. The experimental results obtained and the analytical predictions provide a direction for further reduction of residual silicon and enhancement of ballistic performance.

ACKNOWLEDGEMENT

This work was funded by the US Office of Naval Research (ONR) Contract No. W911-QY-08-C-0093.

REFERENCES

[1] A. H. Heyroth, "Silicon carbide articles and method of making the same," US Patent No. 2,431,326 (1947).

[2] K. M. Taylor, "Cold molded dense silicon carbide articles and methods of making the same," U. S. Patent No. 3,275,722 (1965).

[3] K. M. Taylor and R. J. Palicka, "Dense carbide composite for armor and abrasives," U. S. Patent No. 3,765,300 (1973).

[4] M. Waggoner et al., "Silicon carbide composites and methods for making same," US Patent No. 6,503,572 (2003)

[5] M. Aghajanian, B. Morgan, J. Singh, J. Mears and B. Wolffe, "A new family of reaction bonded ceramics for armor applications," in Ceramic Armor Materials by Design, Ceramic Transactions, Vol. 134., J. W. McCauley et al editors, (2002) 527-540.

[6] M. K. Aghajanian et al., "Boron carbide composite bodies and methods for making same," U. S. Patent No. 6,862,970 (2005).

[7] P. Karandikar, M. Aghajanian, and B. Morgan, "Complex, net-shape ceramic composite components for structural, lithography, mirror and armor applications," Ceramic Engineering and Science Proceedings (CESP)Vol. 24 [4], (2003)561-566.

[9] F. Thevenot, "Boron carbide – A comprehensive review," in Euro Ceramics, Vol. 2:Properties of Ceramics, Elsevier Applied Science Publishers, London/New York, (1989) 2.1-2.25.

[10] P. G. Karandikar, S. Wong, G. Evans, and M. K. Aghajanian, "Microstructural development and phase changes in reaction bonded B_4C," CESP Vol. 31 [5] (2010) 251-259.

[11] P. G. Karandikar, S. Wong, G. Evans, and M. K. Aghajanian, "Optimization of reaction bonded B_4C for personnel armor applications," Proceedings of Personal Amor System Symposium (PASS), Quebec City, Canada, September 13-17, 2010.

[12] J. Whalen, W. Trela, and J. R. Baer, "Processing of SiC with statistically designed experiments," Ceramic Powder Science III, Ceramic Transactions #12, (1990) 903-909.

FABRICATION OF HIGH VOLUME FRACTION SiCp / Al METAL MATRIX COMPOSITES

Ryan McCuiston
Department of Tool and Materials Engineering
King Mongkut's University of Technology Thonburi
Bangmod, Thungkru, Bangkok 10140, Thailand

Sukunthakan Ngernbamrung, Kannigar Dateraksa, Kuljira Sujirote
National Metal and Materials Technology Center
National Science and Technology Development Center
Klong Luang, Pathumthani 12120, Thailand

Jessada Wannasin, Trinnamet Sungkapun
Department of Mining and Materials Engineering
Prince of Songkla University
Hat Yai, Songkla 90112, Thailand

ABSTRACT

High volume fraction silicon carbide particle (SiCp) - aluminum metal matrix composites were fabricated for use in a domestic armor application. A SiC powder mixture, consisting of three different average particle size powders (6.3, 2.2 and 0.68 microns), was designed in order to achieve high packing fractions, on the order of 60 to 70%. Near net shape porous SiC preforms were fabricated by die pressing. The preforms were bisque sintered to give them sufficient strength for infiltration. The bisque sintering was conducted either in air to produce an oxide film or by the conversion of a preceramic polymer polycarbosilane (PCS) under an inert atmosphere. The preforms were infiltrated by squeeze casting using 7075 aluminum in a semi-solid state. The infiltrated preforms were given a T6 heat treatment. Polished cross sections and fracture surfaces were observed using optical and electron microscopy to analyze the microstructure. The infiltrated preforms were found to have low residual porosity. Results on mechanical property evaluation and ballistic performance will be reported.

INTRODUCTION

Commercially available metal matrix composites (MMC) tend to have a low volume percent of a secondary phase, such as a ceramic powder, added to strengthen or reinforce the metal. These MMCs tend to retain their metal like behavior and strength characteristics. The use of an MMC in a ballistic application, where the MMC is used as a strike face, requires a different set of mechanical properties. In a typical composite armor, where a monolithic ceramic is used as the strike face, the combination of high compressive strength and hardness of the ceramic overwhelms the projectile causing it to deform and fragment. The energy from the projectile is spread out within the ceramic and the remnant kinetic energy ultimately absorbed by the ductile backing. In a ballistic MMC, depending on the reinforcement size and content, the projectile may be defeated in several ways.

Karamis et al. studied 20 vol.% Al_2O_3-aluminum composites[1] and 15-45 vol.% SiC-aluminum composites.[2] The size range of the Al_2O_3 powder was 0.5 to 8 mm while the SiC powder was given as 0.25 to 0.5 mm. Both types of composites were produced by mixing the ceramic powder with molten aluminum followed by squeeze casting. Defeat of sub-ordnance velocity 7.62 mm AP was through a combination of projectile deformation without fracture and significant projectile erosion from interactions with the abrasive ceramic phase, which slowed the projectile to complete arrest. For the case of the SiC-aluminum composites, as the volume fraction of SiC was increased from 15-45 vol.%, the penetration depth in the backing was decreased from 15 to 1.5 mm. It was noted that the larger sized ceramic particles tended to fail by cleaving.

Lin et al.[3] produced 25% vol.% SiC-aluminum composites by a combination of cryomilling, hot isostatic compaction and post forging. The microstructure is a combination of submicron SiC particles embedded in a nano-scale aluminum matrix. 50 mm thick plates were tested against 0.50 caliber ball rounds. Defeat of the projectile occurred through significant plastic deformation of the projectile and erosion of the plate, to a depth of 24 mm. The researchers intend this material as a replacement for ballistic steel.

Forquin et al.[4] created sintered SiC preforms with ~17% porosity. The preforms were infiltrated with aluminum by squeeze casting. Defeat of 7.62 mm AP rounds with a velocity of 847 m/s was demonstrated. The defeat occurred at the surface of the composite with minimal erosion. Sub-surface microcracking was observed, similar to the comminuted zone damage which is commonly observed in monolithic SiC that has been ballistically tested.

Pyzik et al.[5] produced B_4C-aluminum composites containing 70-80 vol.% B_4C powder. The ballistic efficiency of the composites was measured against 0.30 caliber AP rounds. The efficiency was found to be 70 to 80% that of monolithic boron carbide. It was determined that the ballistic efficiency of the composites increased with both B_4C content and B_4C matrix connectivity.

The MMCs produced in this research are intended as a strike face in a low-cost composite armor system capable of defeating 7.62 mm ball rounds. This implies that the MMCs should behave more like a ceramic than a metal. It was decided that the MMC would be produced by the infiltration of a presintered preform, in order to provide more connectivity among the ceramic particles. Therefore the ceramic content should be greater than 50 volume percent. As observed by Karamis et al.[1,2] and other researchers, in MMCs containing large diameter ceramic particles, those particles tend to either pullout of the matrix prematurely or fail by cleaving. This is undesirable behavior. On the other hand, while submicron to nano-size particles can produced excellent MMCs, they are cost prohibitive. The SiC powders examined in this research are predominantly abrasive grade and therefore relatively low cost. The particle sizes examined in this research averaged from ~7 to 0.7 microns. To keep the cost low, squeeze casting was used to infiltrate the presintered preforms.

EXPERIMENTAL PROCEDURE

The experimental details are split into fabrication and infiltration of the preforms. The preforms were made from silicon carbide powder. Four size fractions of silicon carbide powder, F1000-W, F1200-FW, 2.5 micron and 0.7 micron were obtained from Electro Abrasives Corp. (Buffalo, NY, USA). Two different powder blends were designed by fitting ternary mixtures to the Funk-Dinger distribution equation.[6,7] The blends were either dry or wet ball mixed. The particle size and particle size distributions of the powders and powder blends were measured using a Malvern Mastersizer 2000S (Malvern Instruments Ltd., UK).

The powder blends were uniaxially die pressed into tiles of various sizes for oxidation bonding via the formation of SiO_2 or mixed with polycarbosilane (Grade SMP-10, Starfire Systems Inc., NY, USA) for subsequent conversion to SiC. To study the oxidation bonding approach, powder was pressed into tiles, 28x28x5 mm. The tiles were heated in air and held at 900, 1100 or 1300°C from 0.5 to 8 hours to determine the amount of oxidation. The PCS was incorporated into the powder via wet mixing in hexane. After mixing was complete, the hexane was evaporated at room temperature under vacuum in order to dry the powder. The powder cake was sieved through a polymer screen to break up agglomerates. After pressing, the PCS treated samples were fired following the recommendations of the manufacture in order to generate crystalline SiC. The porosity of the oxidized and PCS treated preforms was characterized by mercury porosimetry using a Quantachrome Poremaster (Quantachrome Instruments, FL, USA).

Large preforms, 8x8 cm with thicknesses from 9 to 11 mm were infiltrated by squeeze casting with semi-solid 7075 aluminum to produce MMC tiles. The SiC preforms were first preheated to 750°C and then transferred to a steel casting die. The die was heated to a constant 350°C using embedded heating elements. A pre-determined amount of gas induced semi-solid aluminum[8-10] was then transferred to the die and the pressure applied via a 50 ton ram to force the semi-solid aluminum into the preform. The pressure was applied until the aluminum was solidified. The infiltrated preforms were given a T6 heat treatment. Excess aluminum was removed by simple machining. Infiltrated preforms were cross-sectioned and the microstructure observed. Dog-bone specimens were also infiltrated to evaluate the mechanical properties of the MMCs.

Ballistic testing was conducted at the ballistics testing facility of the Royal Thai Police Department, Bangkok. The targets were tested against ball ammunition (7.62x51 mm NATO) with a velocity of ~850 m/s. The distance between the target and the test barrel was 15 meters. The MMC tiles were glued to S2 glass/epoxy backings which were produced in-house.

RESULTS AND DISCUSSION

Silicon Carbide Powder Blends

In order to increase the packing of the silicon carbide powder, ternary powder blends were created. The blends were designed by an iterative approach. A baseline blend was created and the cumulative percent finer than (CPFT) distribution of the blend determined. The CPFT of the blend was then compared to the CPFT of the Funk-Dinger distribution equation with a distribution modulus selected for maximum packing fraction. The volume percent of the different size fractions in the blend were adjusted to produce a fit close to the Funk-Dinger distribution. The Funk-Dinger distribution equation is,

$$\frac{CPFT}{100\%} = \frac{D^n - D_S^n}{D_L^n - D_S^n} \tag{1}$$

where D is the powder diameter, D_S is the smallest powder size in the blend, D_L is the largest powder size in the blend and n is the distribution modulus. To achieve the maximum packing fraction, n should be equal to ~0.37. There are limitations to this approach. Abrasive grades of SiC powder are only available in a limited number of size fractions near or below a 10 micron average. The Funk-Dinger distribution represents a continuous distribution of powder, which is difficult to produce by mixing only three different size fractions together. In addition, physical aspects of the powder such as the morphology, the use of internal lubricants and the actual processing details of the blends themselves, are not accounted for in the Funk-Dinger distribution.

Figure 1(a) shows the CPFT distributions of the four powders used in this research. F1000-W and F1200-FW ar abrasive grade black powders with an average particle size of 6.3 and 4.4 microns, respectively. The 2.5 and 0.7 micron powders are green SiC with average particles sized of 2.2 and 0.9 microns, respectively. While the average particles size of F1000-W, F1200-FW and 2.5 micron are different, the largest and smallest particles sizes are almost identical.

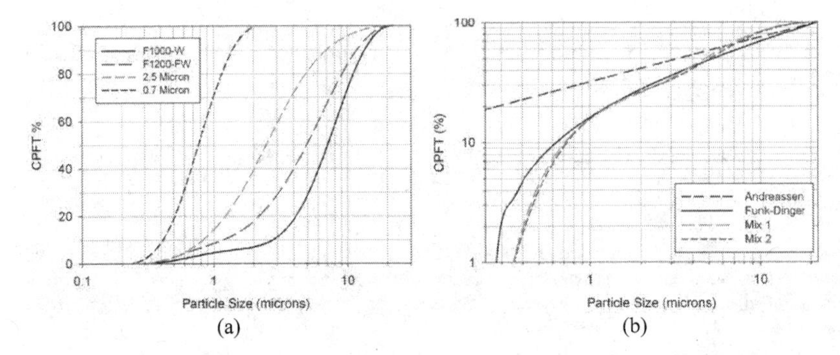

Figure 1. (a) CPFT distributions of the four SiC powders used in this research. (b) CPFT distribution for Mix 1 and 2. They are compared with the Funk-Dinger distribution for D_L=22 microns, D_S=0.24 microns and n=0.37. Please note the CPFT scale is cut-off at 1%.

Table 1. Compositions of Mix 1 and 2 shown in volume percent.

	F1000-W	F1200-FW	2.5 Micron	0.7 Micron
Mix 1	23	64	0	13
Mix 2	66	0	20	14

Two different ternary blends were designed. They are designated as Mix 1 and 2. Their compositions are shown in Table 1. The CPFT distributions for Mix 1 and 2 are shown in Figure 1(b). It is readily seen that both Mix 1 and 2, when compared to the Funk-Dinger distribution, have a large volume of excess particles that are large in size and are missing a volume of particles that are small in size. This is a result of using only three different size fractions, as well as by design. The presence of an extra volume of large particles results in larger porosity, which is advantageous for subsequent infiltration. The plot of the Andreassen distribution for D_L=22 micron and n=0.37 crosses the CPFT axis at ~19 vol.%. This is an indication of the unpacked volume or porosity that can be expected for the Funk-Dinger distribution, which represents a best case scenario. As both Mix 1 and 2 do not fit the Funk-Dinger distribution the actual green porosity is expected to approach 30-40%.

Oxidation of F1000-W Powder

Samples of F1000-W powder were uniaxially pressed at 6000 psi. They were heat treated in air to record the mass gain due to oxidation. Figure 2 shows the results of heat treatment at 900, 1100 and 1300°C for times up to 8 hours. At 900°C for times up to 8 hours, very little oxidation has occurred, less than 1 wt.%. At temperatures of 1100 and 1300°C, the oxidation is much greater after 8 hours, 10.2 and 15.3 wt.%, respectively. As this experiment was conducted only on the F1000-W powder, which has a lower surface area than other 3 powders, it would be expected that the mass increase due to oxidation would be greater for the remaining powders, as well as for the two mixtures. Oxidation bonding conditions of 1100°C for 1 hour, which corresponds to a mass increase of ~5%, were selected for further use.

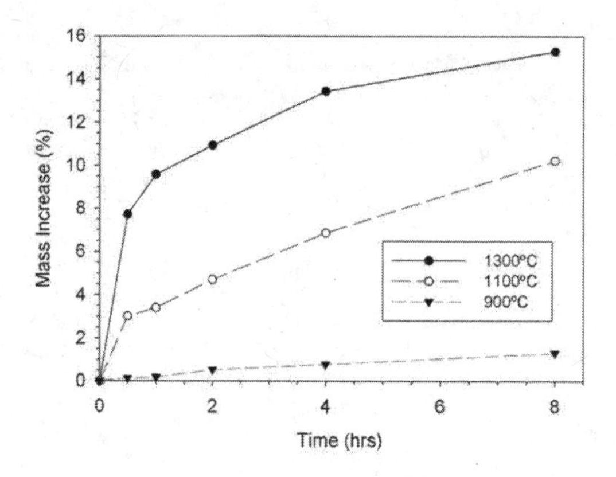

Figure 2. Percent mass increase vs. time for pressed F1000-W samples at 900, 1100 and 1300°C.

Oxidation of Powder Mixes 1 and 2

Samples of Mix 1 and 2 were prepared following the amounts given in Table 1. They were die pressed into 28x28 mm tiles using a 6000 psi applied pressure. As a baseline, samples of the four individual powders were also pressed under the same conditions. No internal lubricants were used. The amount of porosity in the green state was measured. The samples were then oxidized in air at 1100°C for 1 hour. The amount of porosity in the bisque state was then measured. The results are shown in Table 2.

Table 2. Porosity and dimensional changes of tiles pressed at 6000 psi and oxidized at 1100°C for 1 hour.

	0.7 Micron	2.5 Micron	F1200-FW	F1000-W	Mix 1	Mix 2
Green % Porosity	50.70	51.50	49.70	49.70	46.50	45.40
% Bisque Porosity	43.80	45.60	43.50	40.20	38.50	35.60
% Wt. Gain	10.81	8.67	4.31	4.23	6.13	6.8
% Width Increase	5.70	2.70	1.10	0.40	1.10	0.50
% Thick. Increase	Bloated	8.80	2.60	0.70	1.90	1.80

The results in Table 2 show that both Mix 1 and 2 have lower green and bisque porosities than any of the four individual powders, by ~5 vol.%. The decrease in porosity from the bisque to the green state is due to the pore space being filled by SiO_2 resulting from the SiC oxidation. Samples of both Mix 1 and 2 have better green handling strength than pressed samples of the four individual powders as well. It should be noted that the green porosity is can be reduced further using higher pressing pressures, but 6000 psi was selected due to scale-up limitations for larger tiles. Based in part on the results in Table 2, Mix 2 was selected for further research.

Porosimetry

Mercury porosimetry was performed on two samples. Both samples were uniaxially pressed at 6000 psi using the mix 2 powder blend. Sample 1 was bonded by oxidation at 1100°C for 1 hour. Sample 2 had 5% PCS added to it and was heat treated following the manufacturers recommendations. Mercury porosimetry was conducted on the two samples to determine the porosity characteristics. The mode pore diameter for the PCS treated sample was measured to be 1 micron, while the mode pore diameter for the oxidized sample was measured to be 0.21 micron. A plot of the volume intrusion versus pore diameter is shown in Figure 3. It is readily seen that ~50 vol.% of the porosity in the PCS sample is greater than 1 micron in size. For the oxidation sample however, a large fraction of the pore size, ~90 vol.%, is below 1 micron. While both samples were made from mix 2, they experienced significantly different heat treatments. The oxidation sample experienced a maximum temperature of 1100°C, enough to create a nanometer scale coating of SiO_2, but not enough to promote sintering of the SiC powder. The PCS sample however was heat treated in excess of 1800°C to convert the PCS to SiC. SEM images of the microstructure of the PCS sample shows that rounding of the grains has occurred, along with significant necking. As the smallest grains in the microstructure are bonded/consumed, the pore space between them is eliminated. This results in the pore size distribution shifting to a larger average. The creation of a greater percentage of large pores may be the result of coarsening as well.

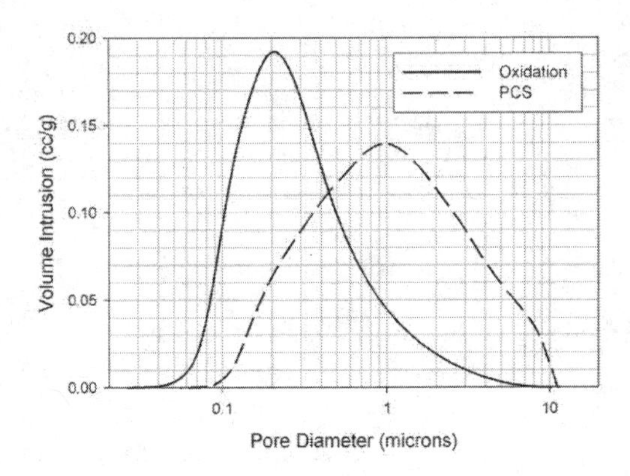

Figure 3. Volume intrusion versus pore diameter for pressed samples of Mix 2 treated with 5% PCS or oxidation at 1100°C.

Semi-solid Aluminum Infiltration

SiC preforms 8x8 cm in size were infiltrated with 7075 aluminum by squeeze casting. The aluminum was cast while in a semi-solid state. This allows for better casting of wrought alloys, such as 7075. Upon infiltration of both oxidation and PCS samples, it was found that the oxidation samples could not be fully infiltrated before solidification of the aluminum. The PCS samples were found to be fully infiltrated. Figure 4 shows cross sections and the corresponding x-ray radiographs for dogbone specimens treated by oxidation and PCS. The infiltration direction of the cross sectioned samples is from top to bottom. The lower right corner of the oxidation sample shows incomplete infiltration of the preform. The x-ray radiograph of the PCS sample shows a uniform color, indicating complete infiltration, without any visible boundaries or cracks. The x-ray radiograph of the oxidation sample shows that the preform is clearly outlined due to incomplete infiltration.

The threshold pressure P_{th}, required to infiltrate a compact of spherical powder with a non-wetting fluid is given by the equation,

$$P_{th} = -6\lambda \cdot \gamma_{lv} \cos\theta \frac{V_p}{D(1-V_p)} \tag{2}$$

where γ_{lv} is the liquid-vapor surface tension, θ is the wetting angle, V_p is the powder packing fraction, D is the average powder diameter and λ is a geometrical factor which accounts for asphericity of the powder and surface roughness.[11,12] This equation does not apply directly to presintered preforms, however several things may be qualitatively inferred from it. Assuming all other variables are held constant, the required threshold pressure for infiltration would increase as either the average powder diameter decreases and/or the packing fraction increases. It may be assumed that as the average powder diameter decreases and the packing fraction increases that the pore diameter decreases as well. Thus it may be inferred that as the pore diameter decreases, the threshold pressure for infiltration increases. The porosimetry results discussed previously show that the oxidation sample has an average pore diameter five times smaller than the PCS sample. The smaller average pore diameter is likely the reason for incomplete infiltration. Due to the difficulty of infiltrating the oxidation samples, it was decided to focus of the PCS approach.

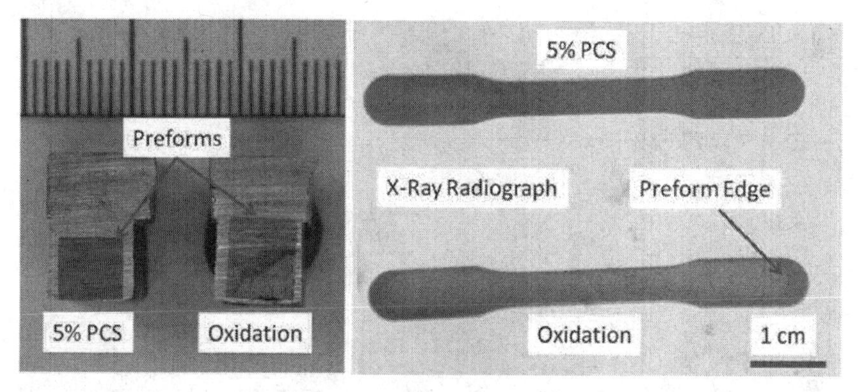

Figure 4. Cross sections and x-ray radiographs of dogbone specimens for PCS and oxidation samples. The PCS dogbone is fully infiltrated while the oxidation dogbone is ~50% infiltrated.

Infiltrated PCS specimens were cross sectioned and polished to examine the microstructure. Optical pictures of the polished specimens are shown in figure 5. At low magnification the presence of powder agglomerates is noticed. These agglomerates were determined to be composed of the 0.7 micron powder, indicating that the initial ball mixing of the blend was insufficient. The wet mixing process used to incorporate the PCS has limited shear action so agglomerates are not expected to be fully eliminated. Improvements to the initial mixing process to remove the agglomerates would likely increase the packing fraction, as well as, improve the connectivity of the matrix. The higher magnification image shows that the microstructure is well infiltrated with minimal residual porosity. Image analysis had determined the porosity to be ~1-2%.

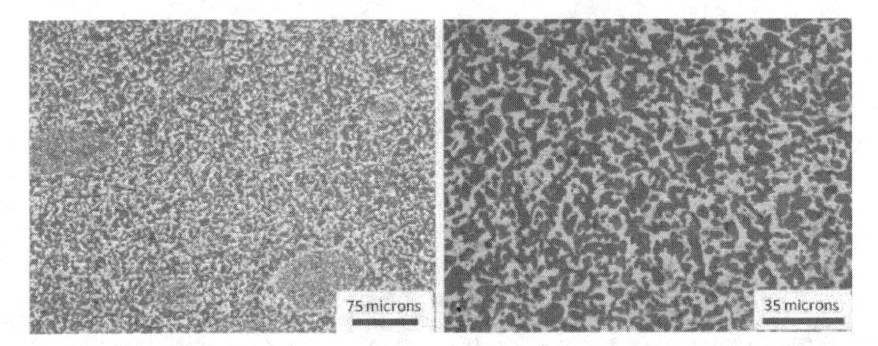

Figure 5. Optical pictures of an infiltrated preform containing 5% PCS. The lower magnification picture shows evidence of agglomeration. The higher magnification picture shows a small amount of residual porosity.

Mechanical Properties

Sic preforms for tensile dogbones and bend bars containing 4, 7, 10 and 14.6% PCS were pressed and infiltrated to produce MMCs. The results of the mechanical property measurements are shown in Table 3. Figure 6 shows a generic fracture surface of a bend bar specimen. The fracture surface is mixture of interfacial failure between the SiC grains and the aluminum matrix, as well as transgranular fracture of the SiC grains.

Table 3. Measured properties for PCS treated tensile and bend specimens.

PCS %	E (GPA)	Tensile Fracture Strength (MPa)	Tensile Fracture Strain (%)	Bend Strength (MPa)
4	141±18	137±65	0.098±0.058	675±28
7	126±36	105±38	0.111±0.061	622±197
10	139±30	93±69	0.070±0.045	681
14.6	143±27	106±13	0.125±0.036	321±82

The large standard deviation of some of the measurements is due to the presence of two main defects. Examination of the fracture surfaces of the tensile and bend bar specimens revealed either large agglomerates due to incomplete mixing, or aluminum filled cracks. The fact that the cracks are infiltrated implies that they were present in the preforms or were formed in-situ during infiltration while the aluminum was still molten. Many of the cracks were so large that the preforms would have failed well before infiltration, so it is likely that the cracks were formed during the infiltration process. One possible explanation is that some of the preforms are not resting perfectly flat on the bottom of the infiltration mold. If a small gap is created between the preform and the mold, tensile forces will be generated in the preform when the infiltration pressure is applied. Due to brittle nature of the preforms, they will simply crack. Sanding of the preforms to remove surface irregularities has eliminated many of the instances of cracking. There was no observable trend in the mechanical properties as a function of PCS content, though that may be due to the presence of defects. Further research has focused on PCS contents in the 5-7 vol. % range.

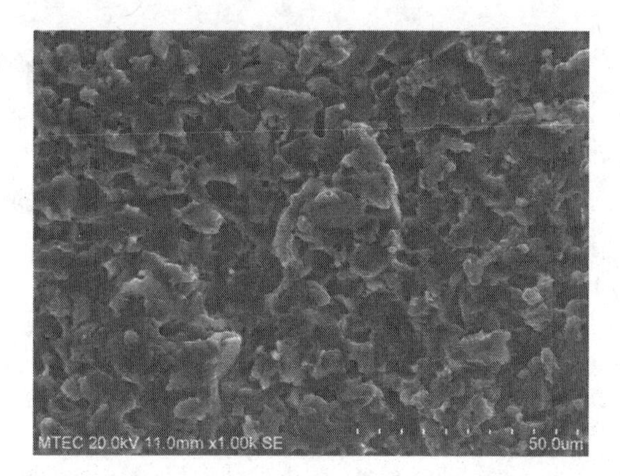

Figure 6. SEM picture of the fracture surface from a bend bar containing 10% PCS. The surface is a mix of SiC/Al interface failure and SiC cleavage.

Ballistic Testing

Fully infiltrated 8x8 cm preforms with thicknesses of 9-11 mm were glued with epoxy resin to S2 Glass /epoxy composite backings to form target assemblies. The backing was composed of 13-15 layers of S2 glass. The backings were produced in-house at the National Metal and Materials Laboratory. The targets were ballistically tested against 7.62 mm NATO ball ammunition. The results are shown in Table 4. There are not enough results to draw statistical conclusions yet, but the MMC tiles do appear capable of defeating the tested round. Five of the six shots were stopped in the backing, with only one complete penetration. The one failure was attributed to the existence of an aluminum infiltrated crack defect. The ballistically tested MMCs fracture/fail in a manner similar to monolithic ceramics, with the formation of a Hertzian cone and radial crack pattern.

Table 4. Ballistic test results for MMC produced with preforms containing 7% PCS.

Preform Thickness (mm)	MMC Thickness (mm)	Measured Velocity (m/s)	Penetration	BFS Depth (mm)	BFS Diameter (mm)
9	11.70	846.24	Complete	--	91.6
9	12.00	856.48	Partial	9.80	60.7
10	11.50	855.40	Partial	24.50	54.2
10	11.70	848.50	Partial	30.30	--
11	11.75	842.07	Partial	13.80	--
11	12.10	857.80	Partial	30.60	--

SUMMARY AND CONCLUSIONS

SiC preforms were created using a ternary blend of SiC powders. The packing fraction was improved over individual powders, but only by ~5 %. This is due in part to processing limitations as well as incomplete mixing. The preforms were subjected to an oxidation or PCS treatment. It was found that the mode pore diameter in the oxidation sample was five times smaller than in the PCS sample. Upon infiltration with semi-solid aluminum it was found that the PCS samples could be fully infiltrated, while the oxidation samples are not capable of being fully infiltrated under the experimental conditions. This is due in part to the smaller average pore size, the brief kinetics of the infiltration process and experimental limitations. Further research was focused on the PCS samples. The measured mechanical properties of the tensile and bend bar specimens show good average properties but the standard deviation in some cases is large. This is due to the presence of two main defects, agglomerates and aluminum filled cracks. The agglomerates may be removed with improved powder processing. The aluminum filled cracks likely result from cracking during the infiltration process. The removal of surface irregularities from the preforms can eliminate the infiltration cracks. Finally, ballistic testing of fully infiltrated preforms was conducted. The MMCs do appear capable of defeating 7.62 mm NATO ball rounds. The one failure was attributed to an aluminum filled crack defect.

ACKNOWLEDGEMENTS

The authors would like to thank the Royal Thai Military and the Royal Thai Police Department for providing funding and access to testing facilities. The authors would also like to thank Dr. Witchuda Daud of MTEC for manufacturing the S2 Glass/Epoxy backings used in the ballistic testing.

REFERENCES

1. M.B. Karamis, A.A. Cerit and F. Nair, "Surface characteristics of projectiles after frictional interaction with metal matrix composites under ballistic condition" *Wear,* **261,** pp 738-45, 2006
2. M.B. Karamis, F. Nair and A. Tasdemirci, "Analyses of metallurgical behavior of Al-SiCp composites after ballistic impacts", *Composite Structures,* **64**, pp 219-26, 2004
3. T. Lin, Q. Yang, C. Tan and A. McDonald, "Processing and ballistic performance of lightweight armors based on ultra-fine grain aluminum composites", *Journal of Material Science,* **43**, pp 7344-48, 2008
4. A.J. Pyzik, P.D. Williams and A. McCombs, "New Low Temperature Processing for Boron Carbide / Aluminum Based Composite Armor", Final Report to U.S. Army Research Office, Report No. ARO-26166-1-MS-A, 1990
5. P. Forquin, L. Tran, P.F. Louvigne, L. Rota and F. Hild, "Effect of aluminum reinforcement on the dynamic fragmentation of SiC ceramics", *International Journal of Impact Engineering*, **28**, pp 1061-76, 2003
6. D.R. Dinger, "One-Dimensional Packing of Spheres, Part I", *Ceramic Bulletin*, **2**, pp 71-76, 2000
7. D.R. Dinger, "One-Dimensional Packing of Spheres, Part II", *Ceramic Bulletin*, **4**, pp 83-91, 2000
8. J. Wannasin, R.A. Martinez and M.C. Flemings, "A novel technique to produce metal slurries for semi-solid metal processing", *Solid State Phenomena*, **116**, pp 366-69, 2006
9. J. Wannasin, S. Janudom, T. Rattanochaikul, R. Canyook, R. Burapa, T. Chucheep, S. Thanabumrungkul, "Research and development of gas induced semi-solid process for industrial applications", *Transactions of Nonferrous Metals Society of China*, **20**, s1010-15, 2010
10. S. Thanabumrungkul, S. Janudom, R. Burapa, P. Dulyapraphant and J. Wannasin, "Industrial development of gas induced semi-solid process", *Transactions of Nonferrous Metals Society of China*, **20**, s1016-21, 2010
11. J. Wannasin and M.C. Flemings, "Threshold pressure for infiltration of ceramic compacts Containing fine powders", *Scripta Materialia*, **53**, pp 657–661, 2005
12. E. Candan, H.V. Atkinson and H. Jones, "Role of surface tension to contact angle in determining threshold pressure for melt infiltration of ceramic powder compacts", *Scripta Materialia*, **38**, pp 999-1002, 1998

DENSIFICATION AND MICROSTRUCTURAL PROPERTIES OF BORON-CARBIDE IN SPARK PLASMA SINTERING

M.F. Toksoy and R. A. Haber
Rutgers University
Materials Science and Engineering
Piscataway, New Jersey, USA

ABSTRACT

Understanding the relationship between microstructural characteristics and mechanical properties are important for enhancing the performance of ceramic armors. This study examines the densification of commercial boron carbide powder using Spark Plasma Sintering (SPS). The particle size, particle size distribution and crystalline purity of the starting powder were examined Spark plasma sintering parameters – temperature, pressure and dwell time were varied. Results will relate hardness, elastic properties, density and grain size to the SPS process parameters.

INTRODUCTION

Boron carbide (B_4C) belongs to the non-metallic hard materials group. This ultra hard material, having a Knoop Hardness Number (KHN) of 2600, is the third hardest material after diamond and cubic boron nitride. Boron carbide also has other extreme properties such as high melting point (2427°C) and low theoretical density (2.52 g/cm^3). Lightweight armors, wear-resistant components and neutron absorbers are some of the many applications for boron carbide based products.[1] Processing of boron carbide is key to achieving these unique properties. There is a direct relationship between microstructure and mechanical properties of materials. There are several manufacturing steps that control the microstructure of final article. Powder attributes, green body properties and sintering methods are the major steps that affect final microstructure.[2]

Powder attributes affect many of the densified boron carbide properties. Boron carbon ratio, particle size and impurities are variables that have to be controlled during component production. Composition of boron carbide, specifically the boron-carbon ratio, can vary over wide range. When the ratio is less than 4, free carbon is present and the material is no longer considered a single phase product. Commercial powders typically have a stoichiometry less than but near $B_{4.3}C$. Although boron carbide exists over a wide compositional range, maximum hardness and indentation values are found at the stoichiometric composition of B_4C.[2]

Boron carbide powders can be produced by several synthesis methods. Carbothermal reduction is one of the methods used to produce boron carbide. This method synthesizes boron carbide via the following reaction.[3]

$$2B_2O_{3(l)} + 7C_{(s)} \rightarrow B_4C_{(s)} + 6CO_{(g)}$$

This process is ideally defined, but experimentally there are challenges to this synthesis method. Reaction occurs above 1600°C and below the melting point of boron carbide. The Arc-melt process is the common synthesis route for commercial powders. The process takes several days to complete and requires extensive size reduction to obtain high quality ceramic powders. Controlling the stoichiometry is difficult and often results in free carbon plus boron rich areas.[1]

Boron carbide is processed using several densification methods including hot pressing, hot isostatic pressing and pressureless sintering. Spark Plasma Sintering (SPS) has also been studied for densification of boron carbide but the resulting properties have not been well examined. Dense boron carbide samples can be readily obtained by hot-pressing. Typical temperature and pressure conditions are 2100-2200°C and 30-40 MPa respectively. Pressing eliminates most of the pores during the cycle by enhancing diffusion in the samples. Typically 15-30 minutes dwelling is sufficient to achieve a densified article. Pressureless sintering is another fabrication method used to produce boron carbide components. This method is preferred for complex shaped articles but achieving full density without additives is very difficult. Additives such as carbon and ferric oxides have been used to increase densification efficiency. Sintering temperature is typically higher than hot-press process temperatures and this procedure is not effective for the applications that need more than 96% of theoretical density.[4,5,6]

SPS is a newly developed process which has several advantages compared to conventional pressure assisted sintering. Rapid and uniform sintering cycles provide fully densified samples. Conventional methods are slow and because of the high operating temperatures controlling grain growth of these materials is very difficult. Grain size and porosity are two important parameters for predicting the performance of densified boron carbide. SPS by applying high electrical current through the sample and the graphite die improves the internal heating in the sample. This allows heating rates to be as much as 500°C/min and also enables more rapid cooling. Densification process is extremely fast compared to the conventional methods.[7]

EXPERIMENTAL

Pure commercial boron carbide powder from UK Abrasives, Inc. was used for current studies. Commercial boron carbide powder was milled adequately to improve homogeneity of the green body. Malvern Mastersizer was used for to measure the particle size distribution of the powder. X-ray diffraction analyses were made prior to sintering.

Table I. SPS trials for boron-carbide densification.

Trial	Temperature (C)	Heating Rate (C/min)	Pressure (MPa)	Dwell (min)
1	1900	300	50	20
2	1925	300	50	20
3	1950	300	50	20
4	1975	300	50	20
5	1850	300	50	20
6	1875	300	50	20
7	1800	300	50	20
8	1825	300	50	20
9	1850	300	50	0
10	1850	300	50	10
11	1850	300	50	30
12	1950	300	50	0
13	1950	300	50	10
14	1850	300	5	20
15	1850	300	20	20
16	1850	300	35	20
17	1850	300	60	20
18	1850	100	50	10
19	1850	200	50	10
20	1850	400	50	10

SPS was used exclusively as the densification method. SPS unit is computer operated and several programs were applied to the sample. Powder was packed in to a 20mm graphite die and 5 MPa pressure was applied on the green body. Each experiment was started under high vacuum (order of 10^{-3}-10^{-5} torr). Chamber was backfilled with argon gas at 600°C and densification trials were continued under argon atmosphere to the sintering temperature and during cooling to room temperature. Argon gas atmosphere was applied to prevent oxidation of the samples and carbon dies.

Various sintering cycles were applied to determine optimum densification procedure for boron carbide. Sintering temperature was varied between 1800-1975°C. The dwell time at sintering temperature ranged from 0-30 minutes. Applied pressure was chosen as third variable on the sintering process and was varied from 5-60 MPa. Heating rate varied from 100-400°C/min were applied to sample. The maximum applied pressures were reached at 1500°C for all samples and pressure was hold until end of the cycle. Table 1 lists the process conditions for the 20 SPS experiments.

Densities of the samples were determined by the Archimedes density method. Samples with 98% and more therotical density were separated and studied further. Mechanical properties of these samples were determined by non-destructive (ultrasound) method. These samples were then cut to the small pieces and polished with the Buehler Polishing Machine.

Microhardness measurements were taken at randomly selected regions. Knoop microhardness test used loads of 500g, 1Kg and 2Kg on the LECO micro hardness tester. Different indentation load were applied to see behavior of the samples with change of indent size. Microstructure were examined using the Zeiss-Sigma FESEM. Indentation from the hardness tests were also examined and photomicrographs were taken.

RESULTS

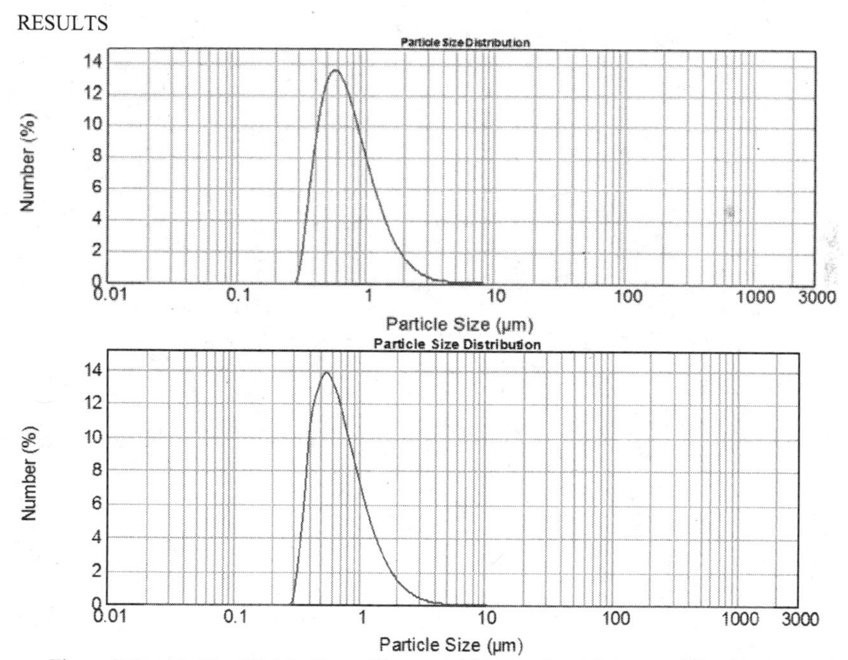

Figure 1. Particle Size Distributions of boron carbide powder a) Prior to milling b)After milling

Figure 1 shows the particle size distribution of boron carbide powder before and after the milling. These two measurements are essentially the same. Short time (20 minutes) milling did not change the particle size of the powder. The powders were mostly submicron materials with all particles less than 5 microns. Average particle size of the boron carbide powder is 0.75 μm and 0.73 μm.

SPS process successfully densified commercial boron carbide powder over a wide temperature range. Figure 2 shows the effects of temperature on density. Boron carbide can be sintered to high density (>2.5 g/cm^3) above 1900°C. Although SPS densified samples at 1900°C, the goal of these trials was to obtain finer microstructures by being able to sinter at lower temperatures since decreasing the sintering temperature should lead to a decrease in grain size. 1850°C was found to be the minimum temperature needed to achieve the desired 2.5 g/cm^3. Additional experiments (trials 14 - 20) focused on the effect of other process parameters while maintaining a constant temperature of 1850°C. The results of these trials are summarized in Figures 3, 4, and 5.

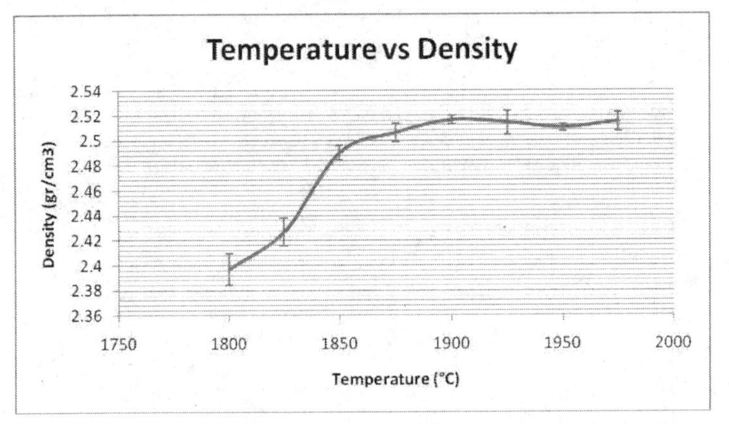

Figure 2. Temperature (°C) vs. Density

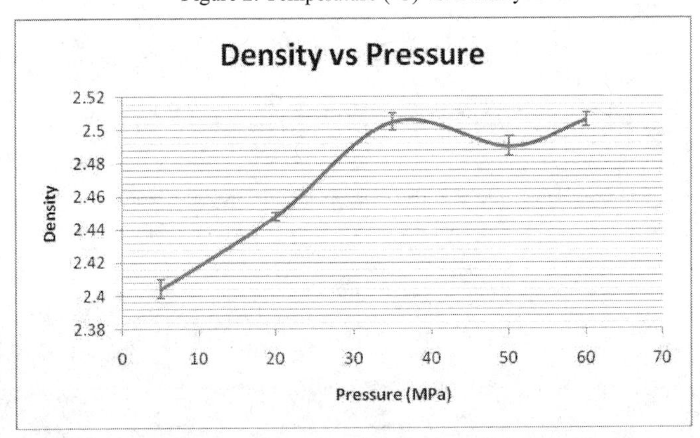

Figure 3. Density vs. Pressure at 1850°C with 300°C heating rate

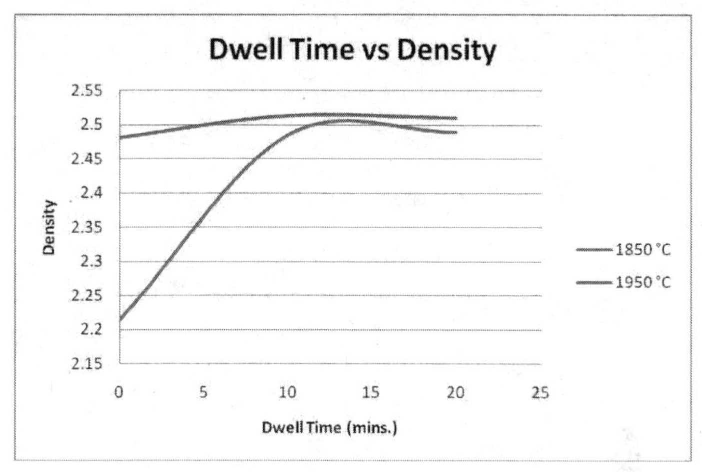

Figure 4. Density vs. Pressure at 1850°C with 300°C heating rate

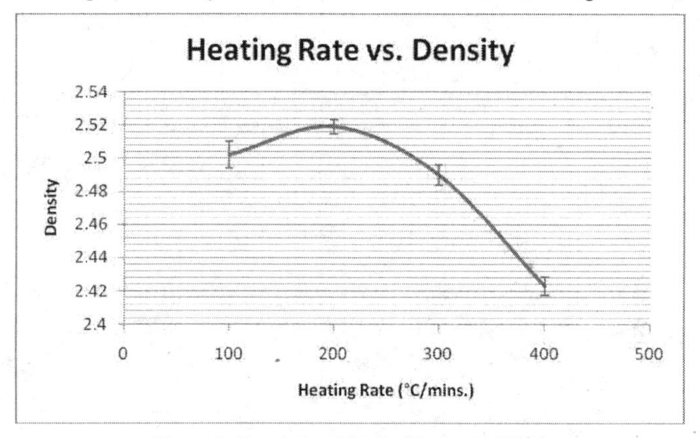

Figure 5. Density vs. Heating Rate at 1850°C

Trials 12 and 13 show that dwell time has a greater effect on density at lower temperatures (1850°C). 2.49 g/cm^3 density is achieved at 1950°C without any dwell time and almost fully dense sample at 1950°C with trials 12 and 13. With no dwell time at 1850°C, the measured density was only 2.42 g/cm^3 but steadily increased as the dwell increased up to 30 minutes. The density increased to 2.48 g/cm^3 after a 10 minute dwell time. Between 10-30 minutes dwell time, density is increased 2.48 to 2.499 gr/cm^3 at 1850°C. These values are below the density measured for the 1950°C sample with 20 minutes dwell.

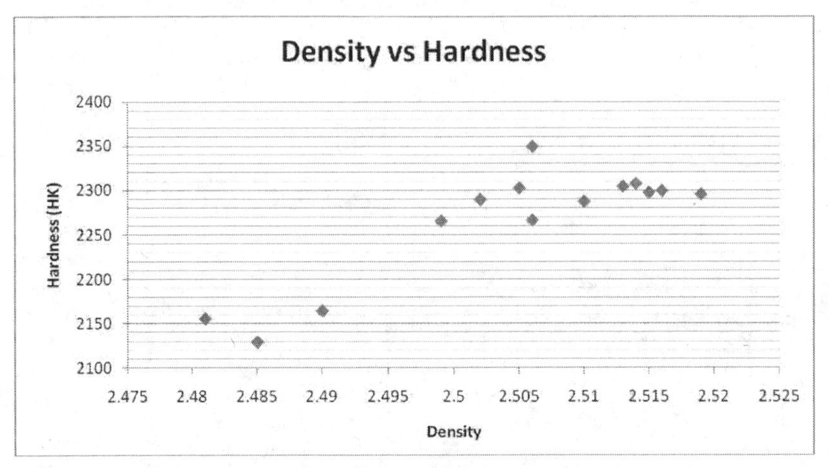

Figure 6. Density vs. Hardness (Knoop Hardness, 1000g. load)

Hardness results show the direct relationship between density and hardness. Sintered samples with over 2.49 densities have over 2250 Knoop (1000 g load). Figure 7 shows temperature effect on the hardness and it shows that the hardness significantly increases over 1875°C. Above this temperature hardness results are essentially unchanged. On the other hand hardness is increased with longer dwell time at 1850°C (Figure 8). Also Hardness values are slightly decreased with longer dwell time at 1950°C.

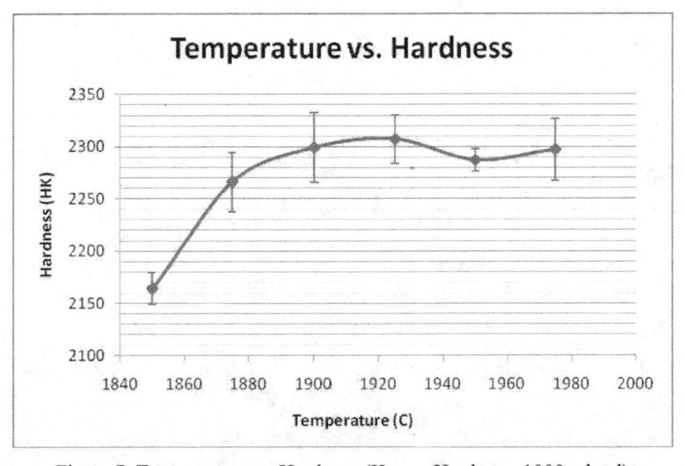

Figure 7. Temperature vs. Hardness (Knoop Hardness, 1000g. load)

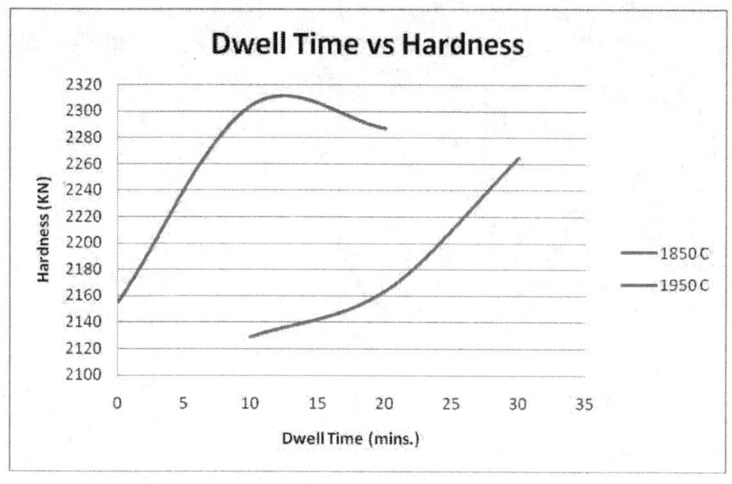

Figure 8. Dwell time vs. Hardness for sintering temperatures 1850°C and 1950°C

Knoop microhardness results were taken at 1 kg and 2 kg loads, but most of the samples exhibit massive crack growth at 2 kg load tests (Figure 9a). As a result 1 kg load values were used to throughout this evaluation. Most of the samples responded without massive crack growth (Figure 9b). As expected these densified boron carbide samples were very hard, but their toughness was not high enough to respond to the applied loads.

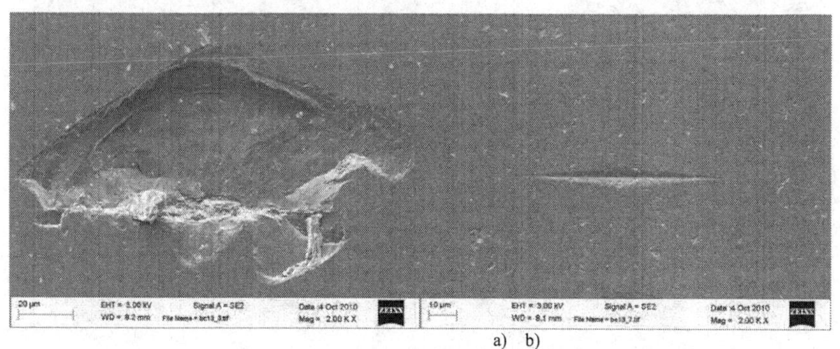

a) b)

Figure 9. a)Massive indent cracks with one of the lowest hard sample with 2000g. load b) Smooth indentation with 1000g laod with same sample

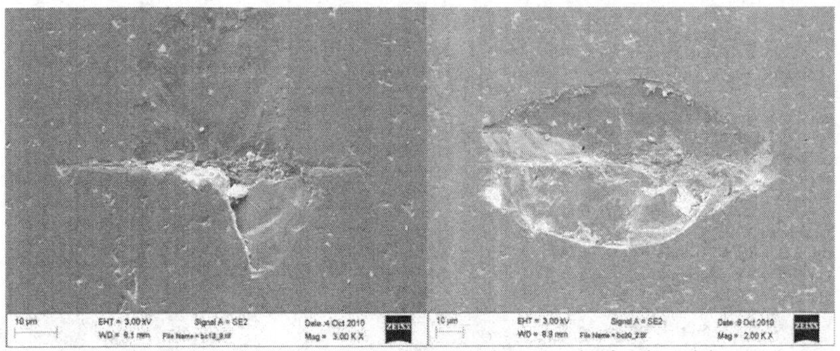

Figure 10. Massive crack growth at 1000 g load as a result of carbon rich area

Carbon rich areas exhibit massive crack growth (Figure 10) due to graphitic inclusions. Although these indents were taken under 1Kg, they exhibit massive crack growth. These massive fracture areas contain graphitic structures. These amorphous phases are not homogonously dispersed throughout the sample and may be responsible for the large scatter in hardness results for some samples. These graphitic regions greatly affect the mechanical behavior of boron carbide samples.

Grain size measurements were done for some of the high hardness samples (Figure 11). Average grain sizes of these three samples were between 1.55 μm-1.75 μm. Also there were some particles observed that were greater than 3-4 μm. The sources of these larger grains have not been investigated.

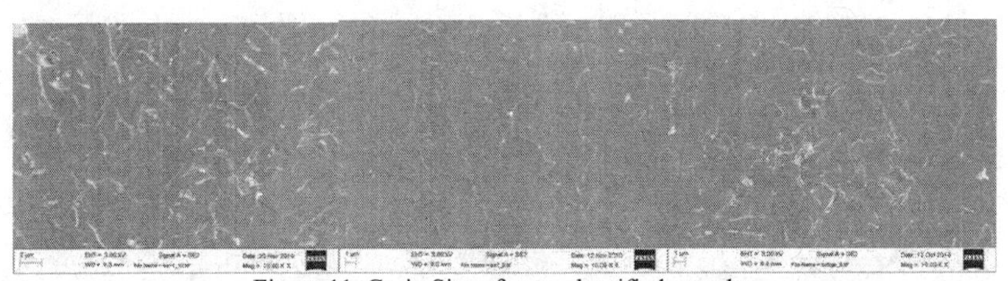
Figure 11. Grain Size of some densified samples

CONCLUSION

Parameters for densifying boron carbide using Spark Plasma sintering were determined. Boron carbide powder was densified extremely fast (around 30-40 minutes) and at lower temperatures. High hardness materials were achieved. Despite high density and high hardness of boron carbide samples, toughness was still too low to response high loads.

REFERENCES

1. Lee H., "Sintering of Boron Carbide", PhD. Thesis, Georgia Institute of Technology, July 2000
2. Cho, "Processing of Boron Carbide", PhD. Thesis, Georgia Institute of Technology, August 2006
3. Weimer, A W et al. "Rapid Carbothermal Reduction of Boron Oxide in a Graphite Transport Reactor", AlChE Journal, 37, 759-767. 1991
4. Speyer R., "Pressureless Sintering of Boron Carbide", Journal of American Ceramic Society, 86,1468-73, 2003
5. Mashhadi M., "Pressureless sintering of boron carbide", Ceramics International, 36, 151-159, 2010
6. Roy T.K., "Pressureless sintering of boron carbide", Ceramics International, 32, 227-233, 2006
7. Van der Biest O., "Modelling of the Field Assisted Sintering Technology (FAST) or Spark Plasma Sintering (SPS)", Katholieke Universiteit Leuven, 1999, retrieved from http://sirius.mtm.kuleuven.ac.be/Research/C2/Modelling%20of%20Field%20Assisted%20Sintering.htm

MODELING HEAT TRANSFER DURING SUBLIMATION GROWTH OF SILICON CARBIDE SINGLE CRYSTALS BY PHYSICAL VAPOR TRANSPORT

J. B. Allen*, C. F. Cornwell*, N. J. Lee*, C. P. Marsh*⁺, J. F. Peters*, and C. R. Welch*
*U.S. Army Engineer Research and Development Center, Vicksburg, MS, 39180-6199.
⁺University of Illinois at Urbana-Champaign, Department of Nuclear Plasma and Radiological Engineering, Urbana, IL 61801-2355.

ABSTRACT

Numerical simulation has been very useful in predicting the temperature distribution and subsequent growth kinetics in various vapor growth processes and can augment difficult or inadequate in situ measurements. Multidimensional models of silicon carbide (SiC) sublimation growth systems of varying levels of complexity have been developed to aid in the design, manufacture, and optimization of these growth systems. Since it is well known that SiC defect density and growth rate are strongly influenced by temperature distribution, an accurate assessment of this temperature distribution is required.

In this work we present a heat transfer model accounting for conduction, radiation, and radio frequency induction heating to investigate numerically the temperature distribution within an axisymmetric apparatus during sublimation growth of SiC bulk single crystals by vapor transport (modified Lely method). Here, the evolution of the magnetic potential vectors and temperature distribution is studied during the heating process and, where feasible, compared with existing numerical experiments found within the literature.

INTRODUCTION

Sublimation growth via the modified Lely method has been one of the most successful and most widely used SiC single crystal growth techniques of recent years[1,3]. The process involves sublimation of SiC from a hot powder source, transport of vapor through an inert gas environment, and condensation on a seed that is colder than the source. The deposition continues until a bulk crystal of reasonable size is grown.

Typically, an SiC growth apparatus (see Figure 1) consists of a Radio Frequency (RF) copper coil, a quartz tube, a graphite susceptor, insulation, and an RF induction-heated graphite crucible containing polycrystalline SiC source powder and a single crystalline SiC seed. The SiC source powder is heated by means of RF induction heating via copper coils that carry a current of approximately 10 kHz and 1200 A. Because of the high electrical conductivity of the graphite susceptor, the time-harmonic electromagnetic field induces eddy currents therein and heat is generated via the Joule effect. In general, the electromagnetic wave penetrates the susceptor with a skin depth of only a few millimeters, but is generally adequate to allow for sufficient heat generation within the inner growth region via conduction and radiation. Convection may also contribute in larger systems[2]. The SiC charge powder is placed within the lower, heated portion of the crucible, and the seed crystal is placed above and is cooled using a blind hole. This temperature difference allows sublimation to occur at the source and crystallization to form at the seed.

The growth rate of the crystal has been shown in numerous studies to be strongly linked to the temperature distribution, particularly with respect to the temperature difference between the source and seed[3-11]. Because SiC dissociates completely into liquid silicon and solid carbon at 3150 K, the growth temperature must remain below this upper limit[12]. The temperature distribution inside the growth system is determined via a complex interaction between the various heat transfer mechanisms, and occurs within materials of different phases, including gases, solids, and porous media. The temperature is further dependent on several configuration control parameters, including coil position, heating voltage, and pressure.

Since favorable growth conditions are critically linked to the temperature distribution, its optimization is an important goal of apparatus design and control. However, because of the high temperatures (approximately 3000 K), experimental verification of the correlation between the properties of the grown crystal, the temperature distribution, and external control is difficult. As a result, accurate numerical simulations have become highly desirable. Despite the various publications relating to numerical simulation of SiC crystal growth, there yet remain several unanswered questions that warrant further investigation.

The primary objective of this work is to develop a coupled, high-resolution, heat transport model to predict the electromagnetic field and temperature variation due to RF heating during SiC crystal synthesis using the modified Lely method. The model proposed here can be used to investigate the effects of variable current intensity and coil position on the temperature distribution within the growth chamber. Additionally, and distinct from the other cited literary sources on the subject, the effects of including radiation effects, and the contributions from including both components of the magnetic flux vector (in computing the electromagnetic field) will be assessed.

ELECTROMAGNETIC FIELD AND RESULTANT HEAT SOURCE

The electromagnetic field produced by RF induction heating can be calculated using Maxwell's equations. For low-frequency ($f < 1$ MHz) induction heating, Maxwell's equations can be simplified using the quasi-steady state approximations. Assuming that the current in the coil is time-harmonic, the magnetic flux density can be expressed as the curl of a magnetic vector potential, $\mathbf{B} = \nabla \times \mathbf{A}$. The Maxwell equations can then be written in terms of the vector potential[13] .

$$\nabla \times \left(\frac{1}{\mu_m} \nabla \times \mathbf{A} \right) + \varepsilon_m \frac{\partial^2 \mathbf{A}}{\partial t^2} + \sigma_c \frac{\partial \mathbf{A}}{\partial t} = \mathbf{J}_{coil} , \tag{1}$$

where μ_m is the magnetic permeability, ε_m is the permittivity, σ_c is the electrical conductivity, and $\mathbf{J}_{coil}$ is the current density in the coil. If the coil and the electromagnetic field are assumed axisymmetric, both the magnetic potential vector $\mathbf{A}$ and the current density $\mathbf{J}_{coil}$ will have only one angular component with an exponential form, such that:

$$\mathbf{A} = \begin{Bmatrix} 0 \\ A_r e^{i\omega t} + cc \\ 0 \end{Bmatrix}; \qquad \mathbf{J}_{coil} = \begin{Bmatrix} 0 \\ J_0 e^{i\omega t} + cc \\ 0 \end{Bmatrix}, \tag{2}$$

where $i = (-1)^{1/2}$, ω is the angular frequency, and cc denotes the complex conjugate. The final equation for vector potential, $\mathbf{A}$, is obtained by substituting Equation 2 into Equation 1:

$$\begin{aligned} &\left(\frac{\partial}{\partial r^2} + \frac{1}{r}\frac{\partial}{\partial r} - \frac{1}{r^2} + \frac{\partial^2}{\partial z^2} \right) \left(\frac{A_r}{\mu_m} \right) + \varepsilon_m \omega^2 A_r + \omega \sigma_c A_i = -J_o \\ &\left(\frac{\partial}{\partial r^2} + \frac{1}{r}\frac{\partial}{\partial r} - \frac{1}{r^2} + \frac{\partial^2}{\partial z^2} \right) \left(\frac{A_i}{\mu_m} \right) + \varepsilon_m \omega^2 A_i - \omega \sigma_c A_r = 0. \end{aligned} \tag{3}$$

The second term on the left side of Equation 3 is negligible. The following boundary conditions can be used to solve Equation 3:

$$A_r = A_i = 0, r \to \infty, z \to \pm\infty\ , \tag{4}$$

which are based on the conditions that magnetic flux density **B** is axisymmetric, and the computational domain is sufficiently large[14]. After Equation 3 is solved for the vector potential **A**, the generated heat power in the graphite susceptor can be obtained using the principles of eddy current:

$$q_{eddy} = \frac{1}{2}\sigma_c \omega^2 (A_r^2 + A_i^2) \tag{5}$$

RADIATION HEAT TRANSFER

Diffuse gray radiation is modeled utilizing heat flux boundary conditions at the relevant heat exchange surfaces. For this work, these include the surfaces pertaining to the SiC source powder and the growth chamber filled with inert gas. Additionally, radiation heat transfer modeling may be applied to the top and bottom cooling holes.

In general, heat flux can consist of idealized radiation:

$$-k\frac{\partial T}{\partial n} = \sigma\varepsilon\left(T^4 - T_{ext}^4\right), \tag{6}$$

where σ is the Stefan-Boltzmann constant, ε is the surface emissivity (considered herein constant for each surface), n is the normal to the surface, and T_{ext} is the external temperature. If the surface k is receiving radiation from another surface j in the system, then the heat flux may be represented[15] as:

$$-k_k\frac{\partial T_k}{\partial n_k} = \sigma\varepsilon_k\left(T_k^4 - \frac{1}{A_k\varepsilon_k}\sum_{j=1}^{N} G_{jk}\varepsilon_j T_j^4 A_j\right), \tag{7}$$

where the parameters A_j and A_k refer to the specific surface areas (N total) for surfaces j and k, respectively, and the factor G_{jk} represents the Gebhardt factors:

$$G_{jk} = \frac{Q_{jk}}{\varepsilon_j A_j \sigma T_j^4}. \tag{8}$$

Here Q_{jk} is the heat transferred from surface j to k, and is computed from the product of the view factor F_{jk} and the radiation leaving surface j (A_jJ_j), where J_j is the radiosity.

$$F_{jk} = \frac{Q_{jk}}{A_j J_j}. \tag{9}$$

The view factors may be computed as

$$F_{jk} = \frac{1}{A_j}\int_{A_j}\int_{A_k}\frac{\cos\theta_j \cos\theta_k}{\pi R^2}\,dA_j\,dA_k \tag{10}$$

where the parameters θ, R, and A are shown in Figure 2.

MATERIAL PROPERTIES

Table I lists values of electrical resistivity and thermal conductivity for the various temperature-dependent components of the growth system. The material properties are shown for temperatures between 773 K and 3273 K. The SiC source material was modeled[16] with an average grain size d_p of 125 μm and was treated as a porous medium with porosity ε_p and effective specific heat:

$$(\rho c_p)_{eff} = (1-\varepsilon_p)(\rho c_p)_{SiC} + \varepsilon_p(\rho c_p)_{gas} \tag{11}$$

where a porosity of 0.4 was selected for this work[2].

Because the heat transfer inside the charge consists of both radiative and conductive components, an expression for the effective thermal conductivity must account for both of these components. From Kansa[17], we have:

$$k_{eff} = (1-\varepsilon_p)k_{SiC} + \varepsilon_p\left(k_{gas} + \frac{32}{3}\varepsilon\sigma T^3 d_p\right). \tag{12}$$

CONVECTIVE HEAT TRANSFER

The relative importance of convective heat transfer (particularly natural convection due to buoyancy effects) within the context of the aforementioned crucible dimensions, as well as the processing temperature and pressure conditions, may be approximated by the dimensionless Rayleigh number:

$$Ra = Gr * \Pr = \frac{g\beta}{\nu\alpha}\Delta T D_c^{\,3} \tag{13}$$

where Gr is the Grashoff number, Pr is the Prandtl number, β is the thermal expansion coefficient, ν is the kinematic viscosity, α is the thermal diffusivity, ΔT is the temperature difference between the crucible wall and the quiescent gas, g is the acceleration due to gravity, and D_c is the diameter of the crucible.

For a typical growth system composed of argon gas, at a pressure of 3.0E4 Pa, a temperature of 2900 K, and temperature difference of 30 K[2], the Rayleigh number may be anywhere between 1.98 and 15.84 for representative (industry standard) crucible diameters between 50 and 100 mm, respectively. Since the critical Raleigh number Ra_{cr} is significantly less than 1708, it is evident that the heat transfer by natural convection can be neglected in these studies[18]. If however, the crucible diameter is increased (i.e., a increase of a factor of four), the relative importance of the buoyancy effects may require including convective transport modeling.

NUMERICAL METHODS

In concert with Equations 3, 5, and 7 of the previous sections, the equation governing the evolution of temperature for the growth system can be stated as

$$(\rho c_p)_{eff}\frac{\partial T}{\partial t} = \nabla \bullet (k_{eff}\nabla T) + q_{eddy}. \tag{14}$$

Additionally, the boundary conditions corresponding to surface heat flux (Equation 7), and magnetic vector potential (Equation 4) are applicable.

The variational form of Equation 14, after integration by parts of the conduction term and the application of the divergence theorem yields

$$\int_{\Omega}\frac{\partial T}{\partial t}v_i d\Omega+\int_{\Omega}\frac{k_{eff}}{(c_p\rho)_{eff}}\nabla T\cdot\nabla v_i d\Omega=\int_{\Omega}\frac{q_{eddy}}{(c_p\rho)_{eff}}v_i d\Omega+\int_{\Gamma}\frac{1}{(c_p\rho)_{eff}}\left(k_{eff}\nabla T\right)\cdot\mathbf{n}v_i d\Gamma \tag{15}$$

where v_i is the basis function, Ω and Γ are the elemental volume and its enclosing surface, respectively, and $\mathbf{n}$ is the unit normal vector to the surface. The linearized partial differential equation for T in the variational formulation is stated as

$$M_{ij}\frac{\partial T_j}{\partial t}+A_{ij}T_j=F_i \tag{16}$$

where M_{ij} is the mass matrix, A_{ij} is the stiffness matrix, and F_i is the force vector. Comparing Equation 15 with Equation 16 leads to the following:

$$\begin{aligned} M_{ij}&=\int_{\Omega}v_j v_i d\Omega \\ A_{ij}&=\int_{\Omega}\frac{k_{eff}}{(c_p\rho)_{eff}}\nabla v_j\cdot\nabla v_i d\Omega \\ F_i&=\int_{\Omega}\frac{q_{eddy}}{(c_p\rho)_{eff}}\nabla v_i d\Omega+\int_{\Gamma}\frac{q_n}{c_p\rho}v_i d\Gamma \end{aligned} \tag{17}$$

where $q_n = k_{eff}\nabla T\cdot\mathbf{n}$ is the heat flux perpendicular to the surface normal.

The numerical integration of Equation 16 over each quadrilateral element is approximated by the sum over all Gauss-point contributions, and solved using the open-source, multiphysics simulation software Elmer[15]. Time-stepping was performed using a second-order Crank-Nicholson scheme and was solved to steady-state convergence with an error tolerance of approximately 1.0E-5.

A schematic of the RF induction heating system for growth of SiC crystals composing the computational domain (0.25 m × 0.4 m) is shown in Figure 1. The domain dimensions were sufficiently sized to allow for a diminishing magnetic potential at the far-field boundaries (see Equation 4). Also shown in Figure 1 are the system component geometries and their respective coordinates in the axisymmetric (r, Z) coordinate system (see Table II). Each component is also shown in Figure 1 along with the appropriate heat transfer mechanisms (conduction (Cond.) and radiation (Rad.)) that occur either within the bulk of the material or at its interface.

A grid composed of 20,916 unstructured elements was utilized, and clustering was performed around the areas of high-temperature gradients (Figure 3).

RESULTS

Figure 4 shows the in-phase (A_r) and out-of-phase, (A_i) contours of magnetic potential for five turns of coil, a current of 1200A, and a frequency of 10 kHz. As indicated, the contours are concentrated along the outer portion of the graphite susceptor, and tend to bend inward along the top and bottom of the susceptor. The in-phase component reaches its maximum strength (~0.0018 Wb/m) around the coil, with its shape being highly influenced by the geometry of the coil. A_i, in contrast, shows a maximum magnitude of approximately 0.00017 Wb/m, and is located both near the coil and within the graphite susceptor. Of particular interest is that both A_i and A_r potentials show comparable strength within the graphite susceptor, and suggest the need to include both components.

Figure 5(a) shows the heat generated per unit volume from eddy currents in accordance with Equation 5. As shown, the heat is generated primarily within a small depth (i.e., skin depth), along the outer portion of the graphite susceptor. Relative to the surrounding insulation, the susceptor's high thermal conductivity (See Table I) functions as a shield to the inner crucible, and does not allow electromagnetic field penetration. An approximate skin depth[14] δ may be computed from

$$\delta = \frac{1}{\sqrt{\pi\mu_0}}\sqrt{\frac{\rho}{\mu_r f}} \tag{18}$$

where $\mu_0 = 4\pi E\text{-}7$ H/m, ρ is the electrical resistivity, μ_r is the relative permeability, and f is the frequency. From Equation 18 and the values from Table I, the approximate skin depth for the graphite susceptor resulted in approximately 1.6 cm.

Figure 5(b) shows the electromagnetic flux (Tesla). As shown, the maximum field strength of approximately 0.03T occurs near the induction coils and diminishes to zero inside the crucible and near the far-field boundaries. The strength is approximately 0.012T along the susceptor wall nearest the coils, and shows an anisotropic preference originating from the coils and extending toward this central portion of the susceptor.

Figure 6 compares eddy current heat generation with results from the existing literature[2]. Specifically, the heat generation within the SiC powder charge, the crucible (modeled as an inert gas), and the graphite susceptor are shown along the radial direction at various heights: $z = 0.0$, $z = 1.3R_s$, and $z = 2.6R_s$ (where R_s is the outside radius of the susceptor, $R_s = 0.07$ m). As indicated, and corroborated in Figure 5(a), the heat power is generated primarily in the outer portion of the graphite susceptor, and reaches an approximate value of 8.7E6 J/m^3. The discontinuity located between $r/R_s = 0.6$ and $r/R_s = 0.8$ occurs as a result of the intermediate inert gas material. The heat generated within the SiC charge at $z = 1.3R_s$ is somewhat negligible compared to that generated in the graphite susceptor at $z = 0$ and $2.6R_s$, showing an approximate difference of two orders of magnitude. Overall, the numerical results of this study compare very well with the results of Chen et al.[2].

Figure 7 shows the temperature distribution for all components minus the coil apparatus. As indicated, the maximum temperature (~3075 K) exists within the graphite susceptor, along the geometric center of the induction coil, and extends radially inward towards the SiC source powder. The coil geometry is thus responsible for the location of the highest temperature distributions within the source powder, and therefore the first instances of SiC sublimation. The effect of coil geometry on the temperature distribution has been conducted in several studies[2], and has been shown to be highly correlated.

The process of sublimation continues as a result of the temperature difference that exists between the source powder and the seed. As shown in Figure 1, the temperature at the top of the growth chamber T_{top} is cooled by a hole exiting to the outside air. This hole helps maintain the positive temperature difference between the source and the seed and thus facilitates the continued sublimation growth at the seed. This temperature difference is approximately 80 K and allows for a sufficiently high growth rate. Further, this hole helps to maintain a proper radial temperature gradient at the seed (i.e., thermal variations of less than 10 K at the seed surface) that ensures low thermal stresses that subsequently assist in minimizing dislocations and micropipes[2].

The effect of radiation modeling on the temperature distribution can be seen in Figure 8. The temperature difference between neighboring isolines is 10 K. As shown, neglecting radiation modeling results in unrealistically large heat gradients inside the growth chamber, where experiments have shown the temperature distribution to be nearly homogeneous. As stated previously, these erroneous predictions can lead to false conclusions not only with respect to the formation of dislocations and micropipes resulting from uncharacteristically high thermal stresses, but also with respect to SiC crystal growth rates[2].

CONCLUSIONS

A comprehensive process model for silicon carbide growth by the modified Lely method has been developed that accounts for axisymmetric geometries, induction heating, radiation, and conductive heat transfer. The model has been successfully applied to the simulation of SiC growth and compared with existing numerical experiments[2]. Computations of the magnetic potential showed the relative importance of including both the in-phase and out-of-phase components, with comparable magnitudes existing within the graphite susceptor. The graphite susceptor with its relatively high conductivity was shown to function as a barrier to the electromagnetic field, allowing only a negligible amount of eddy-induced heating within the inner crucible and growth region. By contrast, a significant amount of eddy current energy was created in the outer portion of the susceptor within a relatively small skin depth. The maximum temperature distribution was found to exist in the graphite susceptor and at the level of the geometric center of the induction coils. Within the SiC charge, the temperature was highest in the central region near the crucible wall. The positive temperature difference between the SiC charge and the ambient-air-cooled seed allows for the sublimation in the charge and deposition on the seed. The effect of radiation modeling was significant, particularly given the high temperature ranges appropriate to this study, and resulted in a substantial decrease in temperature gradients within the growth chamber.

Future simulations will evaluate the effects of including buoyancy effects by convection caused by increasing the growth chamber dimensions. These will necessarily require the coupling of the current model with the compressible Navier-Stokes equations.

ACKNOWLEDGMENTS

This study was funded through the U.S. Army Engineer Research and Development Center Directed Research Program Advanced Material Initiative.

REFERENCES

[1] O. Klein, P. Philip, J. Sprekels, K. Wilmanski, Radiation- and Convection-driven Transient Heat Transfer during Sublimation Growth of Silicon Carbide Single Crystals, *J. Cryst. Growth,* 222(4), 832-851 (2001).

[2] Q.-S. Chen, H. Zhang, V. Prasad, Z. M. Balkas, N. K. Yushin, Modeling of Heat Transfer and Kinetics of Physical Vapor Transport Growth of Silicon Carbide Crystals, *ASME Journal of Heat Transfer,* **123(6),** 1098-1110 (2001).

[3] D. L. Barrett, J. P. McHugh, H. M. Hobgood, R. H. Hopkins, P. G. McMullin, and R. C. Clarke, Growth of Large SiC Single Crystals, *J. Cryst. Growth,* **128**, 358–362 (1993).

[4] T. Kaneko, Growth Kinetics of Vapor-grown SiC, *J. Cryst. Growth,* **128**, 354–357 (1993).

[5] D. Hofmann, M. Heinze, A. Winnacker, F. Durst, L. Kadinski, P. Kaufmann, Y. Makarov, M. Schafer, On the Sublimation Growth of SiC Bulk Crystals: Development of a Numerical Process Model, *J. Cryst. Growth,* **146**, 214–219 (1995).

[6] M. Pons, E. Blanquet, J. M. Dedulle, I. Garcon, R. Madar, and C. Bernard, Thermodynamic Heat Transfer and Mass Transport Modeling of the Sublimation Growth of Silicon Carbide Crystals, *J. Electrochem. Soc.,* **143**(11), 3727–3735 (1996).

[7] N. Bubner, O. Klein, P. Philip, J. Sprekels, and K. Wilmaski, A Transient Model for the Sublimation Growth of Silicon Carbide Single Crystals, *J. Cryst. Growth,* **205**, 294-304 (1999).

[8] R.-H. Ma, Q.-S. Chen, H. Zhang, V. Prasad, C. M. Balkas, and N. K. Yushin, Modeling of Silicon Carbide Crystal Growth by Physical Vapor Transport Method, J. *Cryst. Growth,* **211**, 352–359 (2000).
[9] A. Roy, B. Mackintosh, J. P. Kalejs, Q.-S. Chen, H. Zhang, and V. Prasad, A Numerical Model for Inductively Heated Cylindrical Silicon Tube Growth System, *J. Cryst. Growth,* **211**, 365–371 (2000).
[10] Yu. M. Tairov and V. F Tsvetkov, Investigation of Growth Processes of Ingots of Silicon Carbide Single Crystals, *J. Cryst. Growth*, **43**, 209– 212 (1978).
[11] V.Tsvetkov, R. Glass, D. Henshall, D. Asbury, and C. H. Carter, Jr., SiC Seeded Boule Growth, *Mater. Sci. Forum,* **264–268**, 3–8 (1998).
[12] S. K. Lilov, Study of the Equilibrium Processes in the Gas Phase During Silicon Carbide Sublimation, *Mater. Sci. Eng., B,* **21**, 65–69 (1993)..
[13] J. D. Jackson, *Classical Electrodynamics*, 3rd edition; John Wiley, New York (1998).
[14] B. Wu, V. Noveski, H. Zhang, R. Schlesser, S. Mahajan, S. Beaudoin, and Z. Sitar, Design of an RF-Heated Bulk AlN Growth Reactor: Induction Heating and Heat Transfer Modeling, *J. Cryst. Growth,* **5**(4), 1491-1498 (2005).
[15] Elmer, Finite element software homepage. (2010). http://www.csc.fi/elmer/
[16] M. Sasaki, Y. Nishio, S. Nishina, S. Nakashima, and H. Harima, Defect Formation Mechanism of Bulk SiC, *Mater. Sci. Forum,* **264–268**, 41–44 (1998).
[17] E. J. Kansa, H. E. Perlee, and R. F. Chaiken Mathematical Model of Wood Pyrolysis Including Internal Forced Convection, *Combust. Flame,* **29**, 311–324 (1977).
[18] W. D. Reid, and D. L. Harris, Some Further Results on the Benard Problem, *Phys. Fluids,* **1,** 102-110 (1958).

TABLES AND FIGURES

Table I Material Properties of PVT SiC Synthesis

Component	Electrical resistivity Ωm (773 – 3273 K)	Thermal conductivity W/mK (773 – 3273 K)
Graphite	0.8E-5 to 1.0E-5	50-20
Insulation	~0.001	0.1-3.0
SiC charge	1-0.001	20-5.0
Argon gas	--	0.03-0.05

Table II Geometric Parameters PVT SiC Synthesis

Radial Parameter	Distance (m)	Vertical Parameter	Distance (m)
r1	0.0420	Z1	0.0595
r2	0.0560	Z2	0.0700
r3	0.0700	Z3	0.0847
r4	0.1155	Z4	0.1225
r5	0.1890	Z5	0.1995
--	--	Z6	0.2468
--	--	Z7	0.2590

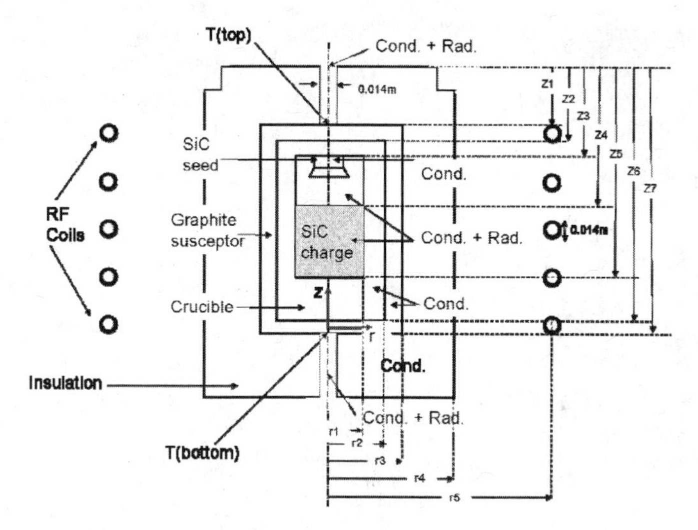

Figure 1 Schematic of the RF induction heating system for growth of SiC crystals. In addition to the primary system components (RF coils, insulation, etc.), the schematic also indicates the heat transfer mechanisms (conduction (Cond.) and radiation (Rad.)) that are modeled in this work. Also shown are the system component geometries in the axisymmetric (r, Z) coordinate system.

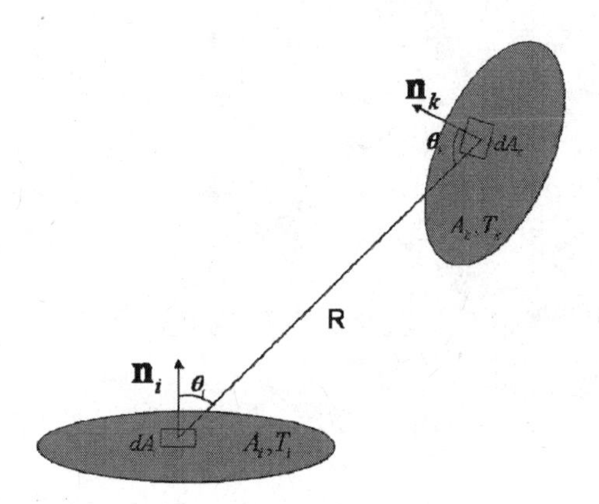

Figure 2 Geometric definition of view factor parameters

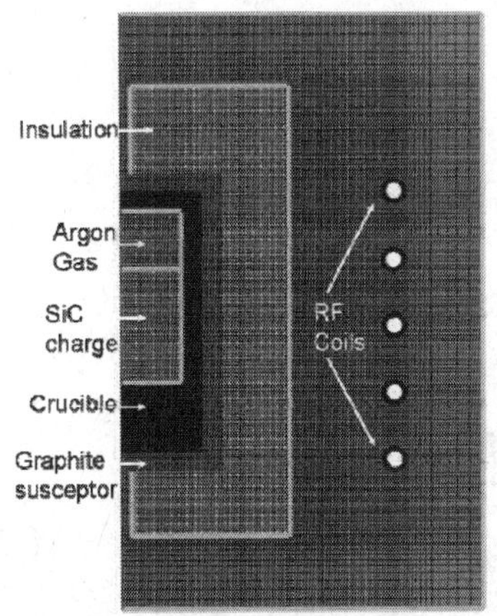

Figure 3 Computational grid for finite element method. Grid lines are highly clustered in regions of large temperature gradients. For purposes of clarity, every second grid line is plotted in each direction.

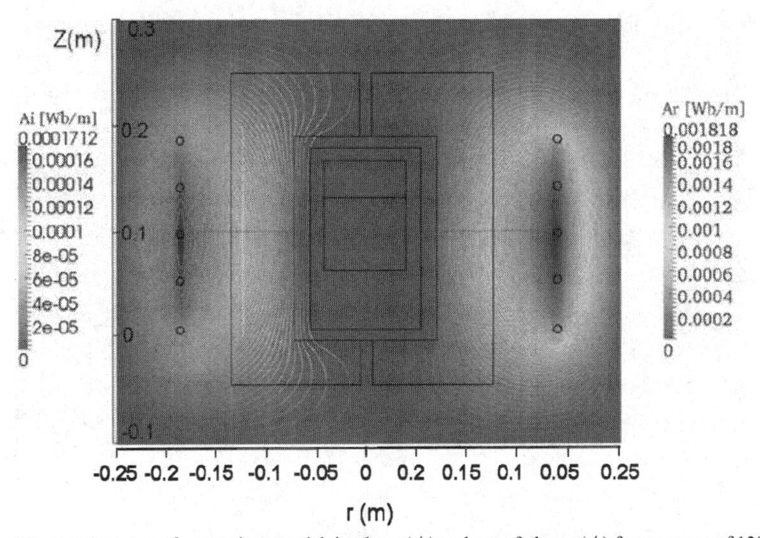

Figure 4 Contours of magnetic potential, in-phase (A_r) and out of phase, (A_i) for a current of 1200A and 10 kHz. The in-phase component reaches its maximum strength around the coil, with its shape being highly influenced by the geometry of the coil. The out-of-phase component, with an average, overall strength of approximately one order of magnitude less than that of the in-phase component, shows a comparable magnitude within the graphite susceptor.

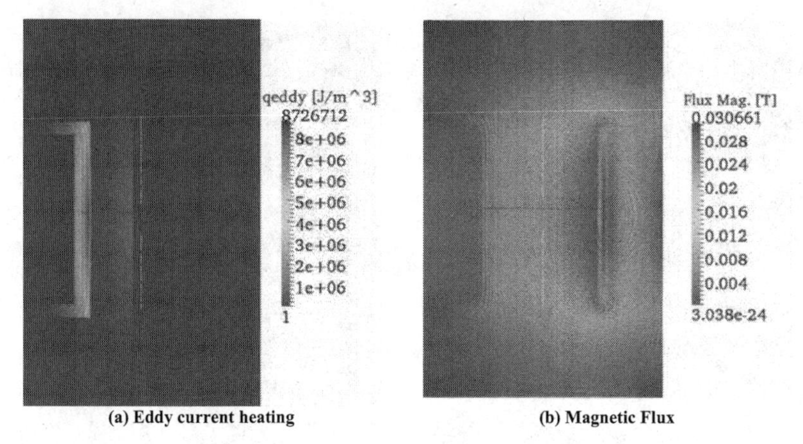

(a) Eddy current heating **(b) Magnetic Flux**

Figure 5 Contours of skin depth and magnetic flux density. The graphite susceptor with its high conductivity functions as a shield to the inner growth region. A significant amount of energy, generated by eddy currents, is created in the susceptor within a relatively small skin depth.

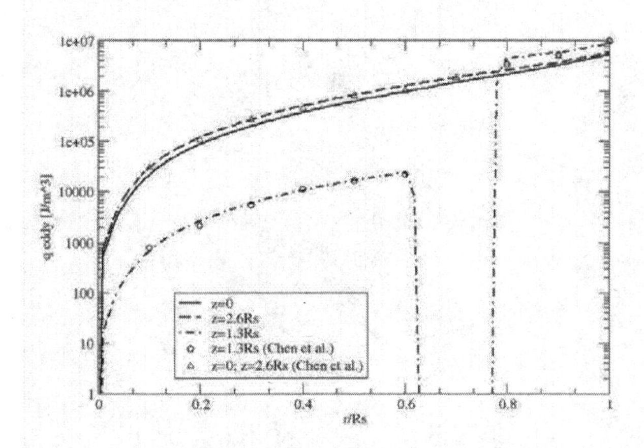

Figure 6 Comparison of generated heat power along the radial direction at different heights with results from Chen et al.[2]. As shown, the heat power is generated mainly in the outer portion of the susceptor. Further, the heat power generated at $z = 1.3R_s$ (within the charge) is negligible compared to that generated at $z = 0$ and $z = 2.6R_s$ (within the susceptor).

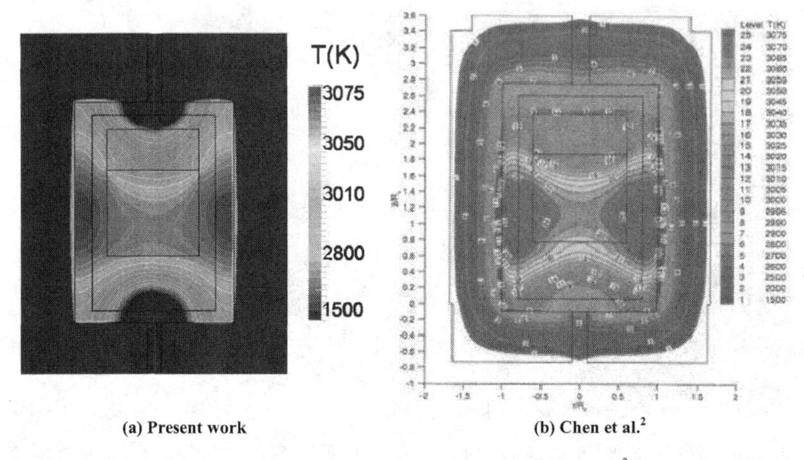

(a) Present work (b) Chen et al.[2]

Figure 7 Comparison of temperature contours between the present work and Chen et al.[2] for a current of 1200A and frequency of 10 kHz. The maximum temperature exists in the graphite susceptor, and at the level of the geometric center of the induction coils. In the charge, the temperature is higherin the middle near the crucible wall than in the bottom and top regions, and it will sublimate there first. The positive temperature difference between the SiC charge and the seed allows for the sublimation in the charge and deposition on the seed.

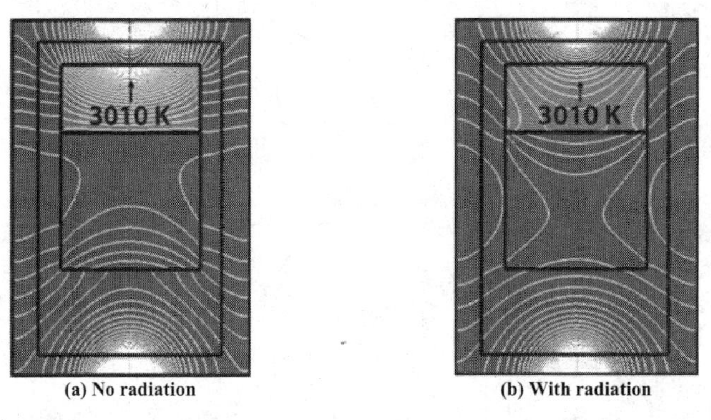

(a) No radiation (b) With radiation

Figure 8 Temperature contours with and without radiation modeling. The temperature difference between neighboring isolines is 10 K. Neglecting radiation results in unphysically large heat gradients inside the growth chamber, where the temperature distribution should actually be almost homogeneous.

DEVELOPMENT OF NANO ZIRCONIA TOUGHENED ALUMINA FOR CERAMIC ARMOR APPLICATIONS

S. Huang[1], J. Binner[1], B. Vaidhyanathan[1], P. Brown[2], C. Hampson[3] and C. Spacie[4]

[1]Department of Materials, Loughborough University, Loughborough, UK
[2]Defence Science and Technology Laboratory (Dstl), Salisbury, UK
[3]Morgan Advanced Ceramics Limited, Stourport-on-Severn, UK
[4]Morganite Electrical Carbon Ltd, Swansea, UK

ABSTRACT

The brittle nature of ceramics can limit their ability to resist multiple ballistic impacts. The present work focuses on investigating the possibility of using a toughened nanostructured ceramic, viz. nano zirconia toughened alumina (ZTA), as a potential ceramic armour. The material comprises of 10 - 20 wt% nano yttria stabilised zirconia (nanoYSZ) in a submicron alumina matrix. Using varied sintering conditions, ZTA samples with a range of alumina and zirconia grain sizes were produced. The effects of the yttria content, amount of zirconia and grain sizes of both YSZ and alumina on the mechanical properties of the ZTA were studied. It was found that 15% 1.5Y-ZTA provided both high hardness and toughness. In addition, the coarsening of zirconia grains benefited toughness enhancement due to the increased transformability of the zirconia grains. Using a modified split Hopkinson pressure bar, the samples were subjected to high strain rate (2×10^{-3} - 4×10^{4} s^{-1}) tests; the ZTA samples with zirconia contents ≥ 15 wt% showed the best performance. A micro-Raman study on the post-tested samples suggested that the transformation toughening absorbed the energy from the impact and protected the samples from breaking.

INTRODUCTION

Over the last 40 years, light weight ceramic armour has been developed for personnel, aircraft and vehicle protection and has proven itself effective in defeating high velocity and armour piercing projectiles. The keys to high performance ceramic armour include good ceramic properties such as high hardness, strength and stiffness as well as the ability to work harden under extreme stress[1]. These properties combine to increase the impact duration allowing the ceramic to erode and blunt or even comminute the projectile and absorb its kinetic energy[2]. Ceramic armour systems are also intended to be able to withstand multiple hits, however, the brittle nature of the material can reduce their multihit capability.

In ductile metals, local plastic deformation occurs under stress, which prevents the fracture of the material. On the other hand, in ceramics, the strong bonds combined with the crystalline structure inhibit dislocation movement and result in these materials displaying brittle properties. Under ballistic impact, ceramic armours fail when localized stresses exceed the material strength. Moreover, the ceramic tiles can be shattered on impact affecting their ability to defeat multiple hits. Many approaches have been investigated with the aim of improving the toughness of ceramics[3,4,5], of which one is the use of zirconia particles to toughen alumina. The zirconia inclusions toughen the alumina matrix via tetragonal to monoclinic transformation when cracks form. The phase transformation causes 4-5 vol% expansion, which provides a compressive stress that acts to reduce the propagation of the crack[6]. Zirconia toughend alumina (ZTA) has already been widely used in several applications, such as cutting tools, wear parts and biomechanical devices[5]. For the ballistic performance, Woodward et al.[7] observed that the toughness of ceramic tiles in an armour system can significantly affect the fragmentation process. With higher material toughness, the volume of fragments decreased in all the size range of the fragment distribution. This suggested that the increased toughness could improve the multihit capability of ceramic armours, since it is possible that exaggerated fragmentation can be prevented.

Toughness can also be improved by decreasing grain size. With the latter, the strength of dislocation pile-ups at the grain boundary increases, which results in a change in the fracture mode from transgranular to intergranular fracture[8]. The latter results in a grater energy requirement for crack propagation and therefore results in higher toughness. When the mean grain size is below 100 nm, i.e. the ceramics are nanocrystalline they can deform plastically and extensively by grain boundary sliding. This 'superplastic deformation' is in sharp contrast to the usual brittle behaviour associated with commercial ceramics[9,10]. Nanosized zirconia is an example of a superplastic ceramics with high toughness[9].

In the present work, the approaches of introducing toughening additives and reducing grain size were combined to produce nano zirconia toughened alumina (nanoZTA). The ceramics were produced through single stage sintering route with a range of sintering and heat treatment conditions to achieve full densities with varying grain sizes. The quasi-static properties and high strain rate performance of samples with 10, 15 and 20 wt% (6, 10 and 14 vol%) nano zirconia additions based on 1.5 and 3 mol% yttria stabilisation were studied and compared with pure alumina materials. The effects of the grain sizes of the alumina and the zirconia in the ZTA on their mechanical and high strain rate properties were also studied.

EXPERIMENTAL

Sample processing

Aqueous zirconia nanosuspensions containing different amounts of yttria, viz. 1.5 and 3 mol%, (MEL Chemicals, Manchester, UK) was mixed with a fine submicron alumina aqueous suspension (Baikowski, Annecy, France) so that the required YSZ content in the final material was 10 wt%, 15 wt% or 20 wt% (6, 10 or 14 vol%). Details of the as-received suspensions are listed in Table I. In order to achieve a high density green body, the as-mixed suspension, which had a solid content of 40-50 wt%, was further concentrated to ~57 wt%. During the concentration, the suspension was subjected to an ultrasound treatment at regular intervals using a ultrasonicator (MSE Scientific Instruments, Manchester, UK) to break up any agglomerates that formed and hence to control the viscosity of the suspension.

Table I The properties of the as-received nanosuspensions

Name	Yttria content /mol%	Solid content /wt%	pH	Particle size (D50) /nm
1.5YSZ	1.5	26.5	2.50	13
3YSZ	3	22.6	2.51	14
Alumina	-	58.8	3.52	160

The high solid content, homogeneous and low viscosity suspension was then spray freeze dryed to yield granules that possessed high flowability, due to their spherical shape, and high crushability, due to their porous structure after the ice sublimed. A foaming agent consisting of 2 vol% Freon was added to the high solid content suspension before the spray freeze drying to improve the crushability of the granules. The efficiency of the spray freeze drying method and the use of Freon has been discussed elsewhere[11,12]. In the spray freezing stage, the suspension was dripped onto an ultrasonic rod using a pipette, so that droplets were formed, which fell into liquid nitrogen and froze into ice crystals. The beaker containing the latter was subsequently connected to a freeze dryer (Virtis®, Benchtop SLC, New York, USA) which was connected in turn to a double stage, oil-sealed, rotary vane vacuum pump (Leybold® D2.5B, Leybold vacuum GmbH, Oberkochen, Germany) with a exhaust filter. The ice crystals were freeze dried under a vacuum of <100 mTorr at -50°C; the process taking about 2 days.

The dry, granulated powder was sieved to achieve a fraction in the range 125 to 250 μm, and the granules were die pressed at 250 MPa followed by cold isostatic pressing (CIP) at the same pressure to form green bodies in the form of discs measuring 10 mm diameter and 6 mm thickness. The pressed pellets reached densities above 55% of theoretical. For sintering, the green bodies were heated at 10°C/min to a temperature in the range 1400-1500°C and held for up to 20 h and then furnace cooled back to room temperature. A post-sintering heat treatment at 1500°C for 20 h was applied to some samples to generate different grain sizes. The details of sintering and heat treatment conditions are listed in Table II, where the sample code indicates the composition and sintering conditions. For example, 10% 3Y-ZTA represents a ZTA with 10wt% 3 mol% Y_2O_3 stabilised zirconia (3YSZ) added in the alumina matrix.

Table II Compositions and sintering conditions of the samples

Sample code	YSZ content / %	Yttria content / %	Sintering T / °C	Heating time / h
Al_2O_3	-	-	1400	2
10% 3Y-ZTA	10	3	1450	10
10% 1.5Y-ZTA	10	1.5	1450	10
15% 1.5Y-ZTA-5	15	1.5	1500	5
15% 1.5Y-ZTA-10	15	1.5	1500	10
15% 1.5Y-ZTA-15	15	1.5	1500	15
15% 1.5Y-ZTA-20	15	1.5	1500	20
15% 1.5Y-ZTA (HT*)	15	1.5	1500	10
20% 1.5Y-ZTA	20	1.5	1500	10

* HT refers to 20 hours heat treatment at 1500°C after sintering

Mechanical properties measurement and High strain rate tests

Densities of sintered samples were determined by Archimedes's method. A FEGSEM (Leo 1530VP FEGSEM, LEO Elektronenskopie GmbH, Oberkochen, Germany) was used for the microstructure observation and the grain size measurement of the sintered bodies. Hardness (*HV*) and fracture toughness (K_{Ic}) were determined by means of Vickers indentation with loads of 9.8 N and 98 N, respectively (10 indentations for each sample). Hardness was calculated as $HV = P/2d^2$, d being the half-diagonal indentation impression and P the indentation load (9.8 N). To calculate fracture toughness, the formula proposed by Anstis et al.[13] was used:

$$K_{Ic} = a \times \left(\frac{E}{H}\right)^{1/2} \times \frac{P}{c^{3/2}}$$

Where E is the Young's modulus of the material, H is the Vickers hardness, P the indentation load, 98 N, and c the crack length. a is a constant: for alumina samples the value is 0.016; for ZTA the value is 0.025[14].

Initial evaluation of the high strain rate characteristics of the ZTA ceramics was carried out using a modified split Hopkinson pressure bar (SHPB) on samples measuring 8 mm in diameter and ~4 mm in thickness. The SHPB setup consisted of a maraging steel elastic striker, incident and transmission bars, each with a diameter of 12 mm, Figure 1. A copper pulse shaper was placed in between the striker and the incident bars, in order to provide a ramp-loading pulse. To prevent the ceramic specimen from indenting into the bar end faces and thus causing stress concentrations, a pair of high stiffness tungsten carbide discs (ø 9 mm × 1 mm) was placed in between the sample and the bars[15]. The stress-strain curves of the ZTA samples were obtained from the analysis of the incident wave, the reflective wave and the transmitted wave.

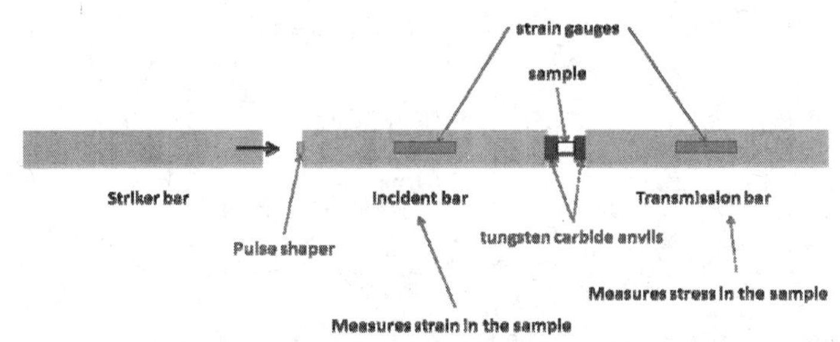

Figure 1 Setup of SHPB for brittle ZTA ceramic testing

Finally, Raman spectra were collected around the Vickers indentations and the top and fracture surfaces of the SHPB tested samples using a Raman spectrometer (Horiba Yvon Raman LabramHR, Horiba Jobin Yvon SAS, Villeneuve d'Ascq, France) equipped with a liquid nitrogen cooled CCD detector and two objective lenses, ×10 and ×50 magnification. The resolution of the instrument is around 1 μm. The phase transformation toughening and the transformability of the ZTA samples were studied and compared using this technique.

RESULTS AND DISCUSSION

1. Effects of microstructure and composition on material properties

Table III listed the densities, Al_2O_3 and YSZ grain sizes, hardness (measured using 9.8 N load) and indentation toughness (measured using 98 N load) of some selected sample used for understanding the effect of microstructure and composition on material properties. In the following paragraphs the effects of stabilizer content, YSZ amount and grain sizes will be discussed.

Table III Grain sizes and mechanical properties of the samples examined

Material	Density / T.D. %	Mean grain size		HV / GPa	K_{Ic} / MPa $m^{1/2}$
		Al_2O_3 / μm	YSZ / nm		
Al_2O_3	99	1.2±0.1	-	18.4±0.4	3.1±0.4
10% 3Y-ZTA	98	0.9±0.2	150±50	17.7±0.3	4.1±0.4
10% 1.5Y-ZTA	98	0.9±0.1	170±60	17.7±0.3	4.7±0.4
15% 1.5Y-ZTA-10	99	1.0±0.1	210±60	17.8±0.3	5.1±0.3
15% 1.5Y-ZTA (HT)	99	1.4±0.2	270±60	17.4±0.8	5.5±0.2
20% 1.5Y-ZTA	98	1.0±0.1	280±100	17.2±0.5	5.2±0.1

1.1 Effect of stabilizer content

The effect of the stabilizer (Y_2O_3) amount was investigated via the comparison of the 10% 3Y-ZTA and 10% 1.5Y-ZTA samples. As shown in Table III, the two samples showed similar hardness but slightly different toughness values. The similarity in the hardness of the two samples is consistent with the linear rule of the composite hardness, since both the matrix and the additive volume percentages, which determine the hardness, are the same between the two samples.

On the other hand, it may be observed that the toughness of the ZTA increased slightly as the yttria content decreases in the YSZ additive; the 10% 3Y-ZTA showed a toughness of 4.1 MPa $m^{1/2}$, whilst the 10% 1.5Y-ZTA had a toughness of ~4.7 MPa $m^{1/2}$. It is postulated that the transformability of YSZ in 1.5Y-ZTA is higher compared to that in 3Y-ZTA which therefore resulted in the increased transformation toughening effect.

To justify this postulation, the Raman spectra of the 10% 3Y and 1.5Y-ZTA samples before (Figure 2 (a)) and after indentation toughness tests (Figure 2 (b)) were compared. In Figure 2 (a), it can be observed that the spectra of both samples showed the distinctive peaks of the tetragonal structure. In Figure 2 (b) it may be seen that the intensities of the two monoclinic peaks at 175 and 188 cm^{-1} were much more significant in the 1.5Y-ZTA sample compared to those in the 3Y-ZTA sample. Therefore, it is evident that the transformability of the YSZ in the 1.5Y-ZTA was higher. This may be explained by the different transformability of the zirconia in the two samples. The critical grain sizes for phase transformation are 1.2 μm for 3YSZ and 110 nm for 1.5YSZ [16,17]. In the present work the mean zirconia grain sizes were 150-170 nm, less then that required for 3Y-ZTA but greater than that required for 1.5Y-ZTA, hence the latter showed a higher transformability than the former.

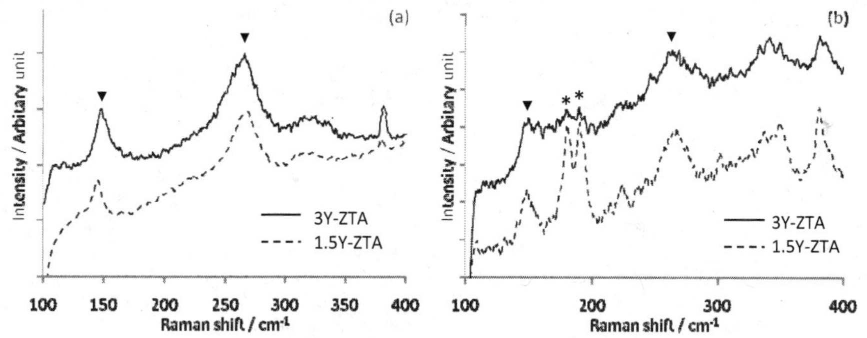

Figure 2 Raman spectra of 10% 3Y-ZTA and 10% 1.5Y-ZTA before (a) and after (b) indentation test, ▼ zirconia tetragonal peaks, * zirconia monoclinic peaks.

1.2 Effect of zirconia content

Comparing the hardness of pure alumina and 1.5Y-ZTA samples with different YSZ contents, the hardness decreased gradually with increasing YSZ content. This is mainly because that the hardness follows a linear relationship with the alumina content[6]. In terms of toughness, increasing the YSZ content from 10 to 20 wt% increased the value from 4.7 to 5.2 MPa $m^{1/2}$, which is as expected. In terms of achieving both high hardness and toughness, 15% 1.5Y-ZTA offered the best properties and hence was studied in the most detail.

1.3 Effect of grain size

The 15% 1.5Y-ZTA samples with alumina grain sizes ranging from 0.8-1.4 μm were produced using various sintering and heat treatment conditions as listed in Table II. The relationship between the alumina grain size and the indentation fracture toughness are shown in Figure 3 (a), whilst the relationship between the alumina grain size and the YSZ grain size is shown in Figure 3 (b). In addition, the grain size-toughness relationship of the pure alumina samples is added to Figure 3 (a) for comparison. Whilst Figure 3(a) shows that there is negligible change in toughness for pure alumina with grain size over the range investigated, whilst the addition of zirconia particles resulted in an increase in toughness with the alumina grain size in the ZTA samples. For the pure alumina samples, although the increasing grain size increased the crack bridging force[18], which could have toughened the material, a change of fracture mode from intergranular to transgranular was observed to be the main effect.

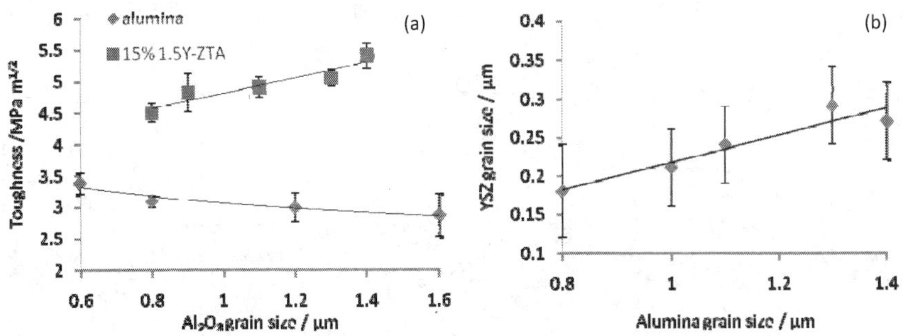

Figure 3 (a) Relationship of the toughness with mean Al_2O_3 grain size for pure alumina and 15% 1.5Y-ZTA samples; (b) relationship between YSZ and alumina grain sizes in 15% 1.5Y-ZTA.

By observing the crack propagation using FEGSEM, it is found that in the 15% 1.5Y-ZTA samples the increasing alumina grain sizes resulted in an increasing proportion of transgranular fracture through the alumina grains from 2% (0.8 µm alumina grain size) to ~17% (1.4 µm alumina grain size). As transgranular fracture requires less energy compared to intergranular fracture, the increasing amount of the former is considered to be harmful to the materials' toughness. Therefore, it is postulated that in the ZTA samples, although the increasing alumina grain size harms the toughness of the material, the addition of YSZ toughens the ZTA effectively when the crack meets the YSZ grains and therefore results in an overall increased toughness with an increase in alumina grain size. This hypothesis is also justified by the YSZ grain size-alumina grain size relationship observed in the ZTA samples (Figure 3 (b)). As the increase in alumina grain size in the ZTA samples is accompanied by a coarsening of the zirconia grains, Figure 3 (b), the transformability of the YSZ grains also increased significantly and therefore resulted in an increased toughness.

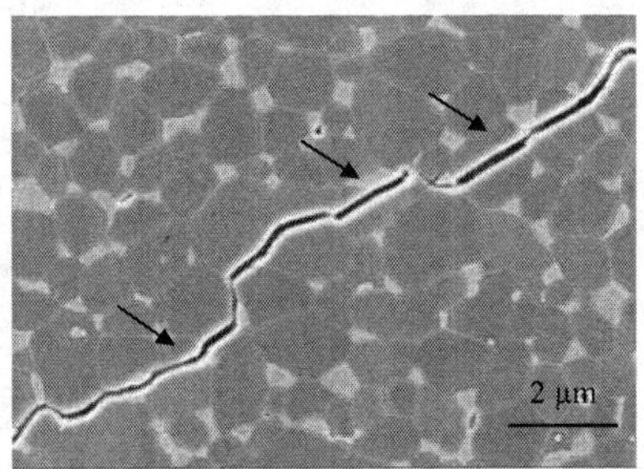

Figure 4 Crack propagation in 15% 1.5Y-ZTA (HT), showing transgranular fracture through alumina grains. The arrows indicate the transgranularly fractured grains.

Using Raman analysis, the transformability of the ZTA was studied according to the equation proposed by McMeeking and Evans[19]:

$$K_{Ic} = K_{Ic}^{m} + \eta V_f \Delta V E \cdot d^{1/2} / (1 - \nu)$$

Where K_{Ic}^{m} is the toughness of the material without tetragonal to monoclinic transformation, η is a constant, ΔV is the lattice dilatation associated to the transformation (4.7%), E is the Young's modulus, v is the poisson's ratio. V_f is the volumetric fraction of the transformable tetragonal phase, and d is the transformation zone size. Because η, ΔV, v and E are constant, fracture toughness is a function of $V_f d^{1/2}$. The latter parameter can be determined by means of Raman analysis following the method proposed by Katagiri et al.[20], which is based on a spatial mapping of the sample performed in the direction perpendicular to the crack, as illustrated in Figure 5. Examples of Raman spectra obtained at different distances from the indentation are shown in Figure 5 as well. The monoclinic fraction has been estimated from the relative intensities of the monoclinic doublet (bands at 181 cm^{-1} and at 192 cm^{-1}) with respect to the tetragonal band at 145 cm^{-1}. The calculated $V_f d^{1/2}$ value with respect to the different grain sizes and toughness of ZTAs are listed in Table IV.

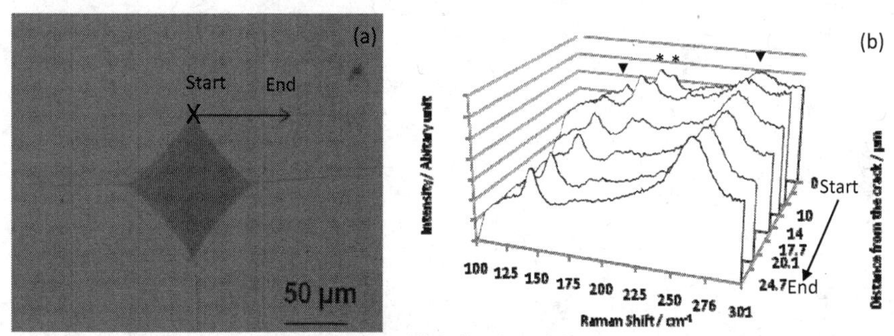

Figure 5 (a) Low magnification image with indication of the mapping paths; (b) Raman spectra acquired at different distances from the crack.

Table IV Transformability parameters and the mechanical properties of 15% 1.5Y-ZTA sintered at different conditions

Material	Mean grain size		HV / GPa	K_{Ic} / MPa $m^{1/2}$	V_f	d / µm	$V_f(d)^{1/2}$ / $(\mu m)^{1/2}$
	Al_2O_3 / µm	YSZ / nm					
15% 1.5Y-ZTA-5	0.8±0.1	180±60	17.8±0.4	5.0±0.4	0.69	8	1.95
15% 1.5Y-ZTA-10	1.0±0.1	210±50	17.7±0.3	5.1±0.3	0.71	8.5	2.07
15% 1.5Y-ZTA-20	1.3±0.1	290±50	17.7±0.4	5.4±0.3	0.67	12	2.32
15% 1.5Y-ZTA(HT)	1.4±0.2	270±60	17.4±0.8	5.5±0.2	0.68	16	2.72

The results in Table IV indicate that the toughness of the ZTA samples increased with $V_f d^{1/2}$ and also with the YSZ grain size. This observation is consistent with Casellas et al.[14], who suggested that the increased grain size of zirconia yielded a lower activation stress, which then lead to both larger transformation zones and more pronounced shielding effects around a propagating crack. Therefore, the current experimental results indicate that the transformation toughening effect induced by the increasing YSZ grain sizes outperforms the detrimental effect of increasing the alumina grain size in the ZTA samples. It is therefore expected that the toughness of the ZTA could be improved further if the alumina grain size could be further limited whilst the large zirconia grain size can be maintained.

2. High strain rate performance

The previous discussion of effects of microstructure and composition on the ZTA properties suggests that the 15% 1.5Y-ZTA samples with large YSZ grain size provided both a reasonably high hardness and toughness. It is important to learn whether, under dynamic compression, this material still shows a superior high strain rate performance. Split Hopkinson pressure bars (SHPB) are commonly used tools to generate families of stress-strain curves at controlled strain rates in laboratories. In the present work, a range of ZTA and pure alumina samples were tested using SHPB, the impact velocity varied from 12 to 22 ms^{-1} (strain rate from 2×10^3 to 4×10^4 s^{-1}) and the results are listed in Table V.

Table V Fracture conditions of the SHPB tested samples

Sample	K_{Ic} / MPa $m^{1/2}$	Impact velocity / ms^{-1}	Degree of fracture
Al_2O_3	3.1±0.4	16	Fractured into pieces
10% 1.5Y-ZTA	4.7±0.4	16	Fractured into pieces
15% 1.5Y-ZTA-10	5.1±0.3	12	Intact
		16	Mainly intact, 2 small fragments
		22	Mainly intact, 2 small fragments
15% 1.5Y-ZTA (HT)	5.5±0.2	12	Intact
		16	Intact
		22	Fractured into 3 large pieces
20% 1.5Y-ZTA	5.2±0.1	16	Mainly intact, 1 small fragment

It was observed that with the 16 ms^{-1} SHPB tests, the pure alumina and 10% 1.5Y-ZTA samples shattered into pieces, whilst the ZTAs with 15% or more zirconia addition resulted in mainly intact samples. The stress-strain relationships of a sample that fractured and one that remains intact are compared in Figure 6. The former sample, 10% 1.5Y-ZTA shows a stepped curve with each step representing a fracture activity in the sample during impact. The fracture process released the stress and hence resulted in the flat steps in the curve. The 15% 1.5Y-ZTA sample showed a much smoother curve without the periodic stress releases. As a consequence, it withstood a much larger strain, ~4% versus <2.5% for the sample that fractured, although the maximum stress was quite similar.

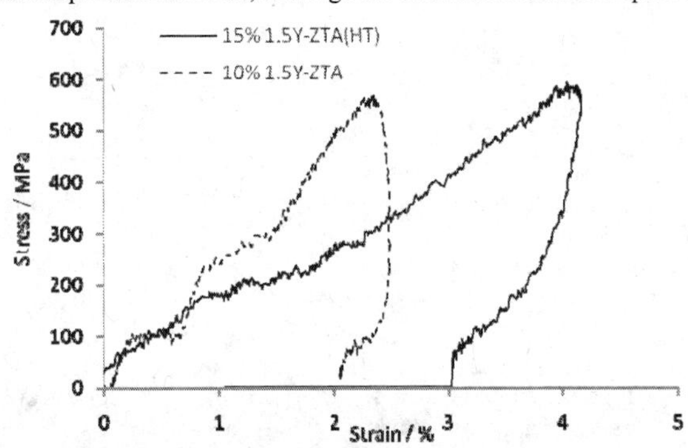

Figure 6 Stress-strain relationships of 10% 1.5Y-ZTA and 15% 1.5Y-ZTA (HT) SHPB tested using an impact velocity of 16 ms^{-1}.

To understand the effect of transformation toughening on the high strain rate performance, Raman analyses on the top surfaces of the 10%, 15% and 20% 1.5Y-ZTA post SHPB tested samples (16 ms^{-1} impact velocity) were performed. To achieve statistically valid data, five Raman analyses were performed on each sample. It was observed from spectra that the 15% and 20% 1.5Y-ZTA samples showed higher monoclinic intensities than the 10% sample, Figure 7. The observation of the monoclinic peaks in the 15% and 20% 1.5Y-ZTA samples confirms the transformation toughening effect of the YSZ particles in the ZTA. In addition, the higher intensity of the monoclinic peaks for the 15% and 20% samples compared to the 10% 1.5Y-ZTA sample may be because of two reasons: firstly, the higher amount of zirconia lead to more transformation toughening effect; secondly, only the 10% sample shattered during the SHPB test, therefore it is possible that the majority of the impact energy was consumed by the sample shattering rather than during the phase transformation.

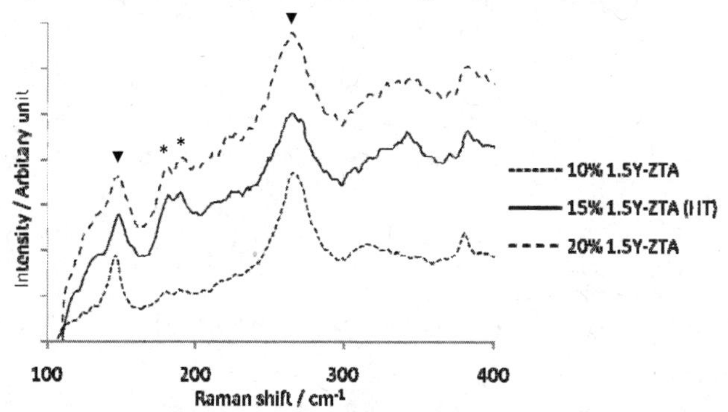

Figure 7 Raman spectra acquired from the SHPB impacted surface of ZTA samples using 16 ms^{-1} impact velocity.

Figure 8 compares typical Raman spectra of the 15% 1.5Y-ZTA samples SHPB tested using 16 and 22 ms^{-1}. The two samples have the same composition and mean grain sizes, however the former sample remained intact whilst the latter sample fractured into three large pieces due to the higher strain rate. By comparing the spectra from the top surfaces of the two, it was observed that the intact sample showed higher monoclinic peak intensity than the other. In addition, comparison between the spectra from the top and the fractured surfaces of the fractured sample showed that the latter showed higher monoclinic peak intensity. The observation of monoclinic peaks on the fracture surface is expected as the crack propagation can promote the tetragonal to monoclinic transformation. However, the lack of monoclinic peaks in the spectrum from the top surface of the fractured sample may be a further indication that the majority of the impact energy was consumed in the fracture process and therefore less was available for phase transformation.

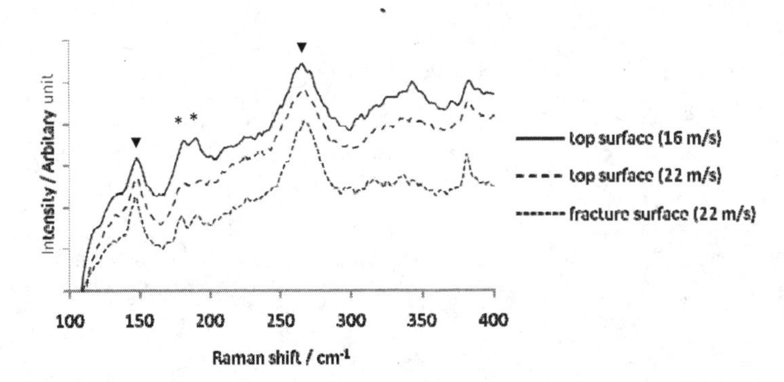

Figure 8 Raman spectra acquired from different surfaces of 15% 1.5Y-ZTA samples SHPB impacted using 16 ms^{-1} and 22 ms^{-1} velocities.

In general, the SHPB results suggest that for the different 1.5Y-ZTA samples, the ability to withstand high impact energy is related to the materials' toughness. In addition, distinctive monoclinic peaks were observed in the Raman spectra from the top surfaces of the intact samples but not the fractured samples. This observation may be an indication that, during the impact, the energy can be dissipated uniformly in the intact samples and therefore result in phase transformation of the tetragonal YSZ grains to monoclinic, which provides a transformation toughening effect to the material.

CONCLUSIONS

In the present work, quasi and high strain rate properties of nano zirconia toughened alumina (ZTA) have been investigated. It has been found that several factors including the stabiliser (yttria) content, zirconia amount and the alumina and zirconia grain sizes in the ZTA affect the materials' hardness, toughness and the high strain rate performance.

The hardness of the ZTA material was found to increase with the alumina content, which is logical since it possesses higher hardness than the zirconia additive. It was also found that the 1.5Y-ZTA sample showed better toughness compared to the 3Y-ZTA samples due to the higher transformability of the zirconia. In addition, a larger zirconia grain size in the ZTA materials is considered to be beneficial to the material toughness, as it promotes the transformability of the zirconia grains.

Preliminary SHPB test results showed that the ZTA samples with 15 wt% or more zirconia addition showed improved high strain rate performances compared to pure alumina and 10 wt% ZTA samples. The toughening effect induced by phase transformation of zirconia grains upon impact were also confirmed by the Raman results.

REFERENCES

1. P.G. Karandikar, G. Evans, S. Wong, and M. K. Aghajanian. A Review of Ceramics for Armor Applications, Presented at the 32nd International Conference on Advanced Ceramics and Composites, Daytona Beach, January, *Ceramic Engineering and Science Proceedings*, **29**, [6], 163-178 (2008).
2. W.W. Chen, A. M. Rajendran, B. Song and X. Nie. Dynamic Fracture of Ceramics in Armor Applications, *J. Am. Ceram. Soc.*, **90**, [4], 1005-1018 (2007).
3. M. Flinders, D. Ray, A. Anderson and R.A. Cutler. High-toughness Silicon Carbide as Armour, *J. Am. Ceram. Soc.*, **88,** [8], 2217-2226 (2005).
4. B.A. Gama, T.A. Bogetti, B.K, Fink, C-J. Yu, T.D. Claar, H.H. Eifert and J.W. Gillespie Jr. Aluminum Foam Integral Armor: A New Dimension in Armor Design, *Composite Structures*, **52**, 381-395 (2001).
5. X.F. Zhang and Y.C. Li, On the Comparison of the Ballistic Performance of 10% Zirconia Toughened Alumina and 95% Alumina Ceramic Target, *Materials and Design*, **31**, 1945-1952 (2010).
6. J. Wang, R. Stevens. Zirconia-toughened Alumina (ZTA) Ceramics, *J. Mat. Sci.*, **24**, 3421-3440 (1989).
7. R. L. Woodward, W. A. Gooch, Jr, R.G. O'donnell, W.J. Perciballi, B. J. Baxter and S.D. Pattie. A Study of Fragmentation in the Ballistic Impact of Ceramics, *Int. J. Impact Engng*, **15**, [5], 605-6180 (1994).
8. A. Muchtar and L.C. Lim. Indentation Fracture Toughness of High Purity Submicron Alumina, *Acta Mater.* **46**, [5], 1683-1690 (1998).
9. J. Karch, R. Brringer and H. Gleiter. Ceramics Ductile at Low Temperature, *Nature*, **330**, [6148], 556-558 (1987).
10. F.Wakai, Y. Kodama, S. Sakaguchi, N. Murayama, K. Izaki and K. Niihara. A Superplastic Covalent Crystal Composite, *Nature*, **344**, [62-64], 421-423 (1990).
11. J. Binner, B. Vaidhyanathan, A. Paul, K. Annaporani and B. Raghupathy., Compositional Effect in Nanostructured Yttria Partially Stabilized Zirconia, *Int. J. Appl. Ceram. Technol.*, (2010) in press.
12. J. G. P., Binner, M. I., Santacruz and K. Annapoorani, Aqueous Nanosuspensions, International patent application Publ. No. WO 2006/136780 A2, Publ. Date 28/12/06.
13. G. R. Anstis, P. Chantikul, B. R. Lawn and D. B. Marshall, A Critical Evaluation of Indentation Techniques for Measuring Fracture Toughness; I, Direct Crack Measurements, *J. Am. Ceram. Soc.*, **64**, [9], 533-538 (1981).
14. D. Casellas, M.M. Nagl, L. Llanes and M. Anglada, Fracture Toughness of Alumina and ZTA Ceramics: Microstructural Coarsening Effects, *J. Mat. Proc. Tech.*, **143-144**, 148-152 (2003).
15. G. Subhash and G. Ravichandran, Split Hopkinson Bar Testing of Ceramics, *ASM Handbook on Mechanical Testing and Evaluation*, ASM International, **8**, 497-504 (2000).
16. H.G. Scott, Phase Relationships in the Zirconia-Yttria System, *J. Mater. Sci.*, **10**, [1], 1527-1535 (1975).
17. J. Wang, M. Rainforth and R. Stevens, The Grain Size Dependence of the Mechanical Properties in TZP Ceramics, *Brit. Ceram. Trans. J.,* **88**, 1-6 (1989).
18. A. Eichl and R.W. Steinbrech, Determination of Crack-Bridging Forces in Alumina, *J. Am. Ceram. Soc.*, **71**, [16], 299-301 (1988).
19. R. M. McMeeking, and Evans, A. G., Mechanics of Transformation Toughening in Brittle Materials, *J. Am. Ceram. Soc*, **65,** [5], 242–246 (1982).
20. G. Katagiri, H. Ishida and A. Ishitani, Direct Determination by a Raman Microprobe of the Transformation Zone Size in Y_2O_3 Containing Tetragonal ZrO_2 Polycrystals, *Advances in Ceramics,* **24**, 537-544 (1988).

MICROSTRUCTURE PROPERTY RELATIONSHIP IN CERAMIC ARMOR MATERIALS

Douglas M. Slusark and R.A. Haber
Rutgers University
Materials Science and Engineering
607 Taylor Road
Piscataway, NJ 08854

ABSTRACT

The presence and spatial distribution of features within the microstructure of armor ceramics have drastic effects upon both the dynamic and quasi-static performance of these materials. Non-destructive evaluation (NDE) techniques such as ultrasound scanning provide a means to examine the bulk of a material without damaging or altering the sample. A study was conducted to correlate the relationship of non-destructive evaluation with microstructural variability and defect distributions for a sintered silicon carbide plate. Ultrasound C-scans were performed at 20MHz to determine the acoustic attenuation coefficient and the elastic modulus at each scanning point. Following non-destructive evaluation, the plate was machined into flexure bars for modulus of rupture determination. The identity and orientation of bars were preserved for comparison of mechanical properties to the NDE maps. Fractography was carried out on fracture surfaces to assess fracture behavior.

INTRODUCTION

Ceramic materials have many attributes that make them attractive candidates for use in armor applications. Among these are low relative density and both high compressive strength and high hardness. One of the most widely utilized armor ceramics is silicon carbide. With a theoretical density of ~3.1 – 3.2 g/cm^3, it is less than half the density of steel (~7.8 – 8.1 g/cm^3) [1]. High hardness and compressive strength as well as high stiffness are due to the strong covalent bonding between silicon and carbon atoms in the matrix. Silicon carbide may be produced by a number of forming and densification methods, including pressureless sintering [2,3].

Mechanical properties of ceramics are influenced by microstructural features such as porosity distribution, grain size, and the presence and distribution of inclusions. The final grain size is dependent on the size of the starting powder as well as the sintering conditions [3]. Sintering conditions such as time, temperature, and applied pressure, as well as the addition of any additives, have an effect on the amount and distribution of residual porosity [3]. Although additives have reduced strength and hardness in comparison to the silicon carbide matrix, they are required to achieve full density during sintering.

The strong covalent bonding present in silicon carbide reduces the mass transport characteristics during sintering, making it challenging to reach full density [4]. Two of the most common additives to increase densification are carbon and boron carbide [3,5]. They act to increase the diffusion rate and help control grain growth during sintering, to reduce porosity, and to remove the SiO_2 passivation layer present on SiC particles [5]. The process used for adding additives and the amount added have an effect on the properties of the material. Adding an excess or not ensuring proper mixing or sintering can result in un-reacted material that forms agglomerates in the finished part. These agglomerates can act as stress concentrators due to the elastic property mismatch between the agglomerate and the silicon carbide matrix [6].

Non-destructive evaluation, including ultrasound scanning, provides the means for interacting with and interpreting the microstructure of a material. Traditional quality control methods involve selecting and testing a part from a production run. In many instances, this leads to the part being destroyed, thereby decreasing yields. A common qualification tool in the production of ceramic armor tiles to measure dynamic performance is ballistic testing. This is a time-consuming and expensive test, and only allows inferences to be made about the remaining tiles in the lot. A fast, reliable method of testing the entire lot would be preferable. While ultrasonic detection of flaws in ceramics has been in use for well over two decades, a further understanding of the microstructural interaction of ultrasound energy in polycrystalline ceramics is required to predict the dynamic performance of a tile [7].

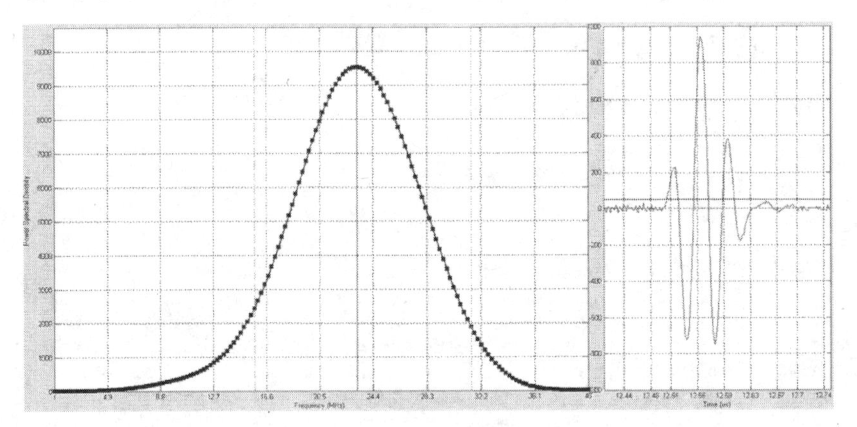

Figure 1. Acoustic spectrum of Olympus ultrasound transducer [8]

EXPERIMENTAL PROCEDURE

Non-destructive evaluation by ultrasound C-scan was performed on a commercially available, fully-dense silicon carbide tile. Scans were performed with an unfocused Olympus planar transducer at 20MHz in pulse-echo configuration, utilizing a 0.1mm lateral step size. As shown in Figure 1, the transducer has a bandwidth of 16-32 MHz when measured at the wavelength range where the output is one quarter the strength at the central emission frequency [8].

At each scanning position, an oscilloscope trace of amplitude vs. time was recorded, as shown in Figure 2. From the position of the corresponding surface reflection peaks, it is possible to determine the longitudinal wave and shear wave times of flight ($TOF_{Longitudinal}$ and TOF_{Shear}). The thickness of the sample (x) at each scanning position was calculated using a method detailed by Bottiglieri et al. [8].

Knowing these, both the longitudinal velocity (C_L) and the shear velocity (C_S) can be calculated [10].

$$c_{L,S} = \frac{2x}{TOF_{Longitudinal,Shear}} \qquad \text{Eqs. 1,2}$$

Poisson's ( ) ratio is calculated according to [10]:

$$= \frac{1 - 2(c_S/c_L)^2}{2 - 2(c_S/c_L)^2} \qquad \text{Eq. 3}$$

The elastic modulus can then be calculated according to the following equation [10]:

$$E = c_L^2 \; \frac{(1-2\;)(1+\;)}{(1-\;)} \qquad \text{Eq. 4}$$

By knowing the thickness of the sample, and taking the ratio of A_2, the amplitude of the second bottom surface reflection peak, to A_1, the amplitude of the first bottom surface reflection peak, it is possible to use an adaptation of the Beer-Lambert Law to determine the acoustic attenuation coefficient, in units of dB/cm, according to the following equation [11]:

$$= \frac{8.686}{2x}\, ln\,(A_2 / A_1)^2 \qquad \text{Eq. 5}$$

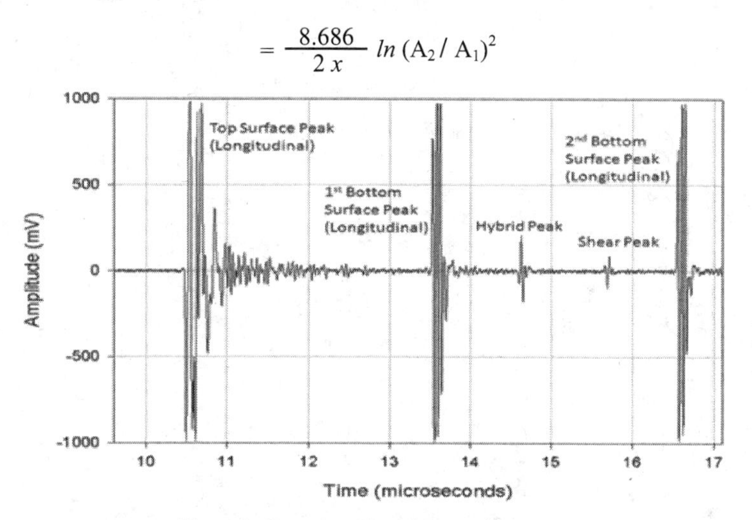

Figure 2. Representative ultrasound trace [8]

After assembling the values of the 20MHz attenuation coefficient and the elastic modulus at each scanning position into ultrasound maps, the tile was machined into ASTM C1161 B-type flexure bars of size 3mm x 4mm x 50mm [12]. One-hundred eight bars were machined from the 100mm x 100m x 14mm tile, such that there were 18 rows, a left and right column, and a top, middle, and bottom layer of bars. The original location, direction, and orientation of each bar relative to the original tile were maintained throughout the machining and testing operation. This was done so that a comparison could be made between the mechanical and acoustic properties of the bend bars. While the ultrasound maps are volumetric in nature, it was hoped that having three layers of bars would increase the likelihood of interacting with features represented in the NDE maps during flexure testing and microstructure evaluation.

The bars were broken by 4-point flexure testing according to ASTM C1161, using a semi-articulating test fixture with a crosshead speed of 0.5 mm/min [12]. Following this, the strength data was analyzed utilizing Weibull statistics to examine the strength distribution throughout the tile, while the primary fracture position was determined for each bar. Optical and field emission scanning electron microscopy were used to characterize the fracture behavior of the test samples, while energy-dispersive x-ray spectroscopy (EDS) was utilized to determine the composition of microstructural features.

RESULTS

Ultrasound Maps

Ultrasound C-scan maps of the 20MHz attenuation coefficient and elastic modulus can be found in Figure 3. The average attenuation coefficient value across the area of the sample was found to be 2.16 dB/cm, with a standard deviation of 0.06 dB/cm. The scale on the attenuation coefficient map varies between 0 and 3.0 dB/cm. Bands of reduced attenuation coefficient values are located along the left and right edges of the map. It is believed that these are caused by the interaction of the ultrasound energy and the glass slides placed underneath the tile at these positions. The average Young's modulus value was found to be 417 GPa, with a standard deviation of 0.6 GPa. The scale used in the elastic modulus map varies from 410 GPa to 424 GPa. The lowest values are located in the lower-center region of the map. A number of acoustic anomalies can be seen in each of the maps. The prominent anomaly that appears in both maps on the horizontal center line towards the left-hand side is believed to have been caused by a feature located in the bulk of the sample. The anomalies that correspond to areas of reduced Young's modulus were most likely caused by pits in the surface of the sample.

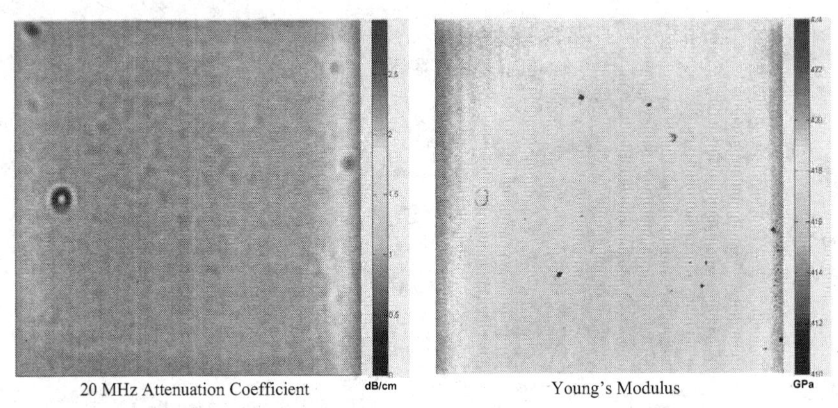

Figure 3. Ultrasound maps

Modulus of Rupture/Weibull Analysis

Bend bar fracture strength was consistent throughout the volume of the tile, although there were outliers, as can been in the Weibull plot in Figure 4. The repeatability of the fracture strength values seen in this study was what is to be expected for bars properly machined from a fully-dense, commercially available tile. The bars from the top and middle layers fractured at an average strength of 452 MPa and 454 MPa, respectively, while the bars from the bottom layer had a lower average strength of 435 MPa. MOR values varied from a maximum of 576 MPa to a minimum of 314 MPa, with an average value of 447 MPa. The majority of the low-strength bars were found in the bottom layer, which is consistent with the lower average fracture strength and lower Weibull modulus for this group of bars, as well as the outliers present in the Weibull plot.

Table I. MOR results / Weibull moduli

Bar Layer	Max (MPa)	Min (MPa)	Average (MPa)	Std dev (MPa)	M
Top	537	367	452	47	11.5
Middle	576	370	454	56	9.7
Bottom	522	314	435	60	8.6
Entire Tile	*576*	*314*	*447*	*55*	*9.9*

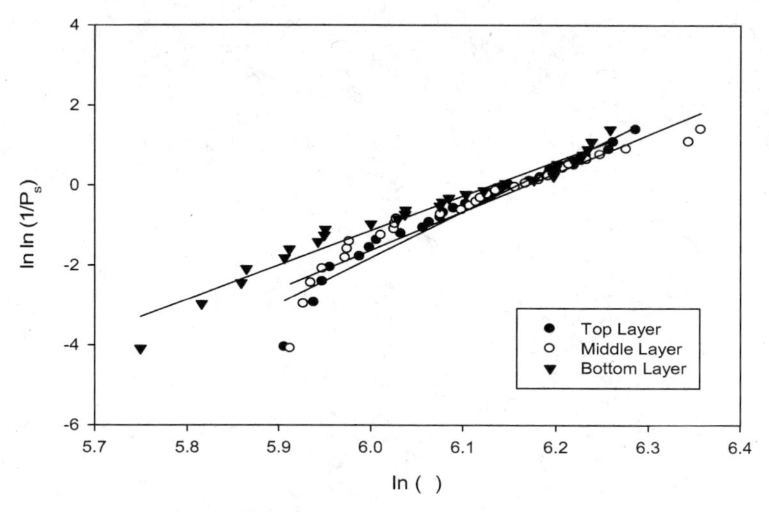

Figure 4. Weibull Analysis – Top, middle, and bottom layers

Primary Fracture Position/Ultrasound Map Overlay

As part of the strength testing analysis, the primary fracture position was determined for each bar. This is a necessary step to confirm the results of the strength testing. A scale diagram of bend bar positions was constructed in order to examine the correlation between features in the ultrasound maps and the fracture position. Fracture positions are indicated on the diagrams as colored marks. Positions from the top layer are shown in black, positions from the middle layer are shown in white, while positions from the bottom layer are shown in red. Where fracture occurred at the same lateral position on more than one layer, the location mark is shown as a combination of both colors. The primary loading zone of the bend test is also indicated in the diagrams by a black rectangle. When testing a B-type bend bar, this is a volume that is approximately 20mm long, 4.0mm wide, and 1.5mm high that experiences the full fracture stress just before rupture.

Fracture position overlay diagrams of the 20Mz attenuation coefficient map and the elastic modulus map can be found in Figures 5 and 6. There was only one instance where fracture occurred at the same lateral position on all three layers, and two instances were fractured occurred at the same position on two layers. The major acoustic anomaly in the attenuation coefficient map did not appear to affect the position of fracture of the surrounding bend bars. This may have been due to a number of reasons, which include that the feature causing the anomaly was located outside of the flexure test primary loading zone. Also, the ultrasound maps are volumetric in nature, and do not contain information as to where in the depth a feature may have been located. A work-around for this has been employed in subsequent ultrasound evaluation testing, where an additional detection gate is defined for the region in the A-scan between the top surface and 1st bottom surface reflection peaks. If a feature with a high enough acoustic impedance mismatch (such as a large pore) is encountered by the ultrasound beam, it will produce a strong reflection. By determining the TOF between the peak and the top or bottom surface reflections, it is possible to determine where in the depth of the sample the feature is located.

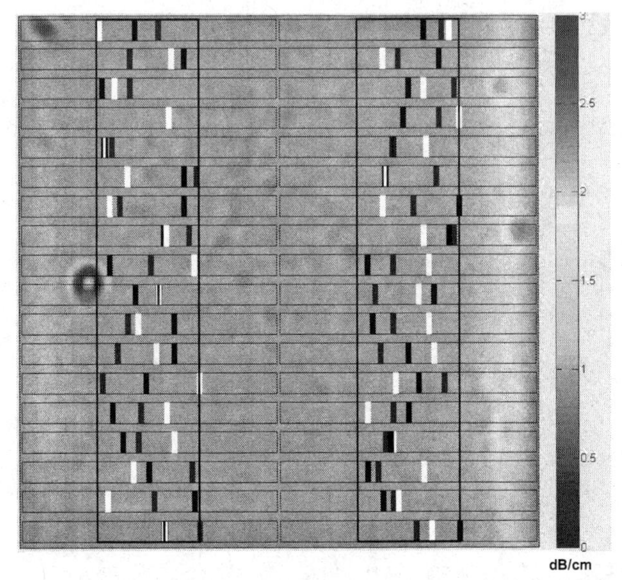

Figure 5. 20MHz Attenuation Coefficient Map / Fracture Position Overlay

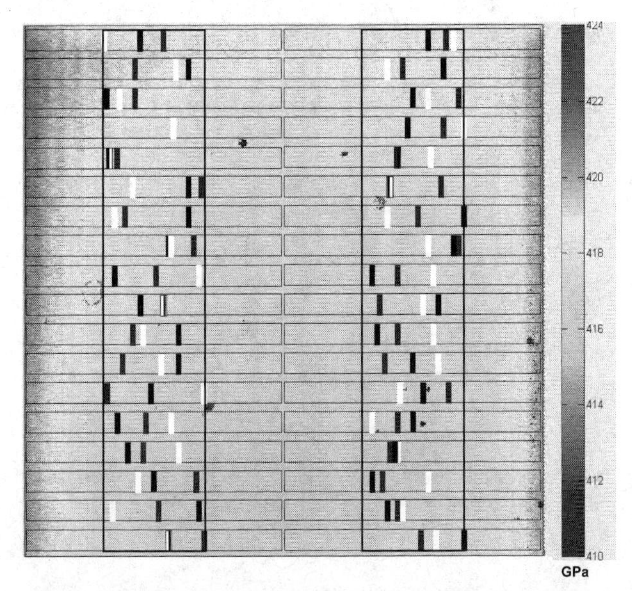

Figure 6. 20MHz Elastic Modulus Map / Fracture Position Overlay

A determination of the variability in the strength data can be made by looking at the relationship between the average strength and standard deviation for each layer of bars. For the top layer of bars, the standard deviation in the strength data was found to be 10.4% of the average strength value, followed by 12.3% and 13.8% for the middle and bottom layers, respectively. There was no discernible pattern as to how the strength values were distributed throughout the tile. When applying the same technique to the attenuation coefficient data, it was found that the standard deviation value of 0.06 dB/cm was only 2.8% of the average value. Furthermore, after a repeatability study performed with this test equipment it was found that the standard deviation of any C-scan attenuation coefficient measurements made was 0.05 dB/cm [8]. Therefore, the standard deviation of the attenuation coefficient values was only marginally higher than the detectable limit. For the Young's modulus map, the standard deviation was found to be less than 1% of the average value for the entire tile. These results show that the variation in strength is far greater than the variation in either attenuation coefficient or Young's modulus.

There are a number of conclusions that may be drawn from these results. One is that the microstructural features affecting strength are not the same ones that cause variation in either of the ultrasound maps, or that the effect of these same features on quasi-static strength and ultrasound energy is different. Another is that the variation of attenuation coefficient or Young's modulus values within the tile is too low to be significant. This study was done on a fully-dense, commercially available tile that did not show enough variation in ultrasound results. The correlation between strength and ultrasound may have been more significant if a purposely engineered sample was used. To produce a lot of variation in attenuation coefficient, a microstructure would need to include features that give rise to a large acoustic impedance mismatch between the feature and the matrix, such as a pore or metallic particle. Large variations in density would likewise cause variation in Young's modulus values.

Fractography

A number of microscopy techniques were employed for the determination of the critical flaw and the assessment of fracture behavior. One of the most useful procedures involved the mapping of fracture surface end faces with a scanning electron microscope. The fracture surfaces shown in Figures 7 and 8 are composite images of the entire primary surface end face. They are each comprised of 9-10 separate SEM micrographs, recorded at 200x magnification, that have been manually stitched together using image processing software.

The results for two bend bars will be discussed here and will be referred to as Bars I and II. These bars were chosen as they broke at similar fracture strengths, showed similar fracture behavior, and broke due to the same type of inclusion. The bars will be discussed in order of decreasing applied fracture stress. It was found on many of the bars examined that one of the sides of the primary fracture surface experienced secondary fractures during flexure testing. This led to remnants of the critical feature and primary fracture path being identified on only one side of the fracture surface end face.

On the composite end face images, the tensile edges of the bars have been indicated as a frame of reference. The left and right faces of the primary fracture surface of Bar I are shown in Figure 7. The results for this example were different than the other in that evidence of the primary fracture path and fracture initiating feature can be seen on both sides of the end face. Texture remaining in the fracture surface indicates that fracture likely initiated near the center of the bar. Bar I was originally from the second row from the top of the right hand column in the middle layer of the starting tile, and broke at a fracture stress of 383 MPa, which placed it in the lower fifteenth-percentile of fracture strengths.

Figure 7. Bar I. Fracture Surface End Faces – Composite [200x magnification]

Figure 8. Bar II. Fracture Surface End Face – Composite [200x magnification]

The image of the left side of the primary fracture end face of Bar II can be found in Figure 8. Bar II broke at a fracture stress of 378 MPa, placing it within the lower tenth-percentile of fracture strengths, and was originally from the eleventh row of the right hand column in the middle layer of the starting tile. Only one portion of the primary end face was preserved after fracture, which appears to have begun in the lower-right portion of the end face.

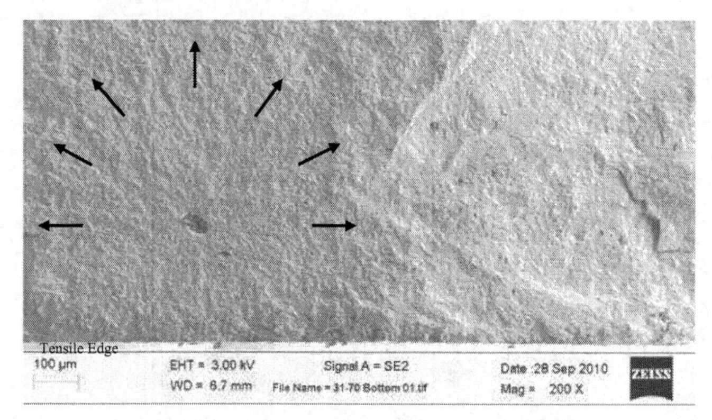

Figure 9. Bar I. Fracture Path [200x magnification]

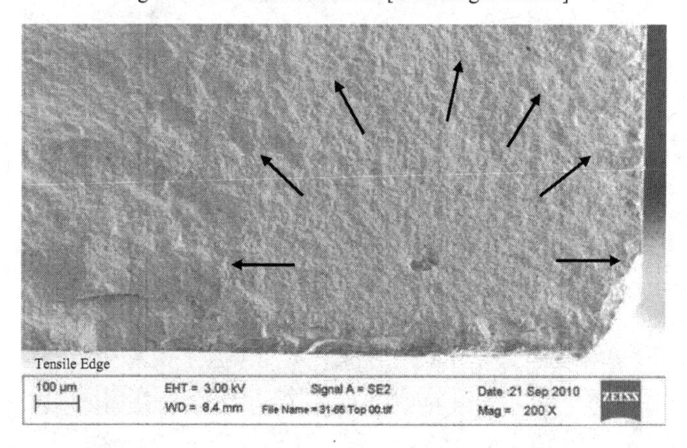

Figure 10. Bar II. Fracture Path [200x magnification]

Figures 9 and 10 each show a close-up view of what are believed to be the fracture initiating features for the two bend bars. Arrows have been added to the images to indicate the likely fracture path. While the path of fracture is very complex, in these images, it appears to radiate out along the direction of the arrows in a smooth pattern before changing direction as the stress field changes as fracture proceeds. Figure 9 contains a SEM micrograph from the central portion of Bar I that shows the region around the fracture initiating feature. A similar region from the lower-right area of Bar II is shown in Figure 10.

Higher magnification images of the likely fracture initiating features can be found in the next group of SEM micrographs. Figures 11a and 11b show the two sides of the likely critical feature from Bar I at 3750x magnification. EDS analysis showed that this feature is comprised of boron carbide, and is most likely an agglomerate of sintering aid added during the mixing process.

The major and minor axes were measured at 59 m and 34 m, respectively, while the minimum distance to the tensile surface of the bend bar is 235 m. Taking this factor into account, along with the height of the bend bar, an estimate of the amount of stress at the feature can be calculated by taking a ratio of the stress at the tensile edge, the location of maximum stress, to the stress at the location of the feature. For this bar, this stress was calculated to be 323 MPa. It is possible to see a plate-like structure that appears on both sides of the fracture surface, as indicated by arrows. Dimpled and raised areas of the feature match-up on either side, indicating that this plate was split during fracture.

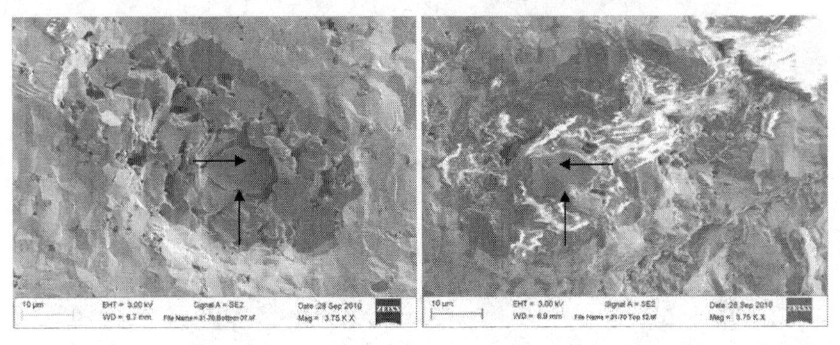

Figures 11a and 11b. Bar I. Boron Carbide Inclusion [3750x magnification]

Figure 12 contains a composite SEM micrograph, recorded at 7150x magnification, of the critical feature from Bar II. The minimum distance between the feature and the tensile edge of the bend bar is 191 m. The major axis was measured at 66 m, while the minor axis was measured at 30 m. In comparison to the fracture stress of 378 MPa, the stress applied at the feature was estimated to be 315 MPa. EDS analysis confirmed that the feature was comprised of boron carbide. In these two high-magnification images, it can be seen that transgranular fracture is the primary fracture behavior. Also of note is the presence of many micron-sized inclusions, which were found across the fracture surfaces. These inclusions were found to be comprised of both carbon and boron carbide.

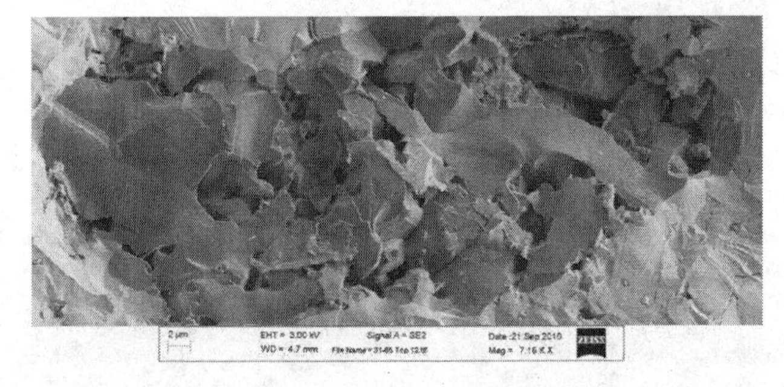

Figure 12. Bar II. Boron Carbide Inclusion [7150x magnification]

SUMMARY

Non-destructive evaluation by ultrasound C-scan @ 20MHz was performed on a fully-dense pressureless-sintered silicon carbide tile. Values of the acoustic attenuation coefficient and elastic modulus were calculated at each scanning position and assembled into two-dimensional maps.

The tile was machined into ASTM B-type bend bars for modulus of rupture testing. The bars were machined such that there were eighteen rows of bars, two columns, and three layers of bars. The position, orientation, and location of each bar relative to the original tile were maintained in order to correlate the mechanical properties of the bend bars to features on the NDE map. Following MOR testing, the primary fracture position for each bar was determined and plotted on a scale diagram of bend bar positions overlaid on the ultrasound maps.

Fractography was carried out by optical and scanning electron microscopy on selected samples. Composite micrographs of the primary fracture end faces were compiled from multiple SEM images that were compiled with image processing software. From analysis of these images, along with energy dispersive spectroscopy, the likely fracture path and critical feature were determined. Fractographic analysis showed that the bars discussed fractured at agglomerates of boron carbide.

Average MOR for all bars was found to be 447 MPa, with a minimum and maximum value of 314 MPa and 576 MPa, respectively. As a percentage of the average, the standard deviation of MOR values for the top, middle, and bottom layers was found to be 10.4%, 12.3%, and 13.8%, respectively. While fracture strength was consistent throughout the tile, the majority of low-strength bars were located on the bottom layer. After analysis of the fracture position overlay diagrams, it was found that fracture position did not correlate with the presence of any acoustic anomalies in the maps, which may have been caused by pits in the surface of the original tile or were located outside of the primary loading zone of the fracture test. It was determined that the variation of the acoustic attenuation coefficient was only marginally greater than the standard deviation of the test setup. Variation in the Young's modulus map were even lower at less than 1%. It was believed that the variation in these values was insufficient to produce a correlation with the strength measurements. The likelihood of determining if a correlation exists would be increased by producing and testing samples that would provide a greater amount of variation in ultrasound results.

ACKNOWLEDGEMENTS

The authors would like to thank the NSF/IUCRC - Ceramic, Composite, and Optical Materials Center (CCOMC) as well as the US Army Research Laboratory - Materials Center of Excellence (MCOE) for funding this research.

REFERENCES

1. Karandikar, P.G., G. Evans, S. Wong, M. K. Aghajanian, and M. Sennett, *A Review of Ceramics for Armor Applications.* Proceedings of the 32nd International Conference on Advanced Ceramics and Composites, 2008.
2. E.R. Maddrell, *Pressureless Sintering of Silicon Carbide.* Journal of Materials Science Letters, 1987. p. 486-488.
3. Williams, R.M., Juterbock, B.N., Peters, C.R., Whalen, T.J., *Forming and Sintering Behavior of B- and C-Doped a- and b-SiC.* Communications of the America Ceramic Society, 1984. **C**: p. 62-64.
4. S. Prochazka and R. M. Scanlan, *Effect of Boron and Carbon on Sintering of Silicon Carbide*, Journal of the American Ceramic Society, 1975. **52**(1–2): p. 72.
5. Stobierski, L., and Gubernat, A., *Sintering of Silicon Carbide I. Effect of carbon.* Ceramics International, 2003. **29:** p. 287-292
6. Bakas, M.P., *Analysis of Inclusion Distributions in Silicon Carbide Armor Ceramics*, 2006. PhD Dissertation, Rutgers, The State University of New Jersey.

7. Rogers, W.P., and Srinivasan, M., *Ultrasonic Detection of Surface Flaws in Sintered Alpha Silicon Carbide.* Engineering and Science Proceedings, Volume 5, 1984. p. 603-613.
8. Portune, A.R., *Nondestructive Ultrasonic Characterization of Armor Grade Silicon Carbide,* 2010. PhD Dissertation, Rutgers, The State University of New Jersey.
9. Bottiglieri, S., Haber, R.A., *Corrective Techniquesfor the Ultrasonic Nondestructive Evaluation of Ceramic Materials.* Advances in Ceramics Armor VI: Ceramic Engineering and Science Proceedings, Volume 31, 2010. p. 57-67
10. Brennan, R., et al., *Elastic Property Mapping Using Ultrasonic Imaging.* Ceramic Engineering and Science Proceedings, 2008. **28**(5): p. 213-222
11. Krautkramer, J., and Hrautkramer, H., *Ultrasonic Testing of Materials.* 4th Fully Revised ed. 1990, Berlin: Springer-Verlag.
12. ASTM Standard C1161, "Standard Test Method for Flexural Strength of Advanced Ceramics at Ambient Temperature," ASTM International, West Conshohocken, PA, DOI: 10.1520/C1161, www.astm.org.

Nondestructive Characterization

ULTRASONIC NONDESTRUCTIVE CHARACTERIZATION AND ITS CORRELATION TO ALUMINA MICROSTRUCTURE

S. Bottiglieri and R. A. Haber
Department of Materials Science and Engineering, Rutgers University
Piscataway, NJ, USA

ABSTRACT

Ultrasonic nondestructive evaluation is currently used to spatially locate heterogeneities within dense ceramics and measure the elastic properties of these materials. The use of acoustic spectroscopy in dense ceramics to measure heterogeneity type, size, and composition has had little investigation. Understanding the acoustic interaction within a ceramic material and the causes of energy loss within different frequency regimes can lead to a predictive method of using acoustic spectroscopy to understand microstructural parameters. This study focuses on the development of acoustic spectroscopy as a tool for microstructural characterization in dense aluminum oxide. As part of this research specific alumina sample sets were fabricated to control and vary different microstructural features such as additive content and grain size. Independently controlling certain aspects of the alumina microstructure lead to a better understanding of how the microstructure causes energy loss over an acoustic attenuation coefficient spectrum.

INTRODUCTION

Acoustic spectroscopy has been proven as a useful tool for measuring particulate size in dilute suspensions, determining average grain size in metallic systems, and understanding differences in mechanical properties such as hardness in metals [1, 2, 3]. When analyzing an acoustic spectrum one must consider the frequency bandwidth emitted by the transducer and which attenuation mechanisms are operable in said frequency ranges. Different materials systems, such as particulate suspensions, metals, glasses, and ceramics will cause different types of ultrasonic loss. Scattering and absorption are the two general regimes of acoustic loss mechanisms which have been studied. Understanding how scattering and absorption occur in different material systems allows for the calculation of microstructural information by the deconvolution of measured attenuation spectra. Particulate size distributions and average grain sizes in metals may be measured through the use of acoustic spectroscopy and knowledge of how scattering and absorption take place in these specific systems [1, 2].

The overarching regimes of absorption and scattering are further broken down into specific types of absorption and scattering; for example, thermoelastic, viscoelastic, thermal conduction, hysteresis, phonon-phonon interaction, and Rayleigh scattering, stochastic scattering, diffuse scattering, respectively [4]. Depending on the frequency range and material studied different absorption and scattering mechanisms will be prevalent. For example, in metallic systems scattering due to grains dominates losses seen in an attenuation spectrum [5]. Using high frequency ultrasound (in the megahertz range) to measure attenuation coefficient spectra in polycrystalline ceramic materials it has been shown that thermoelastic absorption dominates losses seen at approximately 50MHz and below [6]. At frequencies higher than approximately 50MHz scattering begins to overwhelm the attenuation curves and is the prominent loss mechanism. The frequency regimes where specific loss mechanisms are dominant are not well defined and may change for different types of ceramic materials. Frequencies where absorption and scattering strongly overlap require a means of separating the two such that useful information may be obtained. This paper focuses on studying the frequency range where thermoelastic absorption is operable (between 10 and 30MHz) for a set of tailored engineered alumina samples. These samples have been fabricated to have increasing concentrations of 2:1 mullite ($2SiO_2 - Al_2O_3$).

THEORY

Intraparticle thermoelastic absorption is the coupling of the elastic and thermal fields created by the propagating acoustic wave [7]. Local changes in temperature are controlled by the coefficient of thermal expansion and the direction of the acoustic wave. The regions of the material that are in rarefaction will cool slightly while regions in compression will heat up. Heat flow between the compressed and rarefacted regions occurs to seek equilibrium and maintain field continuity between the elastic and thermal fields. This heat flow is irreversible which causes a loss in energy which is in turn measured as acoustic attenuation [7]. In his development on the internal friction in solids, Clarence Zener came up with an expression which relates the frequency of maximum attenuation for a single particle to the size of the particle and its thermal properties (equation 1) [8]. This relationship is used as the basis in the derivation of how much energy a single particle absorbs.

$$f = \frac{\pi\chi}{2\rho C_v a^2} \tag{1}$$

Where f is frequency, a is particle diameter, χ is thermal conductivity, ρ is density and C_v is the specific heat at constant volume.

This paper makes the assumption that energy loss due to thermal elastic absorption is equal to the number of absorbing particles multiplied by the volume of those particles multiplied by a scaling factor as seen in equation 2. This is assumed that more particles will be capable of absorbing more energy and that larger particles bend more in response to the acoustic wave in comparison to smaller particles. As smaller particles have a larger curvature, they are unable to bend as much as larger particles as an acoustic wave passes through. The increase in the size of compression and rarefaction of the particles increases the length of thermal flow to seek equilibrium and thereby increases the amount of energy which must be absorbed. The scaling factor, S, is to account for the nonuniformity of the energy output from the transducer across its bandwidth. The volume of the particle is dictated to be a function of frequency and assumed to be spherical after inverting equation 1 to solve for diameter. This volume can be calculated based on the thermal properties of the material of interest and the frequency range used. The assumption of spherical particles is for convenience and this may deviate strongly for specific materials which have large aspect ratios. Assuming perfect spheres may underestimate the true volume of the particles interacting with the ultrasonic wave which will cause a small overestimate of the number of particles present in the path of the wave.

$$E_{TE}(f) = S(f) \cdot V_{TE}(f) \cdot N_p(f) \tag{2}$$

Where the energy absorbed is in units of joules and volume in m^3; S must have units of Pascals as a result. The second assumption of this derivation is that particles of a specific size only absorb at frequencies as dictated by equation 1. Another way of stating this assumption is that the thermoelastic absorption occurs on a time scale that can be thought of as instantaneous. A final assumption is that only thermoelastic absorption causes energy loss in the frequency range of interest. This assumption may be remedied if one has considerable scattering losses at these frequencies and can include corrections for such losses.

The energy measured is equal the initial acoustic energy introduced into the specimen by the transducer minus the energy lost due to thermoelastic absorption (equation 3).

$$E_m(f) = E_i(f) - E_{TE}(f) \quad (3)$$

The initial energy will be dependent on the frequency bandwidth of transducer output and can be measured by performing a Fourier transform on a transducer delay line peak measured in an oscilloscope. The result is a power spectral density which has units of W·s, or joules. From the Beer-Lambert law it is known that:

$$E_m(f) = E_i(f) \cdot e^{-\alpha(f) \cdot x} \quad (4)$$

Where α is the measured attenuation coefficient (corrected for geometric losses) and x is sample thickness [9]. Substitution and rearrangement yields an equation for the S parameter which is a function of both the transducer output and the measured sample attenuation (equation 5).

$$S(f) = \frac{E_i(f) \cdot (1 - e^{-\alpha(f) \cdot x})}{N_p(f) \cdot V_{TE}(f)} \quad (5)$$

To determine the S parameter E_i was measured (Figure 1) and a 100% alumina sample was created in a spark plasma sintering (SPS) unit to use as a standard for the number of particles. A particle size distribution for the Al_2O_3 sample was measured from SEM micrographs and then normalized to be equal to what volume is contained in the path of the ultrasound beam. This can be seen in Figure 2 below.

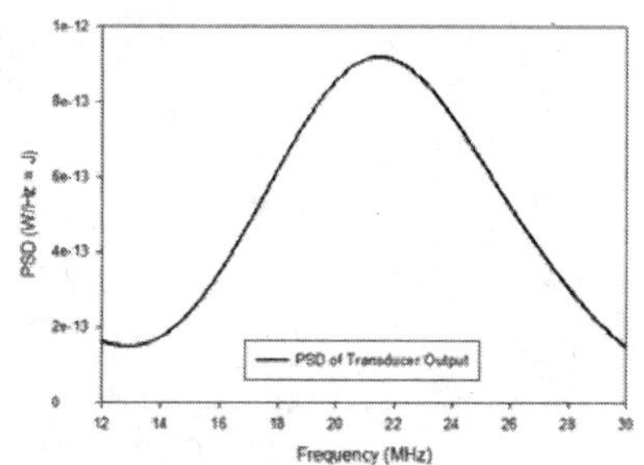

Figure 1. Power spectral density of 20MHz central frequency transducer (units of J).

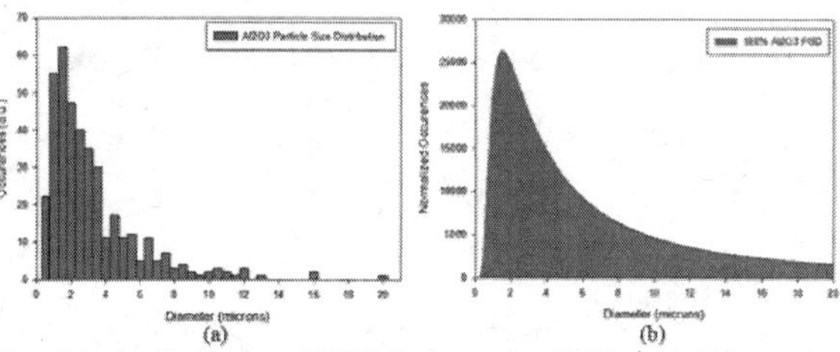

Figure 2. (a) Measured particle size distribution of 100% Al_2O_3 sample. (b) Ultrasound beam volume normalized particle size distribution of 100% Al_2O_3 sample.

Taking the PSD output of the transducer used, the measured attenuation coefficient, and the particle number data from the size range of 0.72 – 1.25μm (dictated by equation 1) a form for the S parameter is obtained. However, this parameter is in units of Pascals and dependent on both the transducer output energy and the volume of the Al_2O_3 grains used as a standard. To obtain a form for the S parameter that is not dependent on the material but purely the transducer, we multiply S by the particle surface area at each frequency to obtain the force output of the transducer (Figure 3). After algebraic rearrangement, equations 6 and 7 give the energy loss due to thermoelastic absorption of any number of particles and a form for the number of particles contained in the path of the ultrasound beam, respectively.

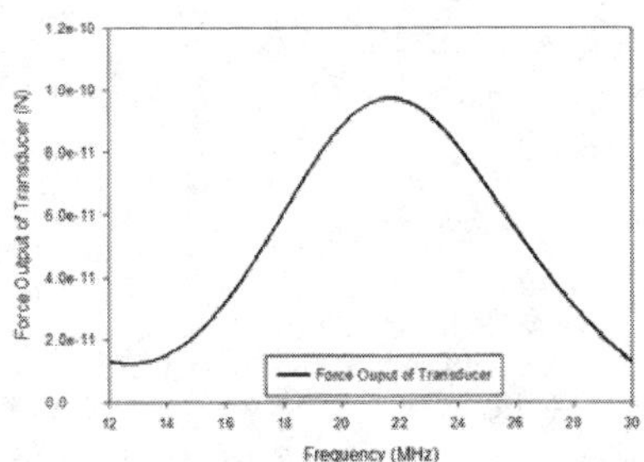

Figure 3. Force output of 20MHz central frequency transducer (units are in N).

$$E_{TE}(f) = \frac{2}{3} F_T(f) \cdot D_{TE}(f) \cdot N_P(f) \tag{6}$$

Where F_T is the force output of the transducer and D_{TE} is the diameter of the particles (as given by equation 1).

$$N_P(f) = \frac{\pi \cdot E_i(f) \cdot (1 - e^{-\propto(f) \cdot x})}{4 \cdot F_T(f) \cdot D_{TE}(f)} \tag{7}$$

Experimental validation of equation 7 involves the creation of tailor-made sample sets as described below.

EXPERIMENTAL PROCEDURE

Degussa Aerosil silica powder, with an average size of 12nm, was added to A16 alumina powder, average size of approximately 1μm, in incremental amounts between 1 and 5 wt%. These five powder mixtures were ball milled in isopronal for three hours and allowed to dry to collect the fully mixed products. An SPS unit was used to densify six grams of each powder mixture. The sintering cycle for each sample used a ramp of 200°C/ min to 1600°C and 3.75MPa/ min to 30MPa and held for 15 minutes. The sintered products were approximately 20mm in diameter and 5mm thick. Each sample had its density measured via Archimedes method shown in Table I. The densities measured were approximately 98% or higher of theoretical based on what was expected to form (from an Al_2O_3-SiO_2 phase diagram – alumina rich mullite) and a linear rule of mixtures.

This study used pulse-echo, immersion-based (distilled, deionized water), automated scanning ultrasound equipment. A 20MHz central frequency transducer with a bandwidth of 10 – 30MHz was used to cover most of the frequency range where thermoelastic absorption occurs. A higher frequency transducer (50MHz central with a bandwidth of 40 – 62MHz) was used to observe scattering behavior. These transducers were acquired commercially through Olympus. The analog to digital converter card (Gage Electronics) has a 3GHz sampling rate which allows for the collection of data points every 0.3ns. The oscilloscope, automated motion, and data acquisition is controlled by a personal computer and software developed at the Rutgers Center for Ceramic Research. The motion control unit and scanning frame were acquired through Techno-Isel and the Pulser-Receiver was furnished by JSR Instruments.

Ultrasonic nondestructive characterization makes use of three modes of data collection; an A-scan which is a point measurement, a B-scan which is a line measurement, and a C-scan which is a fully mapping of a sample piece. Data is measured through the use of electronically gating specific peaks indicative of sample surface reflections. Time of flight and elastic property measurements are made by recording how long it takes the sound wave to travel through the sample and amplitude and attenuation measurements are collected by measuring the relative change of peak intensity from one bottom surface reflection to another [4]. The ultrasonic data collection mode used for the samples in this study was an A-scan. Speed of sound, elastic properties, and attenuation coefficient spectra were recorded for all samples (Table I).

Table I. Density and ultrasound measurements of the tailor engineered alumina sample sets.

Sample (% Al2O3)	Density (g/cc)	cL (m/s)	cS (m/s)	Poisson ratio	Young's Modulus (Gpa)	Shear Modulus (Gpa)	Bulk Modulus (Gpa)
95	3.76	9908	5945	0.219	324	133	192
96	3.77	10212	6028	0.233	338	137	210
97	3.83	10310	6056	0.237	347	140	220
98	3.84	10414	6040	0.247	349	140	230
99	3.92	10576	6197	0.242	374	151	242
100	3.95	10623	6229	0.234	378	153	237

After ultrasonic testing was performed the samples were sectioned for ceramographic preparation and X-ray diffraction (XRD) analysis. Portions of each sample were set aside for polishing, scanning electron microscopy (SEM), and particle size measurements. This process has yet to be performed and results are not shown in this paper. XRD measurements were made on each sample using a Phillip's XRD unit. Scans between 10 and 80 two theta with a step size of 0.03° and dwell time of 2s were taken on each sample. Results show the formation of 2:1 mullite of increasing concentration in the samples containing between 1 and 5 wt% SiO_2. Comparison between ultrasound results and volume percent of mullite show a one-to-one correlation.

RESULTS AND DISCUSSION

Figure 4(a) shows the measured attenuation coefficient spectra on a range of 10 – 62MHz. The disjoint region is due to not having bandwidth overlap between the two transducers used. The lower frequency (10 – 30MHz) data (Figure 4(b)) is the frequency range where it is assumed that only thermoelastic absorption is the dominating loss mechanism. The gap in frequency (30 – 40MHz) contains loss spectra that are a convolution between the thermoelastic absorption and scattering mechanisms. At these frequencies it is expected that scattering begins to become the dominant loss mechanism. To use information from these frequencies would require a transducer to collect the data and a method of deconvolution to correct for the effects of scattering which get stronger as frequency is increased. The scattering, power law, behavior at higher frequencies is neglected in further analysis. It is assumed that the increase in attenuation from the base 100% Al_2O_3 sample spectra is purely due to thermoelastic absorption. A correlation exists between the amount of the mullite phase present and the increase in attenuation coefficient. The analysis below corrects for the increase in attenuation and decrease in alumina as the concentration of the secondary mullite phase increases. This was done by knowing the volume percent of mullite in each sample from XRD measurements.

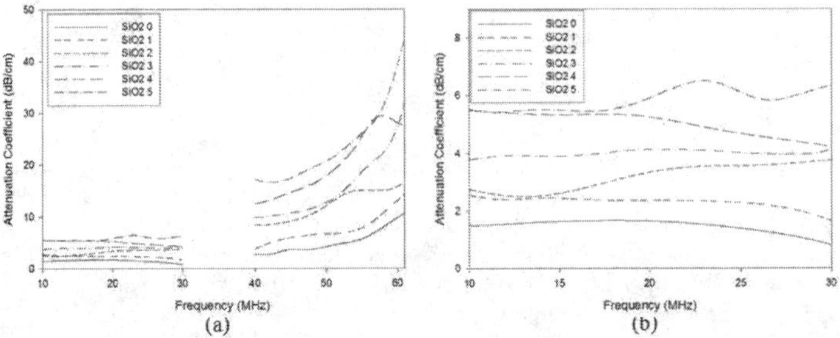

Figure 4. Attenuation coefficient spectra of tailor-made samples: (a) 10 – 62MHz (b) 10 – 30MHz.

XRD spectra are shown in Figure 5 (1wt% SiO_2 – 5wt% SiO_2, (a) – (e) respectively). It can be seen (marked by arrows) that these samples have an increase in the amount of the secondary mullite phase as the SiO_2 concentration increased. This is to be expected and provides a well defined sample set which correlates to every ultrasound property measured. The volume percents, calculated by XRD reference intensity ratio method, of 2:1 mullite which formed in samples having between 1 and 5wt% SiO_2 are 2.1%, 5.6%, 16.3%, 20.6%, and 36.9%, respectively.

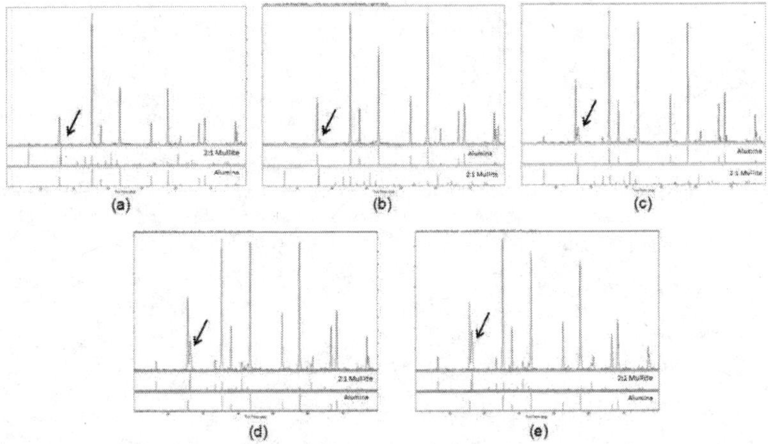

Figure 5. XRD spectra of sample set containing between 1 and 5wt% SiO_2 ((a) – (e), respectively). Arrows mark major 2:1 mullite peak.

Using the derived equation 7, the attenuation coefficient spectra from Figure 4(b), and accounting for the relative percents of mullite in each sample obtained from XRD, ultrasonic measurements of particle size distributions were measured on a size scale corresponding to the frequency range as dictated by equation 1. The ultrasonic particle size distributions are shown below in Figure 6. According to equation 1 and using a frequency range of 10 – 30MHz, a diameter size range for mullite will be between 0.25 – 0.40μm. The five ultrasound particle size distributions were fit to Gaussian distributions to reconstruct the lower and upper bounds of full particle size curves [10]. A lower size limit was set to 0.1μm and the upper limit was allowed to be as large as necessary. The total volume of the mullite present in the path of the ultrasound beam was calculated from XRD measurements. The reconstructed particle size curves for each sample were extended on the lower and upper bounds from the ultrasound particle size curves such that the total area under the curves were approximately equal to the total volume of the mullite in the volume of the ultrasound beam. The reconstructed curves are shown in Figure 7.

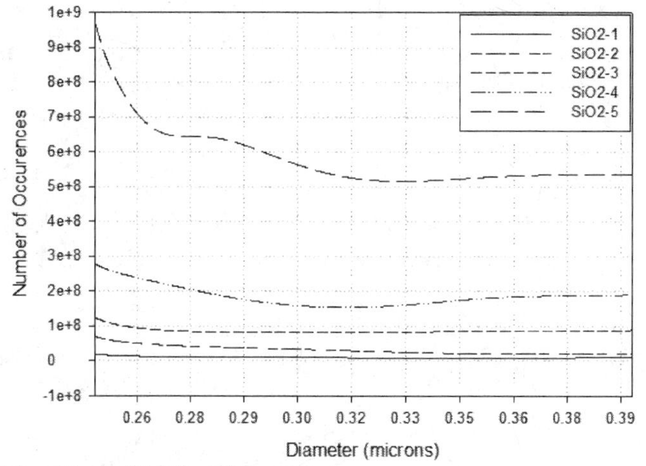

Figure 6. Ultrasonic particle size distribution of SiO_2 sample set.

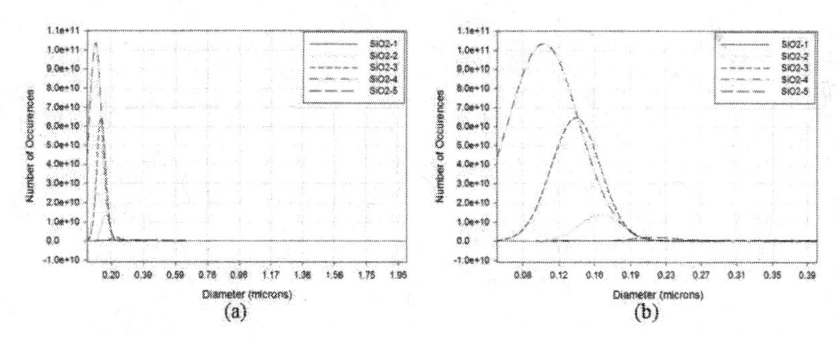

The reconstructed particle size distributions show an increase in mullite grains as there is an increase between 1 – 4wt% SiO_2. The average particle size gets smaller during this increase. For sample SiO_2-5wt% SiO_2 a different trend is observed; the average particle size increases to approximately 0.21μm and there are more occurrences of mullite grains up to about 2μm. This general increase in particle size causes a decrease in the total number of grains at each diameter while maintaining the total volume of mullite in this particular sample. Work is currently being performed to reveal the microstructures of these samples containing varying concentrations of mullite which will aid in verifying the theory put forth in this paper. This includes polishing to a mirror finish, chemical etching with hydrofluoric acid, and SEM imaging for grain size analysis.

CONCLUSION

The theory of thermoelastic absorption has been studied and extended in this paper. Utilizing acoustic spectroscopy as a method of particle sizing has been tested on five well-defined samples containing incremental amounts of the secondary phase mullite. All ultrasonic measured properties show a one-to-one correspondence with the volume concentration of mullite present in each sample. An analytical expression has been derived relating the number of grains to the thermal properties of a specific phase and the measured acoustic attenuation coefficient. An ultrasonic particle size distribution can be measured using this expression and a full scale ultrasonic particle size distribution can be reconstructed by knowing the total volume percent of the secondary phase present via XRD. Future work will verify the reconstructed ultrasound particle size distributions by obtaining measured grain size distributions from SEM micrographs.

ACKNOWLEDGEMENTS

The authors would like to thank the NSF IUCRC Ceramic and Composite Materials Center and the Army Research Laboratory Materials Center of Excellence for Lightweight Vehicular Armor.

REFERENCES

[1] Dukhin, A. and Goetz, P., *Ultrasound for Characterizing Colloids: Particle Sizing and Zeta Potential Rheology*, Publisher: Elsevier Science, Amsterdam, Netherlands, 2002.

[2] Papadakis, E. (1965) "Ultrasonic attenuation caused by scattering in polycrystalline metals." *The Journal of the Acoustical Society of America*, vol. 19, no.10, pp. 940-946.

[3] Vary, A. (1980) "Ultrasonic Measurements of Material Properties." *IEEE Ultrasonics Symposium*, pp.735-743.

[4] Portune, A. R., "Nondestructive Ultrasonic Characterization of Armor Grade Silicon Carbide", Ph.D. Thesis, Rutgers University (2010).

[5] Blitz, J. and Simpson, G., *Ultrasonic Methods of Non-Destructive Testing*, Publisher: Chapman and Hall, London, U.K., 1996.

[6] Bottiglieri, S. and Haber, R. A., "Ultrasonic Nondestructive Characterization of Transparent Spinel.", QNDE Conference, San Diego CA, July 2010.

[7] Bhatia, A.B., *Ultrasonic Absorption: An introduction to the theory of sound absorption and dispersion in gases, liquids, and solids*, Publisher: Oxford Press, 1967, pp. 266 – 273.

[8] Zener, C., "Internal Friction in Solids." Proceedings of the Physical Society, vol. 52, pp. 152-167, 1940.

[9] Bottiglieri, S and Haber, R. A., "Corrective Techniques for the Ultrasonic Nondestructive Evaluation of Ceramic Materials.", 34th ICACC, Daytona Beach, FL, 2010.

[10] Fredlund, M. D. and Fredlund, D. G., "An equation to represent grain size distribution." *The Canadian Geotechnical Journal*, vol. 37, pp. 817-827, 2000.

LOW VELOCITY IMPACT DAMAGE CHARACTERIZATION OF TRANSPARENT MATERIALS

Raymond E. Brennan and William H. Green
US Army Research Laboratory
Aberdeen Proving Ground, MD, 21005-5066

ABSTRACT

Advanced transparent materials are utilized to improve protection efficiency for lightweight vehicles and warfighters in applications such as face shields, riot gear, and vehicle windows. If any damage occurs, the ability to withstand single or multiple hits from various threats could be compromised. While these issues are most likely to occur due to impacts from high velocity projectiles during combat, they may also be the result of low velocity impacts from collisions, severe environmental conditions, or foreign object debris. In this study, transparent materials will be tested by comparing baseline conditions to experimentally controlled damage states. Destructive testing including air gun and sphere impact testing will be used to simulate low velocity impacts in the field. Characterization of the damaged state will include visual inspection, cross-polarization, x-ray, and ultrasound techniques. The combination of destructive testing and characterization of the resulting damage can help to establish a damage acceptance criterion for transparent materials used in protective systems.

INTRODUCTION

Individual transparent materials used in protective systems typically consist of glass, polymeric and ceramic materials. These material components are often stacked and adhered by polymer interlayers to form transparent laminate protective systems[1]. The presence of potentially harmful internal defects in these individual materials (pores, inclusions, secondary phases, etc.) and interlayer defects in the laminates (disbonds, delaminations, etc.) can reduce material properties but may not be visually detectable if index-matched.

Nondestructive bulk characterization techniques can be utilized in the pre-impacted state to detect material inhomogeneities and improve quality control for transparent materials before they are utilized in the field. They can also be used post-impact to detect resulting damage or to compare baseline and damaged states for determination of critical impact conditions.

Current strike face glasses used in transparent laminate protective systems are limited in how thin they can be fabricated before encountering durability issues. Lower density novel glass compositions have been developed that can be fabricated more than ten times thinner while maintaining their durability. By reducing the thickness and lowering the density of the strike face glass, the overall weight will also be reduced. Weight reduction is desirable for vehicle systems to increase maneuverability and transportability while reducing operational costs[1]. This study will focus on a comparison of current and novel glasses used in transparent laminate systems and their ability to withstand low velocity impact damage. Success will be measured qualitatively by how severe the damage impairs visibility and quantitatively by the damage measured through nondestructive imaging methods.

EXPERIMENTAL

The transparent materials chosen for this study were four 14-inch by 14-inch laminate panels. Identical 740-series glass laminate panels, denoted 740-1 and 740-2, each contained three glass and two adhesive layers. The thickness of each panel was ~16 mm. The layers consisted of a strike face layer (Glass A), an adhesive bonding layer, a second glass layer (Glass C), a second adhesive bonding layer, and a polymer backing layer. Two additional 741-series glass laminate panels were fabricated with four of the same five layers as the 740-series. In these panels, the strike face glass layer was changed to a thinner and lower density Glass B, making the total thickness ~11 mm. The 740 and 741-series panels were subject to the same low velocity impact conditions and compared to one another.

Air gun testing was used on 740-1 and 741-1 as a higher mass/lower velocity technique. This method utilized ~19 mm steel spheres to impact the panels. The average impact velocity for the air gun tests was ~30 m/s. In contrast, sphere impact testing was used on 740-2 and 741-2 as a lower mass/higher velocity technique. Impact testing was conducted using ~5 mm steel spheres. Each sphere was launched from a pneumatic launcher at velocities averaging ~400 m/s. For both techniques, a high-speed camera was set up 90° to the projectile path to measure impact velocity.

Visual characterization was conducted on pre-impacted panels by observing visible defects and on post-impacted panels by identifying visible damage. Each panel was illuminated by white light from a light box and digital images were collected for documentation of the pre and post-impacted states of each panel (Figure 1). Cross-polarization techniques were used to observe residual stress states in the transparent laminate systems. The panels were placed between polarized films oriented 90° to one another and illuminated by white light from a light box. When polarized light passed through each panel, the components of the light wave that were parallel and perpendicular to the direction of the stress propagated at different speeds[2-3]. This effect, referred to as retardation, was proportional to the degree of residual stress in the material. It was visibly displayed as a series of fringes that varied in color and intensity. The degree of stress in the transparent laminates was interpreted based on the color patterns by using the Michel-Levy birefringence chart[4]. Digital cross-polarized images were collected for documentation of the pre and post-impacted states of each panel (Figure 1).

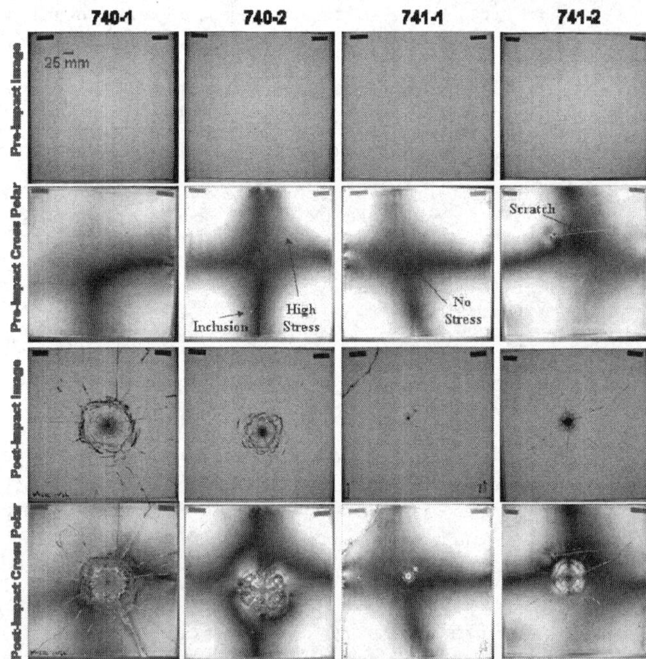

Figure 1. Visual characterization and cross-polarized digital images pre and post-impact.

For radiographic inspection, a beam of penetrating radiation was passed through the transparent laminate systems in a non-invasive manner. While x-ray digital radiography (DR) was used as a 2-D method to produce digital images of the full panels, x-ray computed tomography (XCT) provided densitometric images of thin cross-sections through each panel (Figure 2)[5-6]. Projection DR and XCT were performed through the thickness of the transparent laminate panels using a computed tomography system with a 420 kVP x-ray source and a 512 element linear detector array. The tube energy and current used were 400 keV and 2.0 mA, respectively, and the focal spot was 0.80-mm. The source-to-image-distance and source-to-object-distance were 940.00 mm and 750.00 mm, respectively.

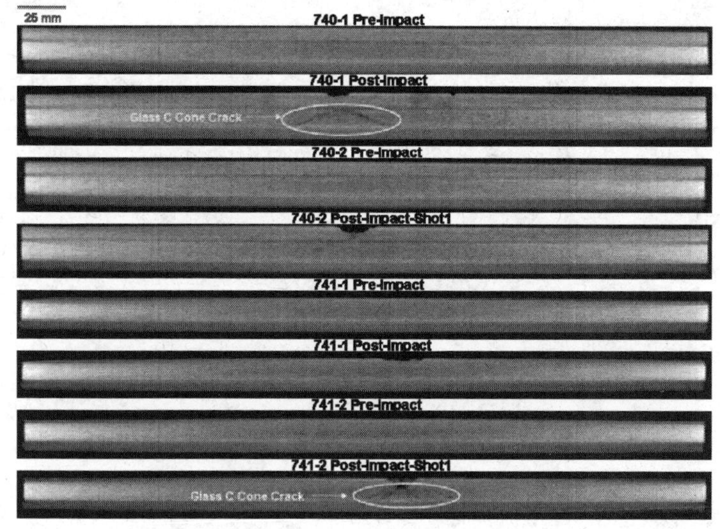

Figure 2. XCT cross-sectional images pre and post-impact.

Ultrasound (UT) characterization was used to detect material variations, defects, and damage in the transparent laminate panels. As the acoustic waves were transmitted into the panels, material changes in individual glass or polymer layers (pores, inclusions, cracks) or laminate interlayers (disbonds, delaminations) resulted in acoustic impedance mismatches that caused reflection of the waves[7-8]. The reflected acoustic waves were selected, gated, and collected as a function of signal amplitude, or intensity. Reflected signals from the top surface of the panels were used to collect surface/near-surface data while bottom surface reflected signals were used to collect bulk data through each panel. Spatial maps, or ultrasound C-scan images, of the gated signals were used to form visual plots of acoustic variations caused by defects and/or damage. Ultrasonic testing was conducted using a 64-element 10 MHz linear phased array transducer. A total of 32 active transducer elements were utilized for each scan, with active area dimensions of 32.0 mm length, 0.5 mm pitch, and 7.0 mm elevation. The transducer was mounted to a scanning bridge for motion control in the x, y, and z axes during setup and testing. A water immersion tank was used to contain the panels during scanning.

VISUAL AND CROSS-POLARIZATION CHARACTERIZATION

Through visual characterization, several millimeter-sized inclusions were found in the transparent laminate panels, numbering ~14 in 740-1, ~18 in 740-2, ~11 in 741-1, and ~36 in 741-2.

In 741-2, two surface scratches and cracking damage at the upper edge of the panel were also visible (Figure 1). Through cross-polarization imaging, variations in stress patterns were observed (Figure 1). According to the Michel-Levy birefringence chart, the black regions represented areas with no stress, the white regions represented areas with minimal stress, the yellow-orange regions represented areas with a significant degree of stress, and the red-blue regions represented areas with a relatively high degree of stress[4]. Despite the panels being fabricated in the same manner, the stress patterns appeared to be drastically different. While 740-1 showed minimal stress with some yellow-orange regions near the edges, 740-2 showed large regions of high stress throughout the center. Panels 741-1 and 741-2 showed some significant yellow-orange stress regions, with 741-2 exhibiting a high stress region near two of the largest inclusions.

After impacting the series 1 panels with the air gun and the series 2 panels using sphere impact testing, the laminates were observed and compared to their pre-impacted states. The first qualitative observation from the digital images was that the degree of damage appeared to be greater for the thicker 740-series laminates. This was especially evident when comparing the impact damage diameter of ~168 mm and extensive radial cracking in 740-1 to the minimal impact damage diameter of ~28 mm and absence of radial cracking in 741-1. The crack in the upper left corner of 741-1 was believed to be caused by the clamp used to hold the panel in place during air gun testing. For the sphere impact tested panels, the ~132 mm diameter damage at the impact region of 740-2 was approximately twice the size of the ~63 mm diameter damage of 741-2. However, for the higher velocity sphere impact testing, the crack formation trend was reversed, with radial cracking present in the thinner laminate panel but not in the thicker laminate. When observing the impacted panels through cross-polarizers, some local stress variations were found, but the overall patterns did not appear to be drastically altered. There was an observable change in 740-2 that appeared to indicate stress relief after impact, in which a blue high stress region changed to a black region where no stress was apparent. In contrast, the impact region of 741-2 appeared to show concentric high stress regions in the same orientation of what was perceived to be a cone crack at the point of impact. Overall, the 740-series panels exhibited a higher degree of damage after impact when compared to the 741-series panels. As the velocity was increased and mass decreased from air gun to sphere impact testing, the magnitude of damage was reduced from 20 times larger (740-1 vs. 741-1) to twice as large (740-2 vs. 741-2). In terms of visibility through the transparent panels, the 741-series panels performed better, as the degree of damage was less pervasive.

DIGITAL RADIOGRAPHY AND X-RAY COMPUTED TOMOGRAPHY CHARACTERIZATION

The DRs of the pre-impacted panels were unable to detect any distinguishable features. All of the images appeared to be homogeneous, as the inclusions and surface scratches could not be resolved. On the other hand, the XCT images detected and contrasted the laminate layers effectively (Figure 2). For the 740-series panels, the Glass A strike face layer, Glass C interlayer, and polymer backing were readily apparent in the cross-sectional image slices. Although the Glass B strike face layers in the 741-series panels were 12 times thinner than the Glass A strike face layers, they could still be resolved in the XCT images, though this was more difficult near the edges.

After impact, the DRs showed damage at the immediate impact point where the sphere struck the panel, but the extent of damage in the surrounding area was unable to be resolved. For 740-1, the radial cracks were apparent upon careful observation, and appeared as hairline cracks extending from the central impact location to the edge of the panel. However, the radial cracks observed in the visual and cross-polarized images for 741-2 could not be resolved in these DRs. The XCT cross-sectional slices captured from the center of each impact were much more revealing (Figure 2). For 740-1, inspection of the strike face showed the amount of material ejected after impact and indicated how deep the damage extended below the surface. The most interesting feature was the cone crack that formed at the top of the Glass C layer and extended all the way through this second layer. While it was difficult to distinguish damage at different layers through visual observation, the XCT images provided a definitive distinction of damage as a function of depth. In contrast to 740-1, the XCT image of 740-2 indicated that the damage at the impact site did not extend all the way through the strike face layer and that the second layer remained undamaged. For 741-1, the damage appeared to occur only in the Glass B strike face layer, while the damage in 741-2 extended into the Glass C layer and formed another cone crack similar to the one observed in 740-1. These results were consistent with the optical images, as panels that restricted the damage to the strike face layer did not result in radial cracking, while panels that allowed penetration into the second layer resulted in radial cracking and more extensive damage. While the DRs provided little information beyond impact location, XCT images provided a wealth of information for effective characterization through the depth of each panel.

ULTRASOUND C-SCAN IMAGING CHARACTERIZATION

For the first set of pre-impact C-scan images, the surface/near-surface reflected signal was gated and the changes in the amplitude of the signal were mapped (Figure 3). Each panel was scanned through the strike face as well as the back face to observe acoustic variations from either side (Figure 3). While the 740-series samples appeared to be relatively homogeneous, surface/near-surface inclusions were detected that were consistent with visual observations.

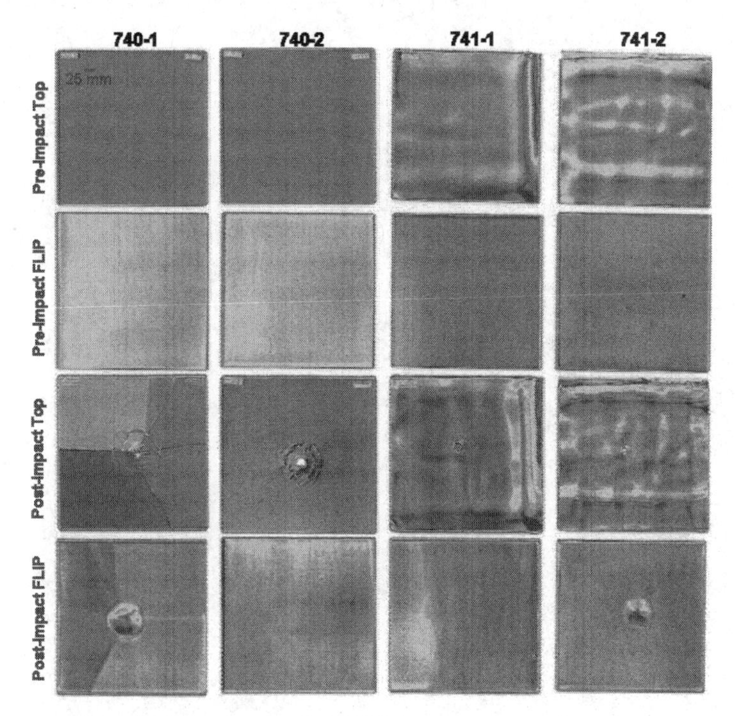

Figure 3. Surface/near-surface UT C-scan images pre and post-impact.

An interesting phenomenon occurred upon observation of the 741-series. Due to the reduced thickness of the Glass B strike face layer, significant amplitude variations caused by surface variations and uneven deposition of adhesive layers in the near-surface were detected. Since the C-scan images collected through the strike face side of these panels were sensitive to adhesive variations, a much greater contrast was apparent. When these same panels were ultrasonically imaged through the back face, these features were not detected, and the results were comparable to those of the 740-series panels scanned in the same orientation. The damage to the upper edge of 741-2 was also distinguished in the C-scan images, as a lower amplitude signal was evident where pre-impact cracking was present.

For the second set of images, the bottom surface reflected signal was gated and the amplitude changes through the entire bulk of each panel were evaluated (Figure 4). The 740-series panel scans were once again homogeneous. The only distinguishable features were inclusions with lower amplitude values than the rest of the panel. The 741-series panel scans showed more intricate adhesive patterns through the thinner strike face glass layers. In this case, the scans through the back face also showed adhesive patterns in which there were acoustic variations. A vertical band on the left side of 741-1 and parallel horizontal bands in the upper and lower regions of 741-2 were indicative of these patterns. As opposed to the previous scans that only showed surface/near-surface amplitude variations, the bottom surface scans represented a greater volume of the sample and enabled similar adhesive features to be observed through both sides of each panel.

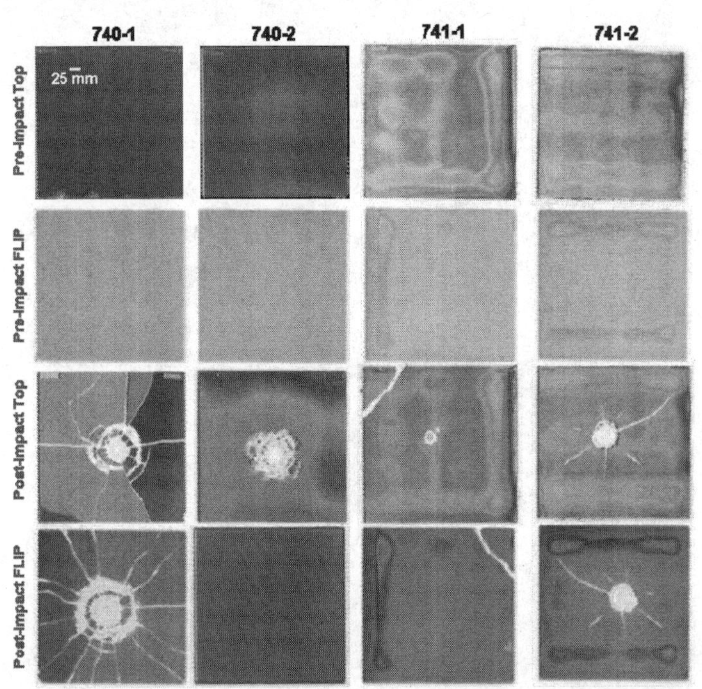

Figure 4. Bottom surface reflected signal UT C-scan images pre and post-impact.

After impact, the ultrasound C-scan images were adept at showing the damage in great detail. For 740-1, the surface/near-surface scans showed the point of impact as a complete loss of amplitude (Figure 3). The cone cracks as well as the radial cracks were also observed in these images, in contrast to the DRs, which were only able to resolve the radial cracks. The C-scan images through the back face of 740-1 showed a circular region representative of the bulge that occurred in the polymer backing layer post-impact. The dimensional change was an observation unique to the C-scan images, and was found in both 740-1 and 741-2, the two samples in which damage to the Glass C layer was identified in the XCT images. For the other two samples, C-scan images through the back face revealed no isolated regions of damage. For the 741-series samples, the adhesive pattern variations appeared to change upon impact to the panels. This could have been an indication of physical change to the adhesive layers, which were designed to mitigate stress and prevent crack propagation into successive layers[1].

The bottom surface reflected signal C-scan images though the bulk of each sample were very effective at contrasting damage, with cracked or damaged regions resulting in a complete loss of signal (Figure 4). For 740-1, scanning through the strike face showed crack patterns within the Glass A layer, while scanning through the back face showed crack patterns in the Glass C layer. While several cracks were able to penetrate both layers, the majority of the 18 radial cracks were unique to either the Glass A or Glass C layer. By scanning through both sides of the panel, the depth locations of these cracks were resolved. For 740-2, the intricate damage pattern was well-represented in the C-scan image and the scan through the back face indicated no penetration into the Glass C layer. For 741-1, the bulk adhesive patterns did not appear to change as much as in the surface/near-surface images. The damage at the point of impact and crack in the corner were clearly visible, and small indications of damage were detected in the back face scan. With the XCT images for 741-2 indicating damage to the Glass C layer, the impact region and resulting radial cracks were as apparent through the strike face images as they were through the back face images. Overall, the C-scan images were very effective at representing the damage to each panel after air gun and sphere impact testing.

QUANTITATIVE DAMAGE THRESHOLD RESULTS

A supplemental technique was developed for quantitatively estimating the percent damage in transparent laminate targets subjected to low velocity impact testing. Ultrasound C-scan imaging was chosen as the basis for obtaining a representative map to contrast damaged and undamaged regions through the bulk of the panels. This method was applied to C-scan images in which the bottom surface signal amplitude was mapped to represent volumetric damage through the bulk of each target. Each C-scan damage map was processed using an inverted grayscale in which the color "white" represented the highest degree of damage. Histograms of the selected C-scan damage maps were plotted as a function of grayscale levels on the x-axis and number of occurrences (or image pixels) on the y-axis. On each histogram, the x-axis ranged from black on the left side to white on the right side. Typically, there was a large curve or series of curves representing the undamaged region of the target. The damaged region was often represented by a small curve at the right side of the histogram where the "white" or very light colors were present. A threshold was chosen on the histogram in which any occurrences to the right represented the damaged regions of the target and any occurrences to the left represented the undamaged regions. The idea was to separate and distinguish the total number of occurrences that represented undamaged and damaged regions of the C-scan image. This was often the most difficult step in the process, since a definitive threshold value was not always clear. Once the threshold value was determined, the histogram data containing the grayscale levels and number of occurrences was referenced. The grayscale level associated with the threshold was located and all occurrences to the right of this value (very light and "white" colors) were selected. The summation of total number of occurrences representing the damaged regions was calculated and this value was divided by the total number of all occurrences in the image and multiplied by 100 to acquire the estimated percent damage in the selected panel.

Estimated percent damage values were calculated for C-scan images representing the strike face and back face of each of the four transparent panels. The inverted grayscale images and corresponding histograms are shown in Figure 5. For panels 740-1 and 741-2, which showed cone cracking in the Glass C layer and significant damage through the back face, the estimated percent damage increased. The damage increased from 10.9% to 15.7% for panel 740-1 due to the higher number of radial cracks present in the Glass C layer. For panel 740-2, in which there was no damage to the Glass C layer, the percent decreased from 8.7% to 0.0% since no damage could be detected in the C-scan image through the back face. For panel 741-1, the percent damage decreased from 1.8% to 1.5%. Despite the absence of a cone crack in the Glass C layer, the crack in the corner of the sample propagated through both the Glass B and Glass C layers, which accounted for the majority of damage through both the strike face and back face images. The quantitative damage threshold technique proved to be a useful means for estimating and comparing volumetric percent damage values for the transparent panels.

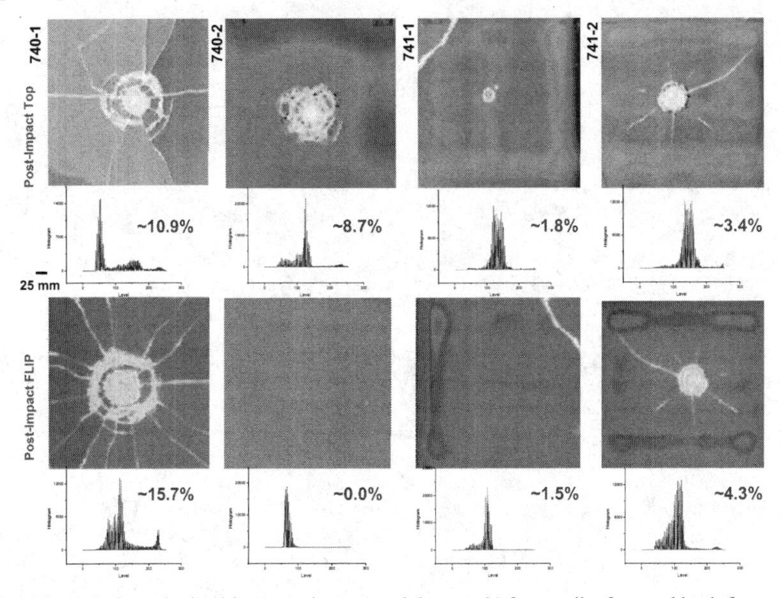

Figure 5. Quantitative histogram images and damage % from strike face and back face.

MULTIPLE IMPACT CUMULATIVE DAMAGE ASSESSMENT

Cumulative damage in a single panel was evaluated by subjecting panel 740-2 to multiple impacts at the same location. The visual and NDE damage results for 740-2 have been presented up to this point for a single impact. These results showed that the damage was contained within the Glass A strike face layer and that no additional damage to the Glass C layer could be detected. The damaged panel was prepared for sphere impact under the same velocity conditions as the first test to induce a succession of three additional impacts to the same location where the first impact occurred. Visual inspection, cross-polarization imaging, and ultrasound C-scan imaging were conducted before and after each impact to study cumulative damage effects, as shown in Figures 6 and 7.

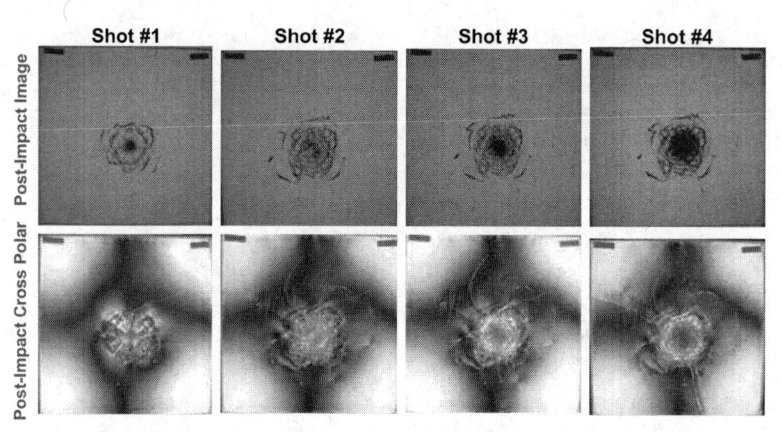

Figure 6. Visual and cross-polarized digital images of 740-2 after each of four sphere impacts.

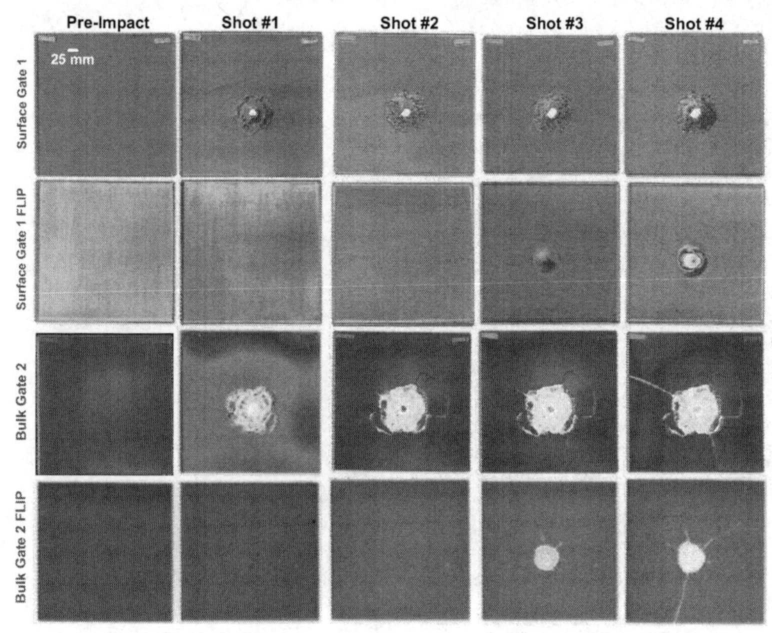

Figure 7. Suface and bulk C-scan images of 740-2 cumulative damage from strike and back face.

Figure 6 shows the progression of damage through digital and cross-polarization images. After the second impact, multiple cracks formed within the original damage zone from the first impact. Several larger cracks also started to propagate, expanding the diameter of the damage zone. The damage to panel 740-2, as observed in the visual and cross-polarized images, did not appear to change significantly after the third impact. After the fourth and final impact, three of the cracks that had started to propagate after the second impact had fully propagated to the edge of the panel.

In Figure 7, C-scan images of the surface/near surface of panel 740-2 after each impact are shown in succession, with the first row representing the strike face images and the second row representing the back face images. The strike face images showed the degree of damage at the point of impact and its slight degree of expansion after each impact. Damage due to cracking was not present until after the final impact, where a radial crack propagated to the lower edge of the panel. The back face images showed no indication of damage to the Glass C layer until after the third impact, at which point there appeared to be a slight physical deformation to the back face at the point of impact. After the fourth impact, a severe degree of damage resulting in complete loss of signal was evident in the surface/near surface C-scan image. The third and fourth rows represented bulk C-scan images through the entire volume of panel 740-2. For the third row containing the strike face images, the damage zone expanded after the second impact, was very similar after the second and third impacts, and resulted in propagation of two radial cracks to the edge of the panel after the fourth impact. For the fourth row containing the back face images, some micro-cracking was evident after the second impact. The third impact revealed the presence of damage and the early formation of radial cracks within the Glass C layer. After the fourth impact, several of these radial cracks began to propagate, with one extending out to the edge of the panel. This demonstrated that out of the three edge cracks that were detected, one of them was present in the Glass C layer. These results showed a large contrast between panel 740-1, which had a large number of radial edge cracks after a single impact, and panel 740-2, in which four impacts to the same location were required to produce a radial edge crack.

The quantitative damage threshold technique was used to determine the increase in damage to the strike face and back face C-scan images after each of the four impacts, as shown in Figure 8.

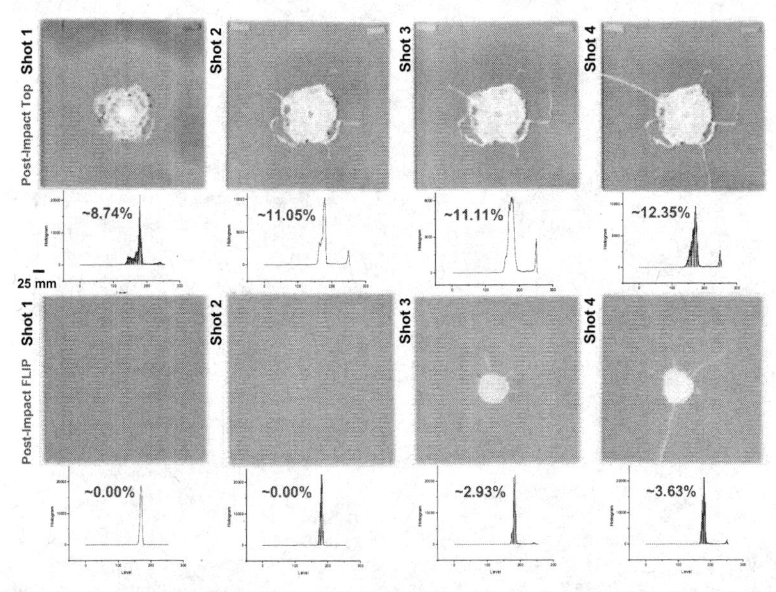

Figure 8. Histogram images and damage % from strike and back face of 740-2 after each impact.

For the strike face images, the quantitative results were consistent with the qualitative observations from visual inspection. The percent damage increased from 8.74% to 11.05% after the second impact with additional cracking to the damage zone and initial crack propagation, from 11.05% to 11.11% after the third impact with a minimal increase in damage, and from 11.11% to 12.35% after the fourth impact with radial crack propagation to the edges. For the back face images, there was no detectable damage after the first or second impacts, as the initial micro-cracking was not distinguished. After the third impact, the damage increased to 2.93% and again to 3.63% as one of the radial cracks propagated to the edge of the panel. All of the quantitative results consistently showed an increase in cumulative damage after each impact test.

CONCLUSIONS

Visual observation methods were effective for addressing one of the critical factors for evaluating transparent material performance – visibility. The simple qualitative tests of looking through the panel post-impact to observe damage or optical distortion, or quantitative tests of measuring the diameter of the impact region, were useful for comparing the utility of one laminate system to another. This method indicated that the 741-series transparent laminate system was the preferred choice due to the reduced average damage diameter. The 740-series panels showed extensive radial cracking that resulted in significant optical distortion. For obtaining a spatial map of stress in these multi-layer stacked laminate systems, cross-polarized imaging was instrumental for checking the integrity of each panel. Post-impact images were useful for determining whether stress was released or enhanced at or near the point of impact. While 740-2 showed the highest degree of pre-impact stress, it also revealed post-impact stress relief. Stress release at the strike face during cracking may have been the reason why there was no penetration into the Glass C layer.

While DR was not an effective characterization technique for mapping pre-impact defects or post-impact damage, XCT cross-sectional imaging was vital for evaluating damage as a function of depth. Not only did the XCT images reveal accurate damage dimensions, but also indicated whether or not the damage extended beyond the strike face. For two of the panels, cone cracks were found in the second glass layer that resulted in additional damage to the overall laminate system. XCT images were more effective than visual and cross-polarization imaging at determining which layer was damaged. Ultrasound C-scan imaging also held several advantages. One was the ability to image changes in the adhesive layers that could not be physically seen. Ultrasound was also capable of detecting inclusions, scratches, and other surface and internal defects that were difficult or impossible to see visually. Dimensional changes to the polymer backing layer were detected through C-scan imaging of the back face, another indication of damage beyond the strike face. For 740-1, in which there were significant differences in the radial crack patterns between the Glass A strike face layer and the Glass C layer, C-scan imaging through different sides of the panel successfully mapped the distinct crack patterns in each layer. In addition to single impact results, cumulative damage was successfully studied by impacting panel 740-2 three additional times under the same conditions. As opposed to panel 740-1, which was subject to Glass C damage after a single impact, the Glass C layer in panel 740-2 remained undamaged until after the third strike. Due to the effectiveness of C-scan imaging for damage detection, a novel quantitative histogram technique was developed to calculate the estimated percent damage based on these images. The results provided an effective means of comparing single impact damage from both sides of the panels as well as cumulative damage from multiple impacts. A combination of C-scan imaging and XCT cross-sectional imaging in addition to quantitative damage assessment from the histogram method provided a complete picture of transparent panel damage.

The authors would like to acknowledge Parimal Patel, Jian Yu, Terrence Taylor, and Dave Spagnuolo of the U.S. Army Research Laboratory for their efforts and support.

REFERENCES

[1]P.J. Patel, G.A. Gilde, P.G. Dehmer, and J.W. McCauley, "Transparent Armor"; *AMPTIAC Newsletter*, **4**, *3* (2000).
[2]J.M. Feingold, "Stress Diagnose It Before It Ruins Your Parts", *Plastics Technology* (2009).
[3]B.R. Hoffman, "How To Measure Stress in Transparent Plastics", *Plastics Technology* (1998).
[4]K.R. Spring, M.J. Parry-Hill, M.W. Davidson, "Michel-Levy Birefringence Chart", *Olympus Microscopy Resource Center* (2010).
[5]J.H. Stanley, "Physical and Mathematical Basis of CT Imaging: ASTM Tutorial Section 3", *ASTM CT Standardization Committee* (1986).
[6]T.H. Newton and D.G. Potts, "Technical Aspects of Computed Tomography", *The C.V. Mosby Company*, **5** (1981).
[7]Mix, P.E., "Introduction to Nondestructive Testing", *John Wiley & Sons* (1987).
[8]Krautkramer, J. and H. Krautkramer, "Ultrasonic Testing of Materials", *Springer-Verlag* (1990).

COMPARISON OF PENETRATION DAMAGE IN NOVEL Mg SPECIMENS VIA COMPUTED TOMOGRAPHY

William H. Green and Kyu C. Cho
U.S. Army Research Laboratory
Weapons and Materials Research Directorate
ATTN: RDRL-WMM-D
Aberdeen Proving Ground, MD, USA

ABSTRACT

X-ray computed tomography (XCT) has been shown to be an important non-destructive evaluation (NDE) technique for revealing the spatial distribution of ballistically-induced damage in metals, ceramics, and encapsulated ceramic structures. Previous and ongoing work in this area includes assessment of ballistically induced damage in relatively lightweight individual ceramic targets and ceramic armor panels. In this paper the ballistic damage in two novel Mg alloy samples was completely scanned and extensively evaluated using XCT 2-D and 3-D analysis. Features of the damage in the samples were compared and contrasted. Some features of the damage were correlated with physical processes of damage initiation and growth. XCT scans and analyses of damage in the samples will be shown and discussed. This will include virtual 3-D solid visualizations and some quantitative analysis of damage features.

INTRODUCTION

Magnesium (Mg) is being studied for use in lightweight protection systems. Lightweight materials are typically used in armor panel structures in order to decrease weight without losing ballistic performance. Mg is the lightest structural and engineering metal at a density of 1.74 g/cm^3 that is approximately 1/5, 2/5, and 2/3 the weight of iron, titanium, and aluminum, respectively [1, 2]. Mg alloys are being considered as extremely attractive lightweight materials for a wide range of the Army's future applications where weight reduction is a critical requirement because of its low density. Furthermore, magnesium has good vibration damping capacity [3] and low acoustic impedance characteristics [4] that could be of additional benefit to vehicle applications. XCT is an effective and important NDE technique for revealing internal fabrication characteristics and spatial distribution of damage in material specimens [5, 6]. Projection digital radiography (DR) is often performed as a precursor to XCT. Previous and ongoing work in the area of XCT evaluation includes assessment of ballistically induced damage in individual ceramic targets, ceramic panels, and metal plates [7]. The purposes of XCT evaluation include characterization and understanding of the detectable fabrication structure and/or damage in the complete 3-D space of the specimen or scanned volume, determination of geometric parameters of fabrication and/or damage features for interpretation and useful engineering data, and correlation of physical structure with fabrication methods and damage features and types with the physical processes of damage initiation and growth. In this paper the ballistic damage in two novel Mg alloy samples was completely scanned, evaluated, and compared using XCT 2-D and 3-D analysis.

DESCRIPTION OF SPECIMENS AND DIGITAL RADIOGRAPHY RESULTS

The first specimen was an approximately 146 mm (5.7") by 186 mm (7.3") rectangular section from a larger impacted test plate with multiple hits, some of which fully penetrated the 39 mm (1.5") thick plate. The specimen included a single complete penetration and the surrounding area as well as undamaged material farther away. The second specimen was an approximately 76 mm (3.0") by 71 mm

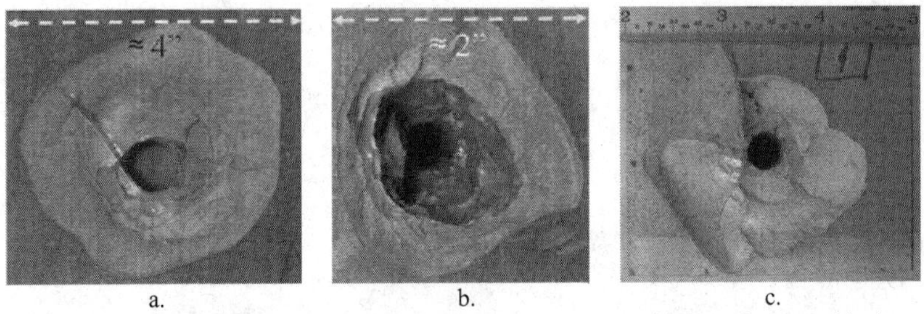

a. b. c.

Figure 1. Photographs of (a) rear (exit) side damage in first specimen, (b) front (impact) side damage in second specimen, and (c) exit side damage in second specimen.

(2.8") rectangular section from a different test plate that was 80 mm (3.1") thick, about twice as thick as the first specimen. This specimen also exhibited a single complete penetration and the surrounding damaged area. Figure 1 shows a close-up photograph of the rear (exit) side of the first specimen and front (impact) and rear side photographs of the second specimen. The first photograph (1a) shows the relatively large amount of material around the main through hole that was pushed out from the rear of the specimen. The second photograph (1b) shows the extensive amount of damage in the entrance region into the second specimen, which was not the case in the first specimen. The third photograph (1c) shows the relatively large amount of material around the main through hole that was pushed out from the rear of the specimen similar to the first specimen. The specimens exhibit similar spall surfaces at their rear sides.

Digital radiographs (DRs) of the first specimen were taken through its thickness and width (edge on) using the 420 keV x-ray tube and linear detector array (LDA) setup in centered rotate-only (RO) mode. The x-ray technique (parameters) of the DRs of the specimen were (400 keV, 2.0 mA) and geometries of source-to-object-distance (SOD) = 750.00 mm and source-to-image-distance (SID) = 940.00 mm. An edge on DR of the second specimen was taken using the 225 keV microfocus x-ray tube and image intensifier (II)/charged coupled device (CCD) camera setup in centered RO mode. The x-ray technique of the DR of the specimen was (190 keV, 0.040 mA) and geometries of SOD = 300.00 mm and SID = 641.50 mm. Figure 2 shows the through thickness (a) and edge on (b) DRs of the first specimen and the edge on DR (c) of the second specimen. In Figure 2a the penetration hole itself is emphasized and shows a darker area to the lower left of the hole due to missing material in the impact side of the specimen. In the edge on (side) image of the first specimen, with the impact side on the left, the main hole is evident, as well as severe damage in the middle and rear regions of the specimen and the missing material that was pushed out of the exit side. This is also the first image that shows some of the nature of the approximately parallel lateral, or "petal-like", damage mode in the middle and towards the rear of the specimen, which is significantly more visible in the cross-sectional XCT images. In the side image of the second specimen (2c), with the impact side (F) on the bottom, the tilted long hole is evident, as well as severe damage in the front region of the specimen and the missing material that was pushed out of the exit side. The horizontal cracking towards the front face of the specimen is not part of the penetration damage.

XCT SCANNING PROCEDURES

The first specimen stood freely on top of a metal plate to raise it up with its exit side facing the x-ray source. Thus, the specimen faces were perpendicular to the horizontal x-ray (collimated) fan beam

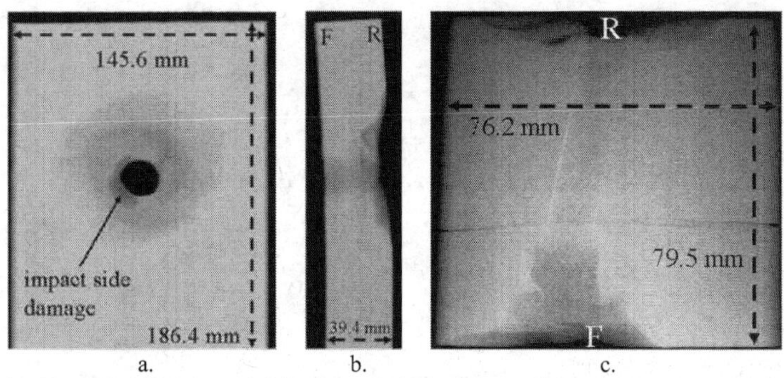

Figure 2. (a) Through thickness and (b) edge on (side) digital radiographs of first specimen. (c) side digital radiograph of second specimen.

resulting in through thickness cross-sectional CT images. The bottom (edge) of the specimen was at a vertical position of about 20 mm and the top was at a position of about 206 mm. The middle of the main penetration hole was at a vertical position of approximately 116 mm. The entire volume of the specimen between the vertical positions of 65.000 mm and 173.900 mm was scanned using the 420 keV x-ray tube and LDA setup in offset RO mode. The scans were vertically overlapping with a slice thickness and increment of 0.500 mm and 0.450 mm, respectively, and each slice was reconstructed to a 1024 by 1024 image matrix. The field of reconstruction (FOR) diameter was 195.00 mm. The tube energy and current used were 400 keV and 2.0 mA, respectively, and the focal spot was 0.80 mm. The SOD and SID were 750.00 mm and 940.00 mm, respectively. The second specimen also stood freely during scanning, but its faces were parallel to the horizontal because geometrically that was a better physical orientation for scanning. The specimen was scanned at regular vertical intervals using the 225 keV microfocus x-ray tube and II/CCD camera setup in offset RO mode. The scans were not vertically overlapping with a slice thickness and increment of 0.596 mm and 1.000 mm, respectively, and each slice was reconstructed to a 1024 by 1024 image matrix. The FOR diameter was 110.00 mm. The tube energy and current used were 180 keV and 0.044 mA, respectively, and the focal spot was 0.005 mm. The SOD and SID were 325.00 mm and 641.50 mm, respectively.

QUALITATIVE AND QUANTITATIVE EVALUATION OF SPECIMENS

Computed Tomography Scans

Figure 3 shows a series of CT scans (images) of the first specimen with the first scan at the vertical position of 116.30 mm (3a), which was within 0.20 mm of the position of the center line of the main penetration hole. The scans in Figures 3b and 3c were taken at vertical positions of 126.20 mm and 131.15 mm, respectively, and the scans in Figures 3d and 3e were taken at positions of 106.40 mm and 101.45 mm, respectively. The impact and exit sides of the specimen are at the bottom and top of the images, respectively, and the thickness is 39.4 mm. The missing material towards the front of the specimen adjacent to the main hole is indicated by the shoulder on the right hand side of the cavity wall in Figure 3a. A series of approximately parallel lateral, or "petal-like", cracks on both sides of the penetration cavity are evident. The cracks have the appearance of starting in one direction from the sides of the cavity and then turning or bending back from the original direction towards the rear of the

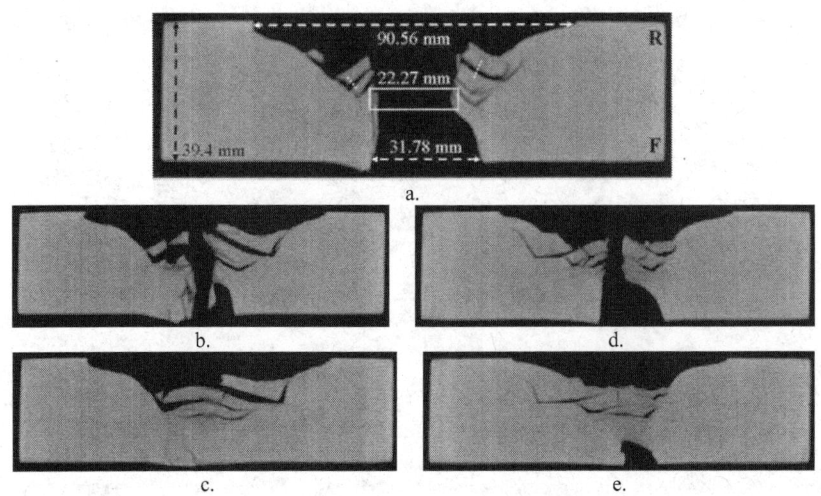

Figure 3. Cross-sectional CT scans (images) of damage in the first specimen. (a) Vertical position of 116.30 mm. (b) 126.20 mm. (c) 131.15 mm. (d) 106.40 mm. (e) 101.45 mm.

specimen. At some depth into the specimen, the lateral damage mode stops and a relatively large amount of material is pushed out of the exit side. The width of the cavity at the very front and rear of the specimen is 31.78 mm and 90.56 mm, respectively. The narrowest width is 22.27 mm, which is in the middle thickness region. The width of the uppermost cracks towards the rear of the specimen on both the left and right hand side of the penetration cavity is about 1.4 mm. The scans in Figures 3b and 3c are 10.10 mm and 15.05 mm above the main hole center line, respectively. The parallel lateral cracking damage mode is evident in both images. These images do not show a high level of cracking oriented approximately in a through thickness direction. The scans in Figures 3d and 3e are 9.70 mm and 14.65 mm below the center line, respectively. The same lateral damage mode is evident in these images, with a similar relative lack of through thickness cracking.

Figure 4 shows a series of CT scans of the second specimen with the first scan (a) located 3.00 mm from its impact face. Due to the orientation of the specimen during scanning the cross-sectional images were parallel to its impact and rear faces, as opposed to perpendicular in the scanning of the first specimen. The scans in Figures 4b, 4c, 4d, and 4e were located 17.00 mm, 50.00 mm, 71.00 mm, and 76.00 mm from the impact face, respectively. Conversely, the scans in Figures 4d and 4e were only about 7 mm and 2 mm from the rear face of the specimen. The small white indications at the edges of the penetration cavity in the first three images are indicative of residual penetrator material. The diameter of the hole in Figure 4c, which was taken at approximately the middle vertical location of the tilted, near constant diameter section, or neck, of the penetration cavity (see Figure 2c), is about 9.6 mm. Multiplanar reconstruction (MPR) visualization, which is a form of three-dimensional (3-D) volume reconstruction visualization, was used to analyze the full set of XCT scans of the specimen to generate individual virtual slices perpendicular to the scanning plane and thus in physically vertical planes from the impact face to the rear face of the specimen. Figures 4f and 4g show two such MPR images, in which the virtual slices are approximately through the center of the neck of the penetration cavity and

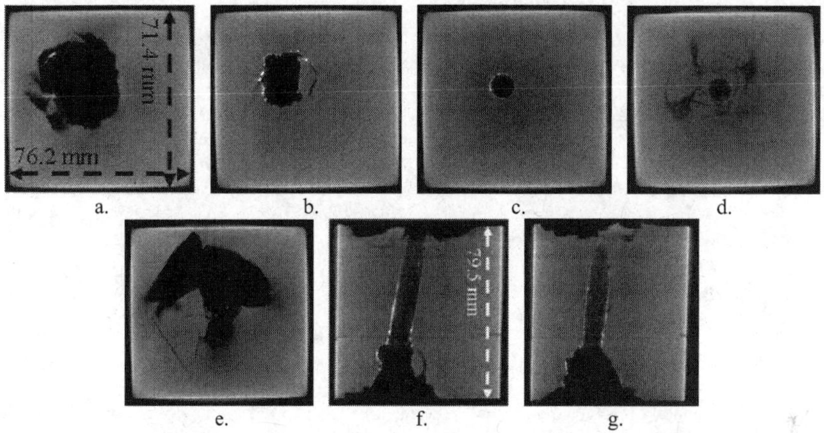

Figure 4. Cross-sectional CT scans and MPR images of damage in the second specimen. (a) 3.00 mm from impact face. (b) 17.00 mm. (c) 50.00 mm. (d) 71.00 mm. (e) 76.00 mm. (f) MPR image through neck region parallel to x-z plane. (g) MPR image through neck region parallel to y-z plane.

parallel to the x-z and y-z planes, respectively. The images show that the neck of the penetration cavity tilts both in the x direction and out of the y-z plane. All of the images in Figure 4 do not show the extensive near parallel lateral crack features that are evident in the first specimen, especially in the through thickness MPR images (4f and 4g). The relatively darker shade of gray in the neck of the penetration cavity in the MPR images does not indicate material in the cavity.

Three-Dimensional Solid Visualization

The excellent dimensional accuracy and the digital nature of XCT images allow the accurate volume reconstruction of multiple adjacent or overlapping slices. A virtual 3-D solid image is created by electronically stacking the XCT images, which have thickness over their cross-sections (i.e., voxels), one on top of the other from the bottom to the top of the specimen, or scanned height, to generate its virtual volume. Figure 5 shows a set of four 3-D solid images of the scanned volume of the first specimen with sections virtually removed in 5c and 5d. The method of virtual sectioning, which is essentially only showing a portion of each scan, allows viewing of generated surfaces anywhere in the scanned volume in 3-D space. Figures 5a and 5b show views of the impact and exit side, respectively, of the entire scanned volume. The very light vertical banding down the middle of the images (top and bottom) is an image artifact from the reconstruction. It is not an indication of a real physical feature in the specimen. The relative shading of lighter and darker gray in the damaged areas is produced by virtual lighting. Physical texture in the surface of the damage in the rear of the specimen is visible in Figure 5b. In Figure 5c, the impact side of the specimen is at the bottom of the image and the sectioned surface is approximately halfway between the top and the bottom of the main penetration hole. In Figure 5d, the impact side of the specimen is on the left side of the image and the sectioned surface is approximately halfway between the left and right sides of the main hole. The relatively lighter surface of the penetration cavity in Figure 5c is due to the angle of the virtual lighting. Both of the surfaces in the sectioned volumes, which are orthogonal to each other, show the parallel lateral cracking damage mode. This is indicative that this damage mode has a significant degree of symmetry about the trajectory of the penetration.

Figure 6 shows a similar set of four 3-D solid images of the second specimen with sections virtually removed in 6c and 6d. Figures 6a and 6b show views of the impact and exit side, respectively, of the specimen. The pitted appearance of some areas of the images is an image artifact from the reconstruction. It is not an indication of a real physical feature in the specimen. In the second specimen the total attenuation of the incident x-ray intensity through it was very high, which resulted in minimal gray level separation between the specimen material and the air immediately around it. This is an understood effect that very highly attenuating materials, whether due to geometry, density, or both, can have on XCT images, resulting in less than desirable gray level separation between different materials and more overall image noise. This produced the pitted appearance in some areas due to the difficulty in finding an optimal separation of the surfaces from neighboring air. The relative shading of lighter and darker gray in the damaged areas is still mainly produced by virtual lighting. The shading in the neck of the penetration cavity in about the middle of the images is also partly due to the tilt of the neck in two directions as it goes through the specimen. In Figures 6c and 6d, the impact (F) side and the rear (R) side of the specimen are at the bottom and top of the image, respectively. In Figure 6c, the sectioned surface is approximately through the center of the neck of the penetration cavity and parallel to the x-z plane. In Figure 6d, the sectioned surface is approximately through the center of the neck of the penetration cavity and parallel to the y-z plane. These images show a more informative depth perspective of the penetration cavity near the impact and rear faces of the specimen. The lighter indications in the area of the sectioned surface closer to the impact face in Figure 6d are indicative of residual penetrator material. The images in Figures 6c and 6d also do not exhibit the extensive near parallel lateral crack features that are evident in the first specimen.

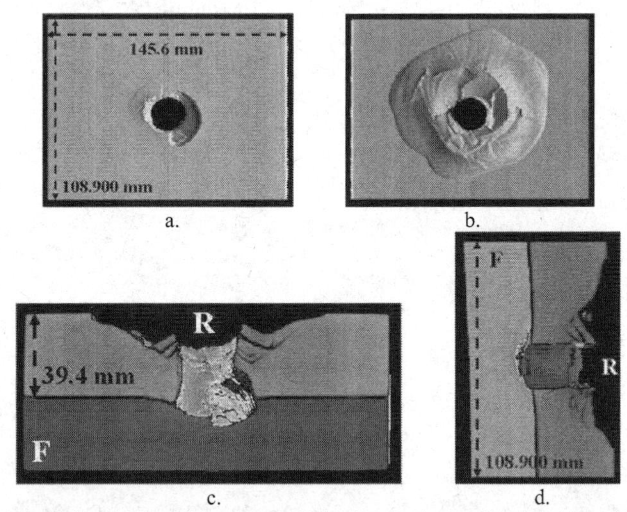

Figure 5. A series of virtual 3-D solid volumes of the damage in the first specimen. (a) front side. (b) rear side. (c) horizontally sectioned through center of penetration cavity. (d) vertically sectioned through center of penetration cavity.

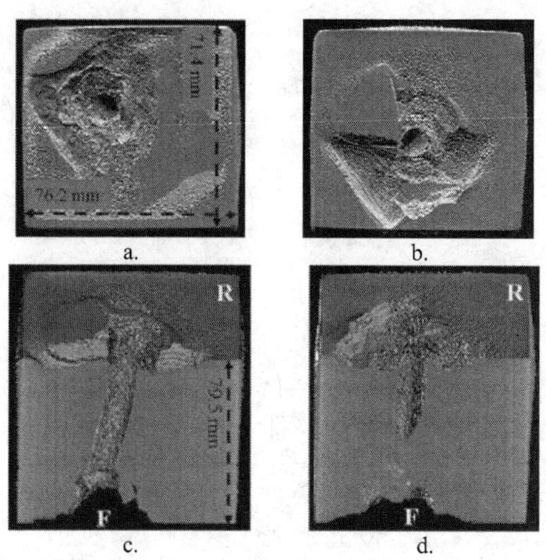

Figure 6. A series of virtual 3-D solid volumes of the damage in the second specimen. (a) front side. (b) rear side. (c) sectioned through center of neck region parallel to x-z plane. (d) sectioned through center of neck region parallel to y-z plane.

Three-Dimensional Point Cloud Visualization

A 3-D point cloud is a set of points in space that define geometrical characteristics (i.e., shape, size, location) of a specimen or scanned volume and features within it. Location of the points is determined by appropriate (image) segmentation of the volume or feature(s) of interest. Figure 7 is a point cloud of the overall damage in the first specimen and its outside surfaces from a top-down view of its thickness with the impact side at the bottom of the image. This view clearly shows that there are three distinct regions of different types of damage. The first region of damage towards the front of the specimen with the shoulder on the right is cylindrical and was produced by the initial penetration of the threat. The diameter of the entrance hole without the shoulder included is 21.04 mm with the center at a height of 116.10 mm. The depth and maximum size (parallel to specimen faces) of this region of damage are about 16 mm and 33 mm, respectively. The middle region of damage, which has clear delineation from the damage towards the front, exhibits multiple parallel lateral cracks with "upturned" ends that go towards the rear of the specimen. The depth of this region to the ends of the cracks closest to the exit face of the specimen is about 16 mm. The maximum distance between the ends (tips) of the upturned cracks is about 68 mm. The last region of damage is the relatively large amount of material that was pushed out of the rear of the specimen, which overlaps with the middle region of damage. The physical morphology of the middle and rear damage regions is closely intertwined, making it difficult to determine a precise boundary between the two regions. The maximum size (parallel to specimen faces) of the rear region of damage is about 100 mm. Figure 8 is an approximate top-down view of the point cloud of the middle region of damage with a small portion of the rear damage to better show the orientation of the lateral cracks relative to the shallow cone of material pushed out the rear of the specimen. The point cloud is tilted backwards a few degrees from a top down view in order to separate the upturned ends of the cracks from the surrounding damage as much as possible.

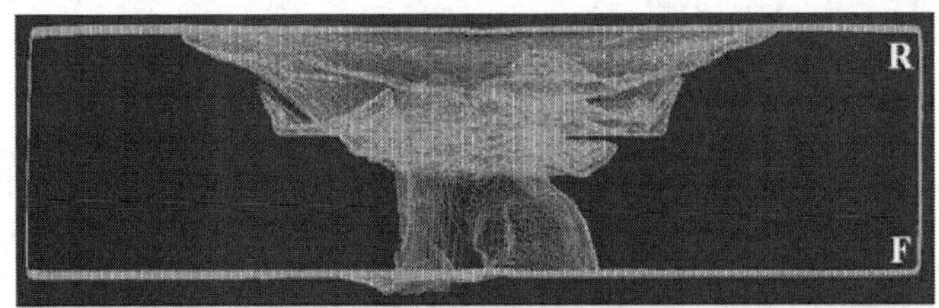

Figure 7. View of 3-D damage point cloud in first specimen, along with faces and sides, looking down at thickness cross section (F indicates front and R indicates rear).

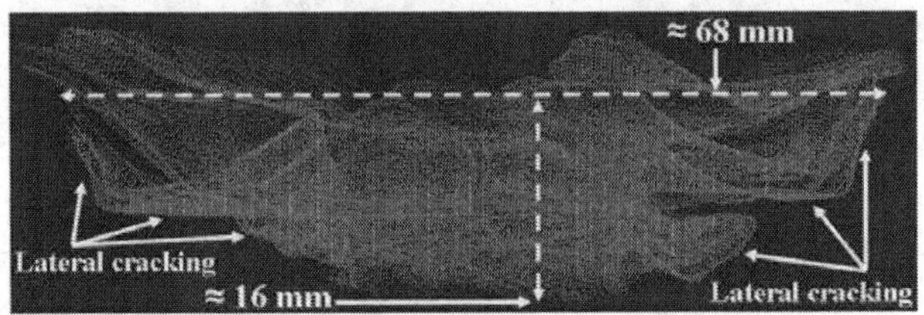

Figure 8. View of middle section (thickness) of damage point cloud only in first specimen. The front side of the plate is below the bottom of the image.

Figures 9a and 9b are point clouds of the overall damage in the second specimen, with its outside boundaries in wireframe mode, tilted forward different degrees from a view looking down at its thickness to better show the entrance and exit regions of the damage. The impact face of the specimen is the top and bottom boundary surfaces in Figures 9a and 9b, respectively. The three regions of damage are distinct and well delineated. The entrance region defined by the inverted cone shape, relatively long neck region, and exit region defined by the material spall are approximately 33 mm, 35 mm, and 11 mm long, respectively. The depth of the entrance region is about 42% of the thickness of the specimen and its beginning at the impact face is larger than in the first specimen. The exit region is only about 14% of the thickness of the specimen. Although there is some damage outside of the neck where it nears the exit region, it appears that this damage is mainly the relatively nondescript precursor to the spall damage and without the same characteristics as the near parallel lateral crack features in the first specimen.

Figures 10a and 10b are isometric views of the damage point cloud only in the first specimen from exit side and impact side perspectives, respectively, with the boundaries of the scanned volume shown in wireframe mode for reference. The face of the specimen that is away from the view perspective is gridded for ease of interpretation. The z axis is vertical and the x axis is parallel to the long dimension of the thickness cross section. Figures 11a and 11b are isometric views of the damage point cloud only in the second specimen from impact side and exit side perspectives, respectively, with the boundaries of the specimen shown in wireframe mode for reference. The z and x axes are shown directly in Figure 11a, indicating the difference in the physical orientation of the specimens in their scans.

Quantitative Damage Evaluation and Discussion

Image segmentation can also be used to produce binary (black and white) images in which a gray level threshold is applied to separate a feature or features of interest from the material around it. The gray level width of the segmented images is set to two in order to replace each pixel gray level in the original images with a new minimum or maximum gray level (e.g., 0 or 255 for an 8-bit binary image). The pixel data of binary images can be statistically evaluated to determine the level or severity of features, such as the amount of physically detectable penetration damage. This process can be performed on a set of XCT scans, as well as sets of parallel MPR images generated from XCT scans.

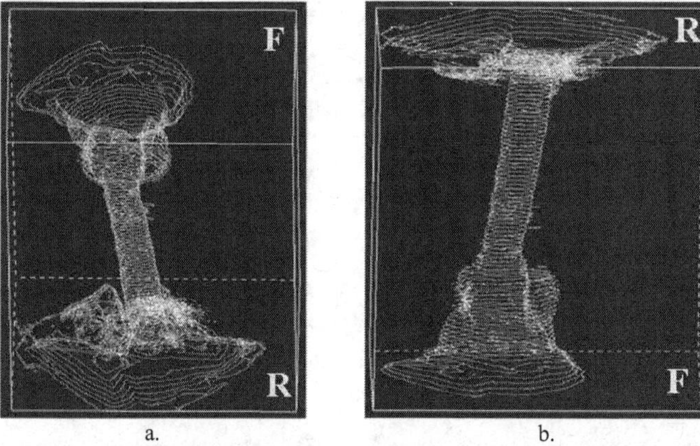

a. b.

Figure 9. Tilted views of 3-D damage point cloud only in second specimen, with boundaries in wireframe mode, looking through thickness cross section (F indicates front and R indicates rear).

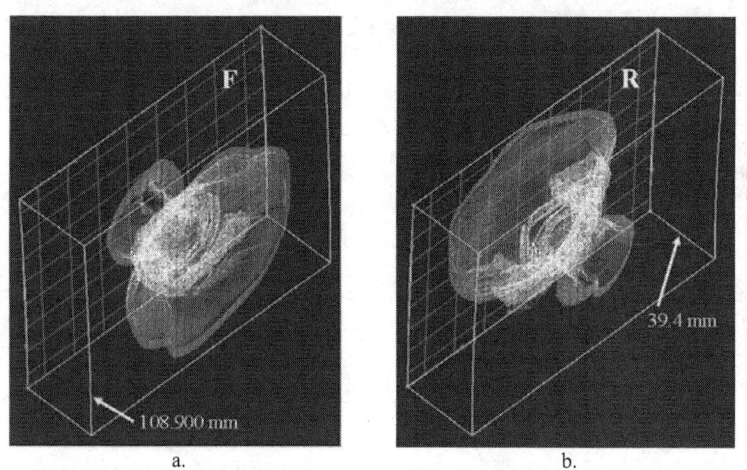

a. b.

Figure 10. Isometric views of 3-D damage point cloud only in first specimen with physical boundaries of scanned volume shown as wireframe and face away from the view gridded (F indicates front and R indicates rear).

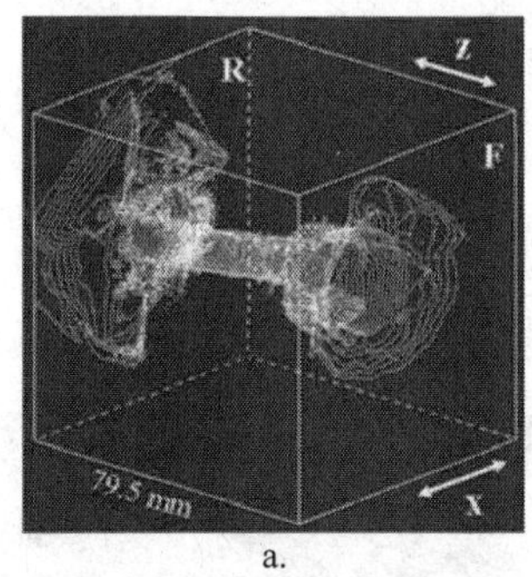

a.

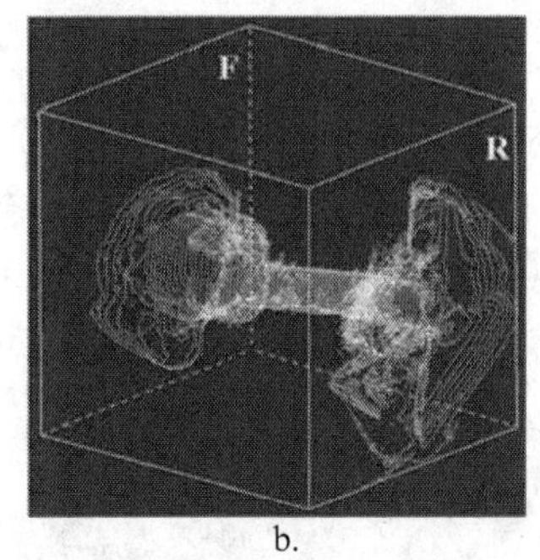

b.

Figure 11. Isometric views of 3-D damage point cloud only in second specimen with boundaries in wireframe mode (F indicates front and R indicates rear).

MPR visualization was used to analyze the set of XCT scans of the first specimen to generate a series of individual virtual slices of the specimen parallel to its faces from its impact face to its exit face, which were segmented to binary images. These virtual parallel slices were vertically oriented in space. The areal pixel density of views in a MPR image, of which there are four (top, side, front, and oblique), is determined differently than XCT images, since each view is not a single reconstructed image to a set matrix size like a CT image. In this case, the physical area of the specimen and the average number of pixels in that area in the vertical slices (front views) of the MPR images was (14.56 cm x 10.89 cm) and 109,114 pixels, respectively. This resulted in an areal pixel density of 688 cm^{-2} and inversely a pixel area of 0.145 mm^2. The number of damaged pixels in each front view was determined using image histogram data and converted to the corresponding amount of damaged area. Figure 12a is a plot of the damaged area parallel to the faces of the specimen versus depth from its impact face. Individual segmented binary vertical slices are overlaid on the plot to show the damage at specific locations. The plot has two local maxima and three local minima. This behavior of local peaks and valleys is due to the parallel lateral cracking damage mode in the middle of the specimen. The local spatial periodicity of the minima to minima (two) and maxima to maxima (one) segments of the damage is about 5 mm. The plot also shows that the area of the penetration hole reaches a minimum at a depth of about 14 mm and goes into a steep rise after a depth of about 26 mm, which is indicative of the large amount of material pushed out the back of the specimen. Figure 12b shows the damaged area plot as well as normalized minimum and normalized damaged area plots. The normalized minimum plot (triangles) is based on the amount of damage that would be present due to a uniform penetration hole through the thickness of the specimen with a diameter equal to the diameter of the base of the threat. In this case, the minimum possible damage area as a function of depth is a constant, so this plot is normalized to one. The normalized damaged area plot (squares) is the damaged area plot (black diamonds) divided by the constant minimum damage area. In this way, the normalized damaged area plot gives a factor difference between the actual damaged area and the minimum possible damage with complete penetration (factor = 1). For example, at a depth of about 14 mm the damaged area is about nine times greater than the minimum possible damage area for a uniform through thickness hole. Similarly, at the two depths of the local maxima, about 18 mm and 22.5 mm, the damaged area is about twenty and thirty seven times greater than the minimum, respectively. The through thickness damaged area is at least approximately one order of magnitude or greater than the minimum possible damage area throughout the penetration cavity.

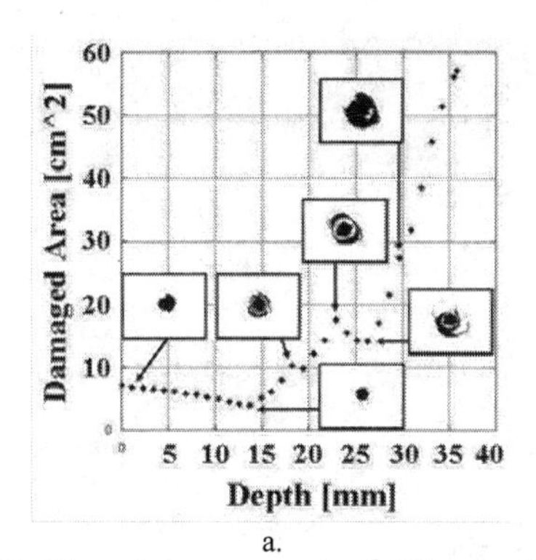

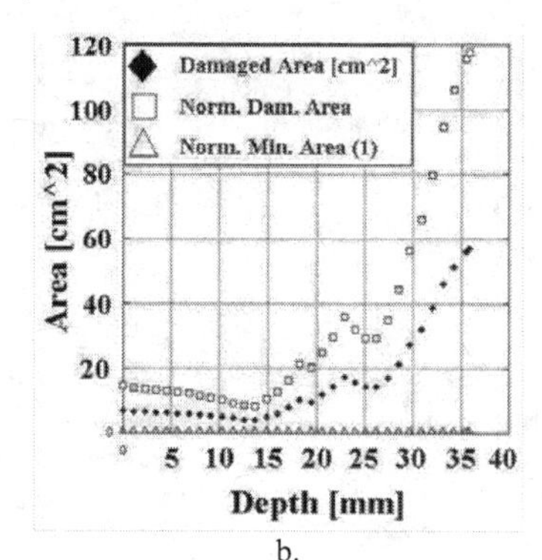

a. b.

Figure 12. Plots of damage vs. depth (distance from front face) in first specimen. (a) damaged area parallel to faces. (b) damaged area, normalized minimum area, and normalized damaged area on same plot for comparison.

Table I. Damaged Area vs. Depth (Distance from Impact Face) in Second Specimen.

Depth [mm]	Damaged Area [mm^2]	Depth [mm]	Damaged Area [mm^2]	Depth [mm]	Damaged Area [mm^2]
0.00	1521.3	30.00	66.0	60.00	74.6
5.00	654.4	35.00	64.1	65.00	75.4
10.00	445.6	40.00	65.1	70.00	137.9
15.00	291.6	45.00	65.6	75.00	1161.2
20.00	258.9	50.00	69.6	78.00	1588.9
25.00	90.2	55.00	71.7		

The image segmentation process was applied directly to the set of XCT scans of the second specimen, since the cross-sectional scanning plane was already parallel to the faces of the specimen. The matrix size and area of each scan was [1024 x 1024] and (11.00 cm x 11.00 cm), respectively, resulting in an areal pixel density of 8666 cm^{-2} and inversely a pixel area of 0.012 mm^2. The number of damaged pixels in each scan was determined using image histogram data and converted to the corresponding amount of damaged area. Table 1 shows the damaged area in the specimen as a function of distance from its impact face (depth) at regular intervals separated by five millimeters. The damaged area decreases significantly in the first 25 mm followed by a near constant, or steady state, amount in the neck of the penetration cavity. The average damaged area between the depths of 30.00 mm and 65.00 mm, inclusive, is 69.0 mm^2. The corresponding normalized area based on the threat is 1.4. The minimum normalized area in the first specimen is about 9 at a depth of about 14 mm, which is significantly higher than the normalized area in the neck region of the second specimen by about a factor of 6.4. The neck region is followed by a steep rise in the damaged area in the exit region of the penetration cavity near the rear face of the specimen, indicative of the relatively sharp transition between the neck and exit regions of the penetration cavity. The data in Table 1, as well as the complete data set for all scans, does not have any local maxima or minima as was the case for the first specimen. The base magnesium material of the two specimens was the same. However, they were subjected to different mechanical and/or heat treatment post processes. Secondly, as noted previously, the second specimen was about twice as thick as the first specimen. It is not clear at this time exactly how these factors affected the behavior of the ballistic threats and the differences in the damage in the specimens. The thickness of the second specimen may have suppressed spall of material out of the exit region of the penetration cavity until relatively late in the penetration event.

CONCLUSIONS

Ballistic damage in two sectioned specimens from different novel Mg plates was scanned and extensively characterized using XCT 2-D cross-sectional (planar) and 3-D volumetric analysis. Damage features in the first specimen including near parallel lateral cracking with "upturned" ends away from the penetration cavity, narrowing and widening cylindrical section of the penetration cavity, asymmetric missing material on one side of the impact face, relatively large area removal of material on the exit side, and three regions of different types of damage and features were captured and discussed. Damage features in the second specimen including an extensive cone shaped entrance region with decreasing taper, a relatively long near constant diameter, or steady state, neck region, relatively large area removal of material on the exit side similar to the first specimen, and three well delineated regions of damage were captured and discussed. Successive application of XCT 2-D evaluation, volumetric solid visualization and analysis, and volumetric point cloud visualization and analysis provided extensive and important qualitative and quantitative data about damage features in both specimens. The amount of detectable damage as a function of the distance from the impact side (depth) was determined for both specimens. The damaged area data for the first specimen was plotted, as well as the normalized data relative to the area of the base of the threat. A subset of the damaged area data for the second specimen was given in tabular form. The plot for the first specimen exhibited features, including local minima and maxima, quantitatively reflecting particular characteristics of the damage. The damaged area data for the second specimen exhibited changes in depth quantitatively reflecting characteristics of the damage, including the extensive entrance cone and steady state neck region. Characteristics of captured damage features provided better understanding of the physical processes of damage initiation and growth in the specimens. Future work is planned to further analyze the damage features and damaged area data in conjunction with microstructural observations and evaluation using relevant approaches and additional specimens taken from those evaluated in this paper or from the respective Mg plates.

REFERENCES

[1]E.F. Emley, *Principles of Magnesium Technology*, Pergamon Press, Oxford, England, (1966).

[2]M.M Avedesian and H. Baker, Ed., *Magnesium and Magnesium Alloys*, ASM International, (1999).

[3]K. Sugimoto, K. Niiya, T. Okamoto, and K. Kishitake, Study of Damping Capacity in Magnesium Alloys, *Trans. JIM*, **18**, 277-288, (1977).

[4]L.P. Martin, D. Orlikowski, and H. Nguyen, Fabrication and Characterization of Graded Impedance Impactors for Gas Gun Experiments from Tape Cast Metal Powders, *Mat. Sci. Eng. A*, **427**, 83-91, (2006).

[5]T. H. Newton and D. G. Potts (editors), *Technical Aspects of Computed Tomography, Radiology of the Skull and Brain*, **5**, The C. V. Mosby Company, St. Louis, MO, (1981).

[6]M. J. Dennis, *Industrial Computed Tomography, Nondestructive Evaluation and Quality Control*, **17**, American Society for Metals (ASM) International, ASM Handbook, USA, (1989).

[7]W. H. Green, R. Brennan, and R. H. Carter, Nondestructive Evaluation of as Fabricated and Damaged Encapsulated Ceramics, *Proceedings of 33rd International Conference on Advanced Ceramics and Composites - Advances in Ceramic Armor V*, **30**, 147-158, (2009).

APPLICATION OF A MINIATURIZED PORTABLE MICROWAVE INTERFERENCE SCANNING SYSTEM FOR NONDESTRUCTIVE TESTING OF COMPOSITE CERAMIC ARMOR

K. F. Schmidt, Jr., J. R. Little, Jr.
Evisive, Inc.
Baton Rouge, Louisiana USA

W. A. Ellingson
Argonne National Laboratory
Argonne, Illinois USA

Lisa Prokurat Franks, Thomas J. Meitzler
US Army RDC TARDEC
Warren, Michigan, USA

W. Green
US Army Research Laboratory
Aberdeen Proving Ground, Maryland, USA

ABSTRACT

A microwave interferometry system has been miniaturized and configured for flexible field use to determine the "status" of composite ceramic armor. The system utilizes Evisive Scan microwave interference scanning technique and has been demonstrated to detect damage on composite ceramic test specimens as well as composite ceramic surrogates with engineered features. The microwave interference scanning technique has demonstrated detection of cracks, interior laminar features and variations in material properties such as density. It requires access to only one surface, and no coupling medium. Data are not affected by separation of layers of dielectric material, such as outer over-wrap. Other methods, including through-transmission x-ray, x-ray Computed Tomography, and destructive examination, have been used to corroborate the microwave data and establish quantitative performance.

Test panels used in this work were provided by commercial manufacturers, the US Army Research Laboratory, US Army Tank-Automotive Research, Development and Engineering Center (TARDEC) and by the Ballistics Testing Station through Argonne National Laboratory. This paper will describe the system and present current results. This work is supported by US Army Tank-Automotive Research, Development and Engineering Center (TARDEC) and US Army Research Laboratory.

INTRODUCTION

The microwave interference scanning technique has been successfully demonstrated on armor panels constructed of high-performance technical ceramics. The ceramic armor is employed in the form of plate inserts in garments and seats; in panels in vehicles, aircraft and vessels; and as an appliqué in armored vehicles. Ceramic armor provides effective and efficient erosion of and defeat of ballistic threats. Effectiveness of ceramic armor can be degraded by defects present from production and by operational damage resulting from handling or impact with objects in the environment, other than projectiles. In normal use, ceramic armor is routinely exposed to the possibility of such damage[1].

A means to detect damage and manufacturing defects which are not visually apparent is needed to determine the integrity of the ceramic armor so that appropriate replacement can be made. Recently, a microwave-based method, having US and international patents[2, 3, 4, 5, 6], has been developed and demonstrated that is as applicable to ceramic armor systems[7]. Applications development has included optimization of antenna – material interaction, and miniaturization of the microwave interferometry system. The method permits real time evaluation by inspection from one surface only, through non-contacting encapsulation, with panels hung in place.

DESCRIPTION OF THE METHOD

The interference scanning method requires access to only one side of a part. The microwave interference pattern is created by bathing the part in microwave energy as illustrated in Figure 1. The probe (transmitter and receiver antenna) is moved over the part, bathing it in microwave energy. Some energy is reflected and transmitted at every interface of changing dielectric constant in the field of the transmitter. This includes the front and back surfaces of the part, and every "feature" in the part that has a discontinuity in dielectric properties. A microwave interference pattern is created when the reflected energy is combined with the transmitted signal to create the measured detector voltage at each of the receivers. The voltage values for both receivers are saved with the associated X-Y position on the object.

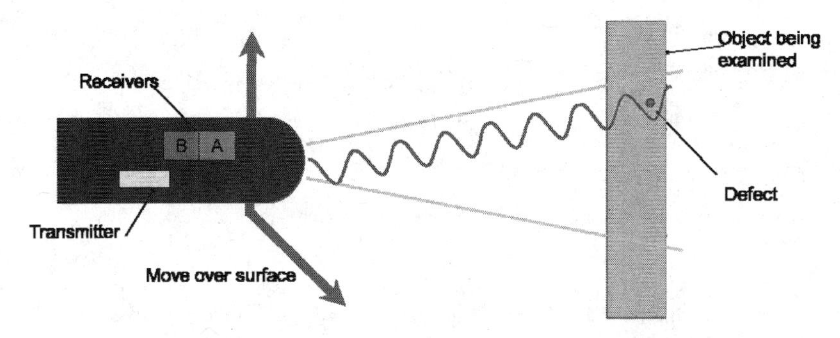

Figure 1. Diagram of transmitter and receivers positioned above the surface of a part under inspection.

The combination of dielectric constants of the engineered ceramic materials and the microwave frequency used in the tests yields wavelengths in the material of about 8 mm (0.33 inches) to 20 mm (0.83 inches). The magnitude of the phase difference between the emitted signal and reflected signal determines the voltage of the signal. This interference pattern is depicted theoretically in Figure 2 for a point reflector such as shown as a defect in in Figure 1. The actual scan image of a 2.5 mm (0.10 inch) spherical conductive reflector is shown in Figure 3, for both channels. The scan image confirms the theoretical image and illustrates the phase difference associated with difference in position of a quarter wavelength ($\lambda/4$) in the wave propagation dimension, Z.

Because the hardware Channels A and B are separated by ($\lambda/4$), the reflection of features at every depth is near a maximum phase gradient. In any "image" data, the rate of change of the detected signal value impacts the "clarity" of that image. This is true for detected Z axis features as well. Thus the "image" data of a feature is optimized visually at a Z dimension associated with maximum rate of change of the signal in the Z dimension. This is achieved for each channel by moving the emitter (and receiver) within a quarter wave length in the Z direction. This position is referred to as the "Stand-Off" distance.

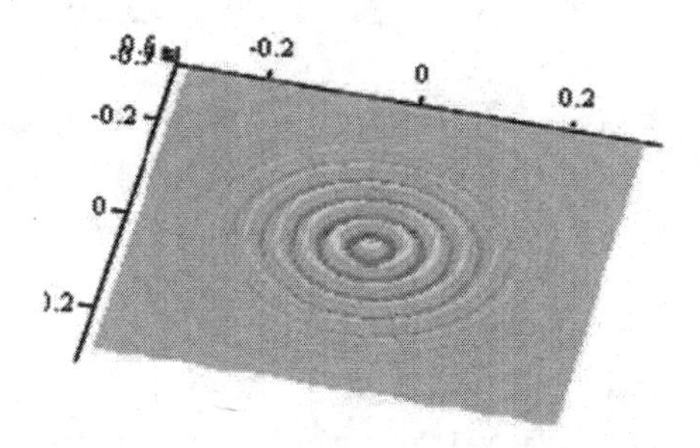

Figure 2. Theoretical image of a point reflector.

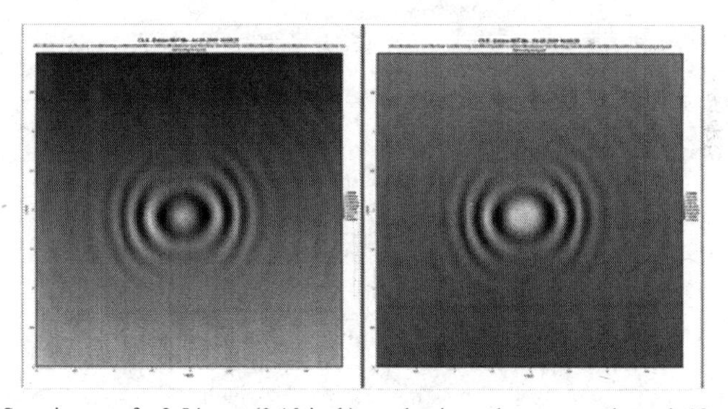

Figure 3. Scan image of a 2.54 mm (0.10 inch) conductive sphere target through 19 mm (0.75 inches) of glass. The image confirms the theoretical prediction and illustrates the phase difference between channels. Channel A is on left and B on right for a flanged wave guide

EQUIPMENT CONFIGURATION

The portable version of the microwave interferometry equipment is shown in Figure 4. This shows the laptop-computer with the driving electronics and the scan head. The same computer and electronics can be coupled in the laboratory to an XY Positioning Table as shown in Figure 5. Operating, interface and display software resides on the interface and display computer.

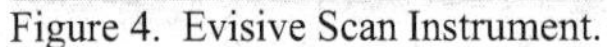

Figure 4. Evisive Scan Instrument.

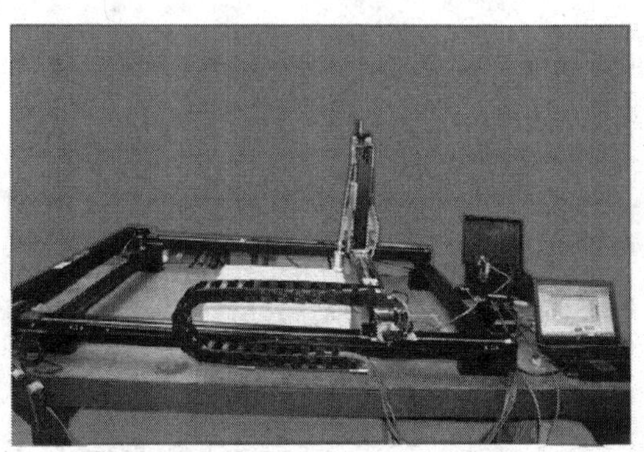

Figure 5. Evisive Scan Instrument interfaced to an XY Positioning system.

Data are collected via an X-Y raster scan over the surface of the part. The data rate is sufficiently high that mechanical positioning or position feedback for manual positioning is the only limitation in scan speed. The scan data are available in near real time. This scanning technology has been applied in the laboratory, with X-Y planar, X-Y cylindrical and r-θ positioning, and in the field with surface X-Y and multi-degree of freedom positioning devices.

With the exception of the infrared tracking position system, the images presented here were acquired on the X-Y positioning table. Datum spacing in the scan direction was 0.003 inches and raster increment was 0.05 inches unless otherwise stated. Scan speed on the X-Y positioning table reach 3 inches per second and ramp to start and stop. Scan speed with the hand held infrared tracking system varies and may exceed 10 inches per second.

PORTABLE FIELD CONFIGURATION AND WIRELESS HAND-HELD DEVICE:

A number of portable configurations have been applied to field use: The instrument and control system has been interfaced to mechanized pipe scanners and with a multi-axis position system for free-form manual positioning, and with a variety of manual position encoders.

A portable system has been interfaced to an infrared camera for correlation in other related studies. The equipment is shown in the field in Figure 6. The probe is manipulated manually, position tracked and presented in real-time (Typical before and after position tracking images are at the lower left). The tracking display facilitates control of coverage and scan density. Before and after images are automatically saved with each scan.

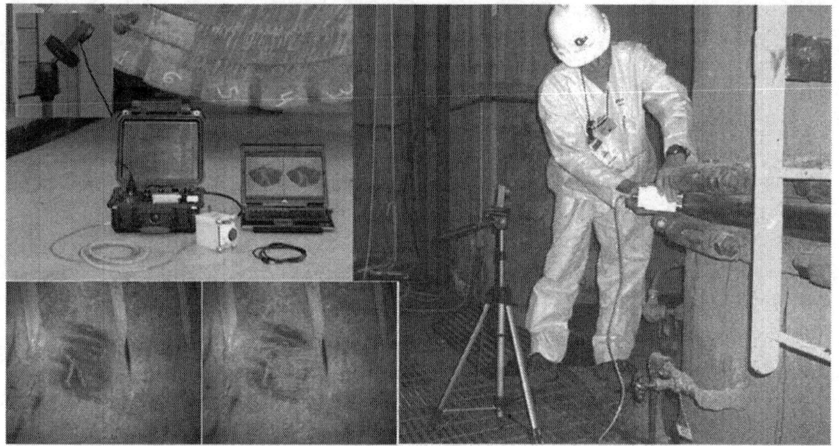

Figure 6. Portable microwave interferometry equipment configured with infrared position tracking

The system electronics have been incorporated into the hand held probe housing and a wireless interface of the probe and control computer has been developed. This required miniaturization of the signal processing components, and development of an additional communications protocol. Bluetooth was used for its efficient application. The self-contained power system benefitted from the low power requirements of the Gunn diode, receivers and signal processing components.

This significantly reduces the system size, as well as improving field applicability. The Wireless Hand Held Evisive Scan system is shown in Figures 7 and 8.

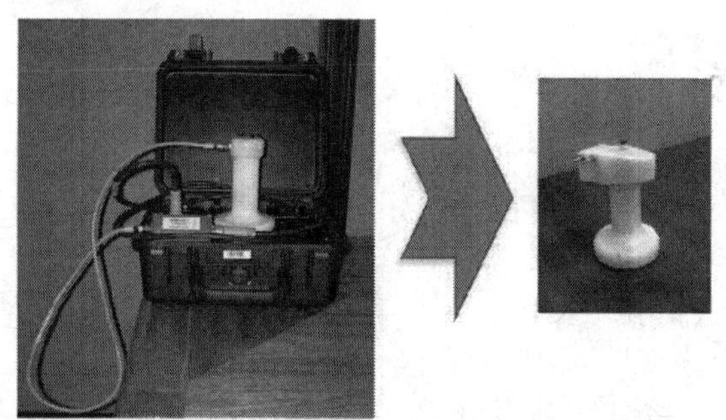

Figure 7. In the Wireless Handheld System, the microwave interferometry instrument electronics and controls have been incorporated into the Probe Assembly, which communicates by Bluetooth with the User Interface Computer. A miniature computer or display device functions as the operator interface.

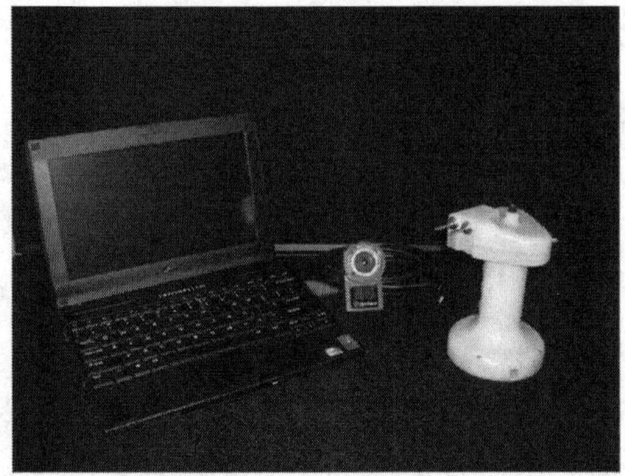

Figure 8. Prototype Wireless Handheld System, using a netbook computer as the operator interface.

IMAGES OF ARTIFICIAL LAMINAR DEFECTS

A multi-layer sample panel of ceramic composite armor was examined. The panel has multiple layers of fiber reinforced resin, ceramic tiles of two compositions, two conductive material layers and one elastomeric layer. The panel has six artificial laminar features, arranged at two depths: above and below the elastomeric layer. The scan image in Figure 9 shows that the microwave interferometry system detected all six artificial laminar features. Similar artificial features were detected below (left in the image) and above (right in the image) the elastomeric layer in the part. The difference in geometry of the left and right presentation of these features relates to contours of layers above the deeper features which are on the left in the image. The difference in depth of the features is indicated by the difference in gray scale (voltage) which relates to their relative phase positions.

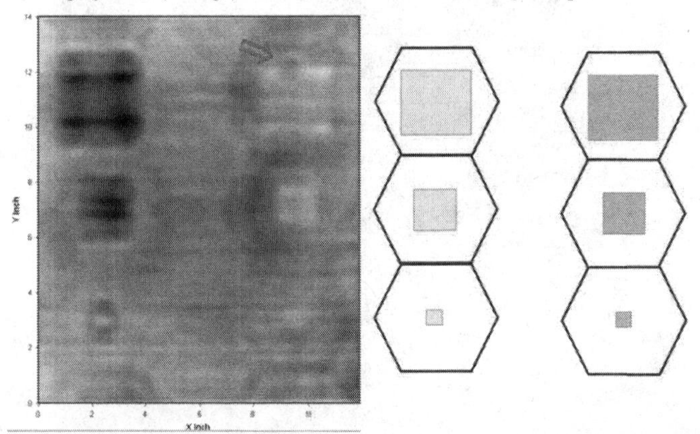

Figure 9. Scanning microwave image and cartoon of a test sample with six square artificial delaminations, and an unintended round feature.

The phase position of laminar features can be adjusted to "focus" the acquired image data at specific depths in the material, or to minimize the effect of specific laminar features. This is particularly beneficial in optimizing the technique for detection of laminar features at a specific layer in the complex material structure.

The small circular indication at (9.5, 12.5) in the upper right in Figure 9 is an unintended laminar feature very near the artificial feature.

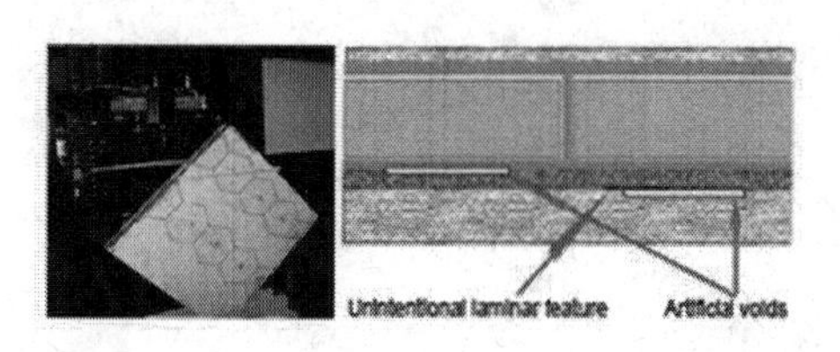

Figure 10. Photograph of the test panel in the x-ray imaging system, and cartoon of the placement of artificial voids and unintended laminar feature. The upper corner is marked off using a small diameter welding rod to validate the position.

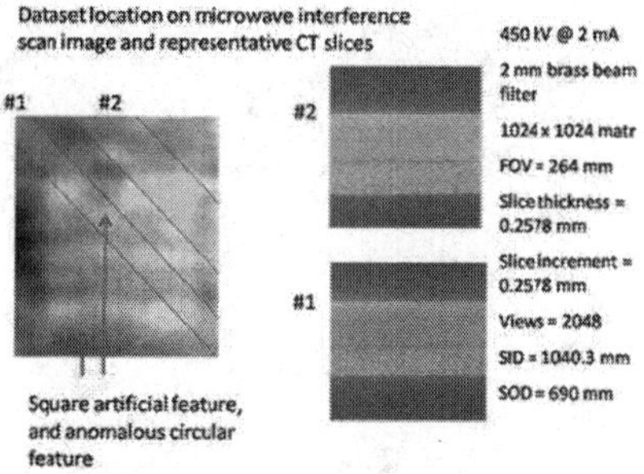

Figure 11. Correlation between high spatial resolution x-ray computed tomographic images and scanning microwave data. Location of x-ray CT images is shown on the microwave scan image (left), and x-ray CT sections are shown at right.

The artificial feature and anomalous feature at (9.5, 12.5) in Figure 9 were examined by through transmission x-ray and x-ray computed tomography (CT). The set up for x-ray tomography is shown in Figure 10. A welding rod was placed across a corner to create a temporary position registration. Examples of the acquired CT images are presented in Figure 11 along with the locations of the CT images relative to the microwave interferometry image data. The complex cross section structure of the specimen is clearly visible in the CT images. While the data was not available at the time of the test, it seems that the thicknesses of the very thin artificial laminar features are below the spatial resolution of the CT image. The anomalous laminar feature is also demonstrated to have a through wall dimension less than the minimum resolution of the CT image (smaller than about 0.25 mm (0.01 inches)).

These experiments demonstrate that the microwave interference scanning method is applicable to complex ceramic armor samples, and that the method seems capable of detecting very thin delaminations at various depths.

DETECTION OF CRACKED ARMOR TILE

The microwave interferometry system has been demonstrated to detect cracked ceramic armor tile in a typical ceramic armor layered configuration. Figure 12 shows a photograph and the corelation between a through-transmission x-ray image and the microwave scan of the same cracked tile. The wide, light gray patterns shown in the scan follow the crack centerlines. The microwave interferometry data has sufficient dynamic range, (8 bit resoultion), to identify the centerlines and edges of features within the data position precision.

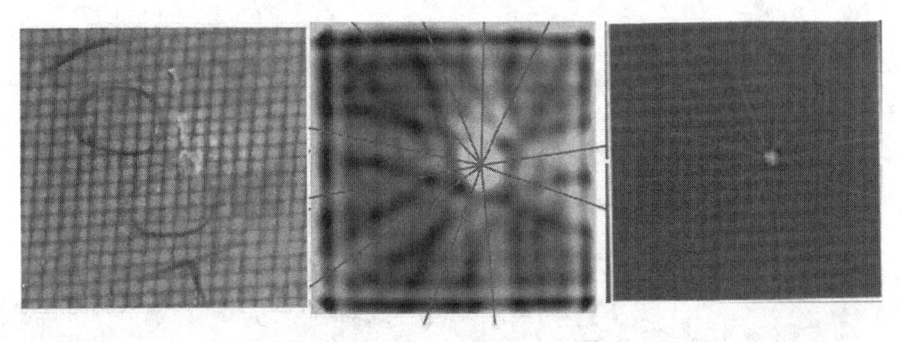

Figure 12. Photograph (left), microwave interference scan image (center) and x-ray image (right) of an artificially cracked tile in an armor panel.

An image of the same cracked tile is presented in Figure 13 to illustrate the detailed nature of the information presented in the scan image. For clarity, the scan image gray scale in Figure 13 (shown at th left) is reversed from that in the scan image show in Figure 12. Thus, the crack features in Figure 13 are dark, instead of light as in Figure 12, for the same values. The plot of voltage versus position for a single X value (right in Figure 13) illustrates the range of values which make up the scan image.

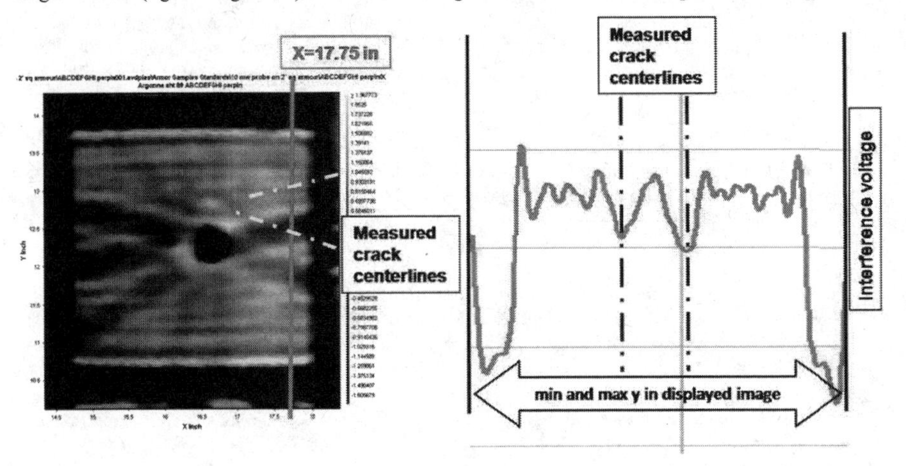

Figure 13. Scan image and cross section of voltage values associated with cracks and gaps to adjacent tiles.

The data is rich with information and can easily be made convenient for interpretation.

CONCLUSIONS

A portable microwave-based system has been developed and demonstrated on layered ceramic armor to detect cracks and delaminations within ceramic armor systems. Examination requires access from one side only and is effective in applications with metal backing. The capability of the method allows determination of size, depth and orientation of features within the dielectric solid. The laboratory instrument has been successfully coupled to X-Y positioning systems as well as multi-axis scanning systems and free-motion position tracking systems.

The system has been miniaturized and wireless communication incorporated facilitating application in field environments.

Further laboratory testing including destructive analysis of samples will establish scan and data interpretation protocols and qualify the technique for field nondestructive testing applications. The equipment will be further optimized and hardened for field use in the Phase II SBIR project.

The system will provide means for warfighters to verifying the integrity of ceramic armor in-theater. For ceramic armored vehicles, this permits confirmation of armor integrity following non-ballistic challenges, and avoids down time for inspection of armor.

ACKNOWLEDGMENTS

Evisive, Inc. expresses its sincere appreciation to the US Army Small Business Innovative Research Program, and US Army Research Laboratory and US Army Tank Automotive Engineering Research and Development Command who have made this program possible.

REFERENCES

[1]Salem, J., Zhu, D., 2007: Edited by L. Prokurat Franks, "Advances in Ceramic Armor III", *Ceramic Engineering and Science Proceedings,* Vol. 28, Issue 5

[2]Little, J., 2002: United States Patent 6,359,446, "Apparatus and Method for Nondestructive Testing of Dielectric Materials", March 19, 2002

[3]Little, J., 2003: United States Patent 6,653,847, "Interferometric Localization of Irregularities", Nov. 25, 2003

[4]Little, J., 2005: International Patent PCT/US2005/026974, "High-Resolution, Nondestructive Imaging of Dielectric Materials", International Filing Date 1 August, 2005

[5]Little, J., 2007: Canadian Patent 2,304,782, "Nondestructive Testing of Dielectric Materials", Mar. 27, 2007

[6]Little, J., 2005: New Zealand Patent 503733, "Nondestructive Testing of Dielectric Materials", PCT/US2005/026974, International Filing Date 1 August, 2005

[7]Little, J., 2005: Australian Patent 746997, "Nondestructive Testing of Dielectric Materials", PCT/US2005/026974, International Filing Date 1 August, 2005

[8]Schmidt, K., Little, J., Ellingson, W., 2008: "A Portable Microwave Scanning Technique for Nondestructive Testing of Multilayered Dielectric Materials", *Proceedings of the 32nd International Conference & Exposition on Advanced Ceramics and Composites*, 2008

[9]Schmidt, K., Little, J., Ellingson, W., Green, W., 2009: "Optimizing a Portable Microwave Interference Scanning System for Nondestructive Testing of Multi-Layered Dielectric Materials", *Review of Progress in Quantitative NDE*, 2009

STATISTICAL QUANTIFICATION AND SENSITIVITY PREDICTION OF PHASED-ARRAY ULTRASONIC DATA IN COMPOSITE CERAMIC ARMOR

J. S. Steckenrider
Illinois College
Jacksonville, IL

T.J. Meitzler and Lisa Prokurat Franks
US Army, TARDEC
Warren, MI

W. A. Ellingson
Argonne National Laboratory
Argonne, Illinois USA

ABSTRACT

A series of 16-inch square by 2-inch thick, multi-layered ceramic composite armor specimens, some of which had intentional design defects inserted between the layers, were inspected using a 128 element, 10MHz immersion phased array ultrasound system. To overcome some of the issues associated with the acoustic wave propagation in layered media, two digital signal processing methods (Fast Fourier Transform (FFT) and Wiener filtering) were employed. While previous work has been presented on the significant improvement in defect detection associated with these methods, the authors present a detailed and quantitative statistical analysis of these results. This analysis suggests that these intentional defects were a) not detectable when the defect was in a particular configuration, b) readily detectable in all cases for alternate defect position configurations, and c) clearly identifiable in most cases for those configurations. However, even in the configuration where intentional defects were not detected (owing to inherent design issues in the armor structure), significant variation in interfacial quality was observed and quantified, and these results will also be presented.

INTRODUCTION

Although ceramic vehicular armor offers significant potential improvement over historical materials by providing a greater capacity for energy absorption and dissipation per unit mass (i.e., very high fracture toughness), they are more susceptible to manufacturing defects that may reduce that high toughness. Thus, an efficient non-destructive evaluation (NDE) method which can both identify and quantify these defects before the armor is placed into service, and perhaps after with portable inspection units, is critical to their effectiveness[1]. Conventional ultrasonic techniques have been used to both locate and characterize such defects in the monolithic ceramic tiles that make up the "backbone" of these armor panels[2,3]. Furthermore, phased-array ultrasound[4] (PA-UT) has demonstrated significant improvement over these methods[5] as it offers both enhanced sensitivity and improved throughput[67].

PA-UT has also demonstrated its performance with regard to the actual implementation of ceramic armor that incorporates these monolithic tiles into a thick, multi-layered ceramic composite structure[8]. In the current and previous work, a layered structure is used in the assembly of composite armor panels. The panels were made up of tessellated hexagonal high-toughness monolithic ceramic tiles, a carbon-based matrix to encase these tiles, an elastomeric layer to distribute and attenuate mechanical stresses transmitted by the ballistic impact, and a glassy layer which provides a monolithic substrate to support the composite armor. These were arranged as shown in the cross-section and topography of Figure 1. To evaluate defect detectability, planar inclusions of a range of sizes (0.5", 1.5" and 2.5" diameters) were intentionally inserted at the two most critical boundaries (i.e., on either side of the elastomeric bonding layer) to simulate a "disbond" at the locations where it would have the greatest effect in reducing the ballistic performance of the panel. While the results of the previous effort showed that the use of FFT and Weiner filtering methods significantly improved defect detection, the current work expands upon those results by quantifying the associated detection limits for various inspection configurations.

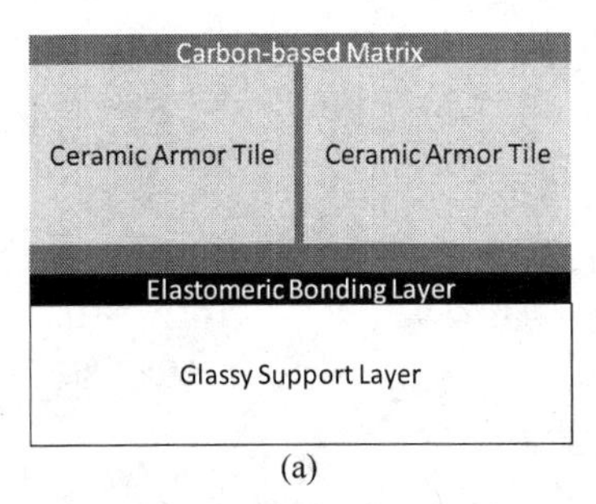

(a)

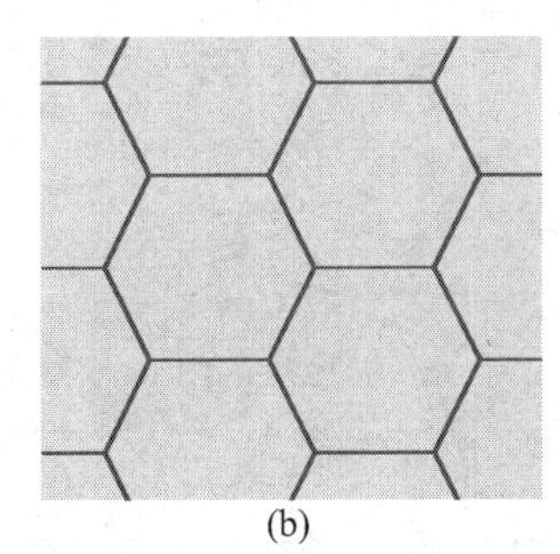

(b)

Figure 1. Schematic diagram of composite layered armor panel: a) cross-section and b) top view.

THEORY

In order to effectively evaluate the ability of Phased-Array Ultrasound (PA-UT) to detect, identify and quantify defects and their severity in composite ceramic armor, it is necessary to evaluate the Signal-to-Noise Ratio (SNR) of the inspection methodology. To do this requires that the noise level of the defect-free inspection be quantified. Since the "data" resulting from a PA-UT inspection is usually an image, it is often necessary to do some type of image processing for best results. In image processing, the SNR is defined as the ratio of the signal mean (in this case, the average pixel intensity in the region known or presumed to be defective) to the standard deviation of the background[9] (the non-defective region of the image), as

$$SNR = \frac{\mu_{sig}}{\sigma_{bg}} \tag{1}$$

However, this assumes that the mean of the background is zero. In situations where this is not the case (i.e., where a low-level response is expected in the background of the image), a more accurate definition would be

$$SNR = \frac{\left(\mu_{sig} - \mu_{bg}\right)}{\sigma_{bg}} \tag{2}$$

As a result, the SNR is simply a measure of how many standard deviations a particular pixel or group of pixels is from the mean of the remainder of the image. This presents two difficulties for PA-UT relative to composite armor inspection. First, this definition of SNR assumes that the pixels of the image (in this case an ultrasonic c-scan image) are normally distributed. While this assumption is valid for shot-noise limited systems (as is the case for ideal optical images)[10], it is not valid for PA-UT images, as will be shown below. Second, the significance of the SNR value is dependent upon the fraction of the total image occupied by the signal pixel(s). For instance, in a 1024 x 1024 image, a single pixel must attain a SNR greater than 4.76 in order to be considered statistically meaningful (i.e., in order to be definitively distinguished from the normal background variation), as this would represent a confidence of roughly 99.9999% (i.e., 1 part in 1024^2), but a 10 x 10 (100 pixel) region within that image would only need an average SNR of greater than 3.73 in order to be considered statistically meaningful, as this would now represent a confidence interval of only 99.99% (i.e., 100 parts in 1024^2). Thus, it is actually the relationship between the area of the defect and the distribution of pixel intensities in that image that is of greatest interest in determining the detectability of a particular defect. In any case where the confidence interval for the mean intensity of a nominally defective region of the image exceeds the fraction of the image not occupied by the nominally defective region, that region is determined to indeed be defective, as it could not have attained the intensity value it has within the distribution of intensities due to regular fluctuations in the background.

To address the first concern, the distributions of pixel intensities were measured for a number of different composite ceramic armor panels using c-scan data and it was found that these data were indeed not normally distributed. Figure 2a shows a c-scan of panel B3, a defect-free panel used for establishing baseline parameters for the PA-UT methodology, and Figure 2b shows a histogram of the pixel values for the central bright rectangular region of that c-scan. (Note that, for this initial multi-facility study, all panels were sealed with a composite wrap to maintain the integrity of the panel as it was shipped between facilities, and that to insure the integrity of the seal this wrap overlapped the edges of the panel, as shown by the dark frame around the edges of Figure 2a. Thus, the panel surface was uniform only in the central bright region, so only this region was used for quantitative analyses.) Although the Normal Distribution fit in Figure 2b is somewhat close, the χ^2 fit (for which the probability distribution is equally well known as

$$f(x,k) = \frac{1}{2^{(k/2)}(k/2-1)!}x^{(k/2-1)}e^{-\frac{x}{2}} \tag{3}$$

where x is the pixel intensity value and k is the degrees of freedom) is clearly superior, and would be expected to provide a much more accurate assessment of the confidence interval for each pixel intensity value in the image. A χ^2 fit was used for each panel in order to determine the probability for each pixel intensity. Any pixels that exceeded this probability determined if and where any defects were present. However, because of the significant variation present in this feasibility study (panels were made in batches of four, with sometimes significant variation between batches, were wrapped individually, with significant variation in wrap between panels, and were inspected intermittently over a period of several months through the round-robin protocol established) it was not reasonable or feasible to establish a single baseline intensity distribution for all panels (as would be used in a conventional manufacturing process). Therefore, the intensity distribution for each c-scan was measured and the χ^2 probability density function was individually fit for each panel.

(a)

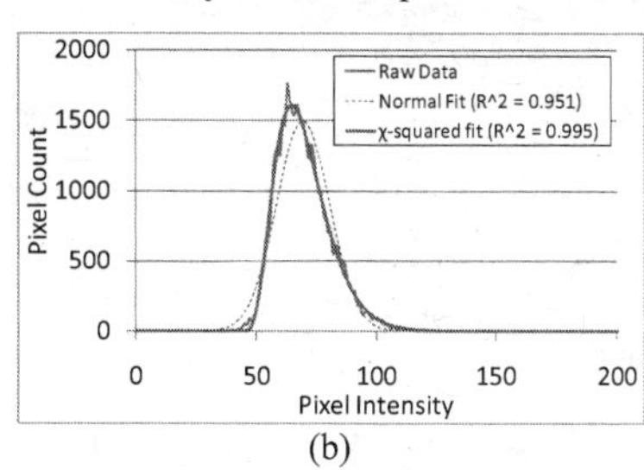

(b)

Figure 2. Example of determination of statistical parameters: a) C-scan image and b) Histogram of intensity values showing Normal and χ^2 distribution fits taken from the central bright rectangular region for panel B3, a defect-free ceramic armor panel.

To address the second concern (establishing the significance of a particular SNR as a function of defect size), intensity values were directly translated into confidence intervals. This yielded a non-linear intensity distribution that was directly correlated to the statistical likelihood that the intensity value would have occurred spontaneously without the presence of a defect. Details of this approach are provided in the Intensity Distribution section below.

PROCEDURE

Although the locations of the intentional defects were known for the panels examined in this study, the analytical methodology must work autonomously without user intervention. The analysis procedure prescribed below is therefore designed to operate automatically. However, given the nature of this initial feasibility study, there were some additional steps added to the procedure to compensate for variation present here that would not be included in an ultimate in-process implementation.

Spatial Filtering

The first accommodation that must be made for these preliminary study panels relates to the panel overwrap (which would not be present in full-scale manufacturing evaluations). As mentioned above, these ceramic armor panels were encapsulated in a woven composite wrap to maintain panel integrity throughout the evaluation process. Unfortunately, this wrap altered the ultrasonic coupling efficiency at the surface, so that regions between the fabric fibers transmitted less acoustic energy into the panel than the regions along the fibers. Thus, the raw c-scan data was superimposed with this fiber pattern, as shown in Figure 3a. To reduce the impact of this artificial surface pattern, which carries no information about the bulk properties of the armor panel, a 5-pixel diameter median filter was applied to the image, with the result shown in Figure 3b. Although the effects of the composite weave have been virtually eliminated in Figure 3b, all other features (which exhibit a spatial dimension greater than 5mm) have been retained, and thus are available for further investigation.

(a)

(b)

Figure 3. Example of use of median filtering: a)-PA-UT c-scan of panel B3 as inspected and b) PA-UT c-scan after 5x median filtering.

ROI selection

The second accommodation that must be made for these test panels derives from the fact that only a central portion of the panel, where ballistic impact would ultimately be applied, was constructed according to the actual in-service armor design. The remainder of the panel was built with "filler" material of a lesser quality and cost. Thus, for the feasibility study, a smaller region-of-interest (ROI) in the center of each panel where the ultimate armor materials were used was examined. Because the location of this region varied from panel to panel, the ROI had to be manually selected for each panel. However, the shape of the ROI was known, so only three different selections were used, and these are shown in Figure 4 for inspections from a) the support side (where the 190 mm x 220 mm ROI was limited by the overlapping encapsulating overwrap around the edges), b) the ceramic side (in the cases where a single high-value 102 mm height hexagonal tile was used, for a 7613 mm^2 ROI) and c) the ceramic side (in the cases where three high-value hexagonal tiles were used, for a 25,308 mm^2 ROI).

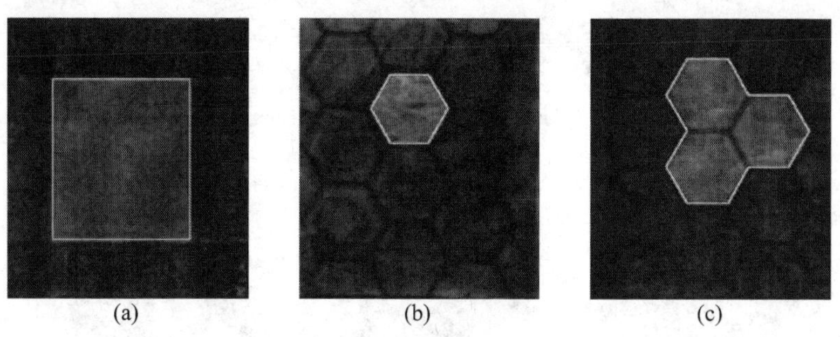

Figure 4. Selection of Region-of-interest (ROI) definition: a) support-side inspection, b) ceramic-side inspection with a single high-performance tile and c) ceramic-side inspection with three such tiles.

Table I. Example Log LUT values for panel B3

Input Intensity	Cumulative Probability	Log Output Intensity
≤ 46	≤ 0.001%	0
47	0.01%	26
48	0.1%	51
51	1%	77
57	10%	102
68	50%	128
85	10%	153
104	1%	179
119	0.1%	204
135	0.01%	230
≥ 148	≤ 0.001%	255

Intensity Distribution

A histogram analysis was applied to each panel to determine the distribution of pixel intensities within the ROI. The histogram output was then fit using the distribution of (3) as shown in Figure 2b above to assign a specific probability value to each pixel intensity in the c-scan. A lookup table (LUT) was then created to map the input pixel intensity to its associated cumulative probability. This LUT was a logarithmic one in which input intensities were mapped to orders of magnitude of probability of random occurrence, as shown in Table I. Here, the cumulative probability shown corresponds to the area of tail of the distribution beyond the pixel value given (i.e., the values peak at 50% for the median pixel value and decrease toward 0% on either side of the median). The limits of the log scale were chosen so that an output intensity of 0 represented the single-pixel limit on the low-amplitude end of the distribution (i.e., the point where the tail of the distribution would represent a total area of less than 0.5 pixels for the ROI used) and 255 represented the single-pixel limit on the high-amplitude end. Because the defects are likely to be small relative to the ROI inspected, this log-scale LUT provides much greater resolution to differentiate background regions that are just randomly bright from statistical outliers that represent actual defects. It should be noted that, because of the nature of the χ^2 distribution, the ability to resolve meaningful variations on the lower end of this distribution is much worse than the ability to resolve them on the upper end. Fortunately, because defects almost always increase the acoustic impedance discontinuity in the specimen, they will increase the reflected acoustic amplitude, and therefore manifest on the upper end of the distribution, where resolution for detecting and differentiating these defects is greatest.

Finally, this LUT is applied to the raw image, yielding a logarithmic mapping of the probability that each pixel actually represents a defect. This is shown below in Figure 5 for panel B3 (again, in which there were no intentional defects present). Figure 5a shows a photograph of the panel, Figure 5b shows the original c-scan of the panel after median filtration, and Figure 5c shows the logarithmic probability plot.

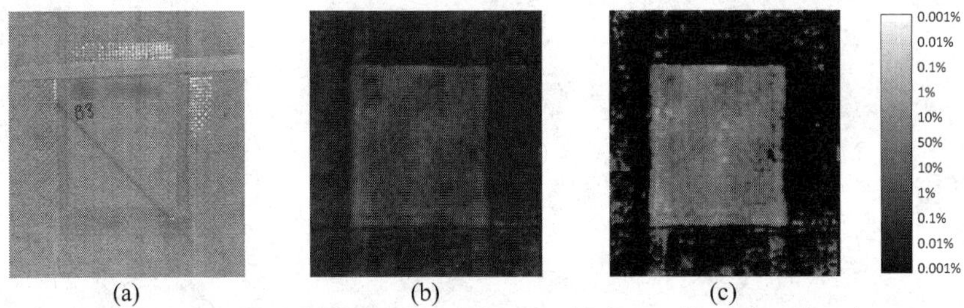

Figure 5. PA-UT Images of panel B3 showing a) photographic image, b) median filtered c-scan, indicating surface contamination, c) logarithmic map of probability. The scale given corresponds to the probability of random occurrence within the background for c).

Defect Identification

The logarithmic map of probability in Figure 5c shows that even the brightest of the raw c-scan areas are only in the 5% probability range, and therefore are not statistically significant. On the other hand, while the log map shows that there are no regions in which the c-scan amplitude (i.e., pixel intensity) is higher than the background with any statistical significance, it does clearly show that there is a single region on the right-hand side of the ROI in which the probability is zero, indicating a statistically significant difference (i.e., a defect) is present. Although, as stated above, the lower tail of the distribution has a significantly reduced resolution, the size of this area warrants investigation, and is seen in Figure 2b as the slight deviation above the χ^2 fit just below pixel values of 50. However, further analysis of Figure 5b shows that this reduction was likely the product of operator-induced damage in that region, as the "defect" region clearly reflects writing (the name of one of the inspection engineers ("Dick") appears above a series of numbers), where someone appears to have either written on one of the panel layers or used the panel as a writing surface beneath some other medium.

In general, in order to determine whether any portion of a particular c-scan differed from ordinary background variation in a statistically significant way, two methods were used. First, each log mapped image was examined to see if any single pixel value was either 0 or 255. Since the original c-scan had already been subjected to a median filter, any single pixel exceeding the threshold for statistical significance (either high or low) was deemed to represent a defect. Second, the brightest and darkest (i.e., highest and lowest probability) 0.5" diameter regions (a 137-pixel area, due to image pixelization) within the ROI were averaged, and if the tail of the χ^2 distribution beyond this average probability represented a smaller area than the 137 pixels measured then this region was also deemed to represent a defect. However, in cases where an intentional defect larger than 0.5" diameter was present, this larger region size was used to determine the detectability of that known defect.

RESULTS

Given the designed character of the central elastomeric layer to absorb mechanical vibrations, inspection of the panels was performed from both sides – the impact or ceramic side (where the armor tiles were located closer to the surface) and the glassy or support side. Because each of these required a different inspection methodology, they also require unique analysis approaches as well, and therefore are presented separately herein.

Support-side Inspection

Figure 6 and Figure 7 show the median filtered c-scans and log probability maps, respectively, for the J-series of panels. In this series, a 0.5"-diameter inclusion was located between the support and elastomer layers during manufacturing. This inclusion is located directly beneath a single high-value ceramic tile (albeit on the opposite side of the panel), and is roughly in the center of the ROI for panels J1, J2, J5 and J6, and near the top of the ROI for panels J3 and J4. Knowing its position, the defect can be identified in all six panels, as indicated by the arrow in Figure 6. Furthermore, all defects demonstrated a SNR greater than 1:1, as tabulated in Table II. However, a more thorough statistical analysis indicates that the only panels in which the defect response differs from the background by a statistically significant amount are panels J1 and J3, as shown in Figure 7. In other words, although the defects show a low probability of random occurrence in all panels (as indicated by a SNR > 1), the predicted area of the tail of the distribution equal to or greater than the average pixel intensity for the defect region (the "Differentiated Area" in the table) is smaller than the area of the actual defect for only these two panels (indicating that, for these c-scans, there are more pixels in the tail of the distribution than would be predicted by the random variation of the background). This means that, if the location of the defect had not been known, it would still have been possible to assign absolute certainty to the presence of defects only in panels J1 and J3.

Figure 8 shows the log probability maps for the N-series of panels. In this series, a 38mm (1.5") -diameter inclusion was located between the support and elastomer layers during manufacturing. This inclusion is located directly beneath the intersection of three hexagonal ceramic tiles (again on the opposite side of the panel), and is located on the left of the ROI for panels N1 and N2, and on the right of the ROI for panels N3 – N6. Note that the outline of the hexagonal tiles in N4 – N6 was the result of surface contamination, which altered the ultrasonic coupling efficiency (i.e., an ink outline of the underlying tiles applied by one of the other inspection facilities) and is not the result of acoustic interaction with these tiles. Again, the SNR for each tile was greater than 1:1, as shown in Table III. However, in this case, because the area of the defect was larger than in the J-series panels, the defects in all six panels differ from the background by a statistically significant amount. In other words, the maximum area meeting or exceeding the average defect intensity, as predicted from the random variation of the background, is smaller than the actual defect size in all panels. This is also demonstrated in the histogram analysis of Figure 9, where the background variation is well aligned with the χ^2 fit, but the defect manifests a significant deviation from that fit at pixel values centered around 152 (the defect's mean intensity) as encircled by the ellipse in the figure.

Table II. Defect characteristics for the J-series panels. Defect area = 137 pixels. Shaded defects were definitively, statistically significantly different from the background. SNR was calculated using (2).

Panel #	χ^2 Fit R^2 value	Defect Mean	Confidence Interval	Differentiated Area (pixels)	Defect SNR
J1	0.994	135.6	99.87%	56	4.01
J2	0.993	71.1	89.01%	4595	1.14
J3	0.994	136.7	99.92%	34	4.04
J4	0.991	142.0	92.95%	2945	1.59
J5	0.992	159.2	98.13%	780	2.69
J6	0.992	89.8	96.86%	1312	2.44

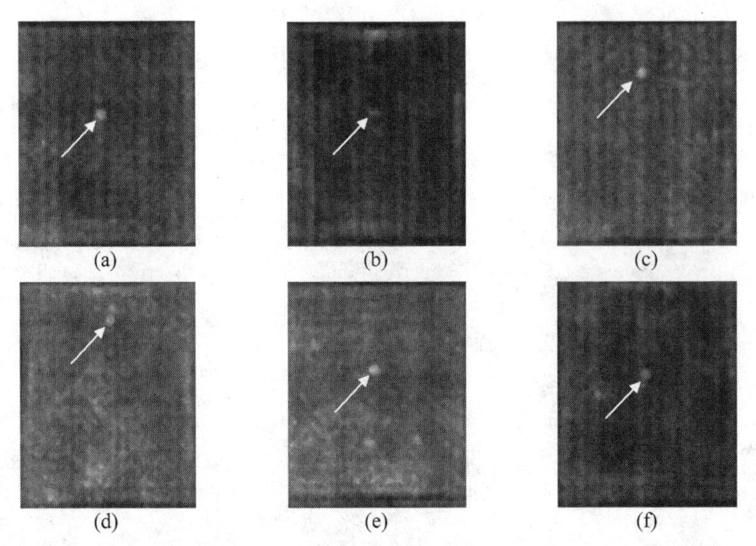

Figure 6. Median filtered support-side inspection c-scan data for panels J1 – J6 (a – f). The intentional defect shown was a 13mm (0.5") diameter inclusion between the support and elastomeric layers, and only the ROI is shown.

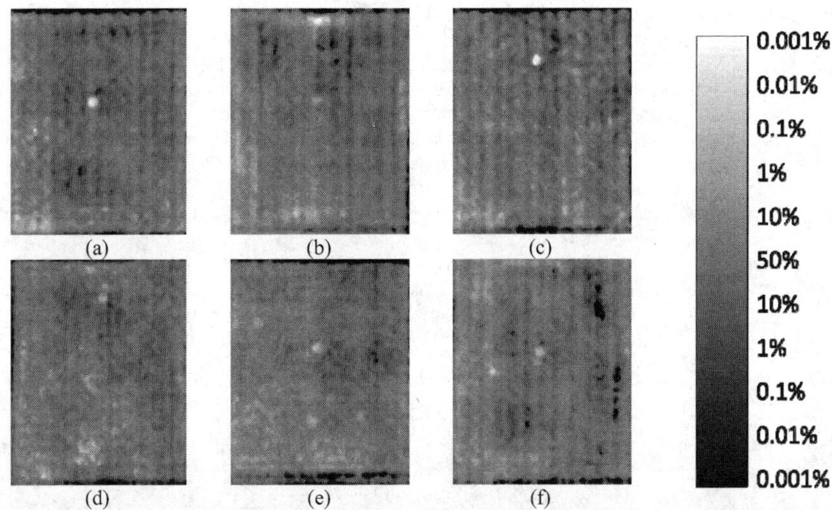

Figure 7. Log probability maps of data from Figure 6 for panels J1 – J6 (a – f). The scale shown references the area of the tail of the distribution beyond that gray-level.

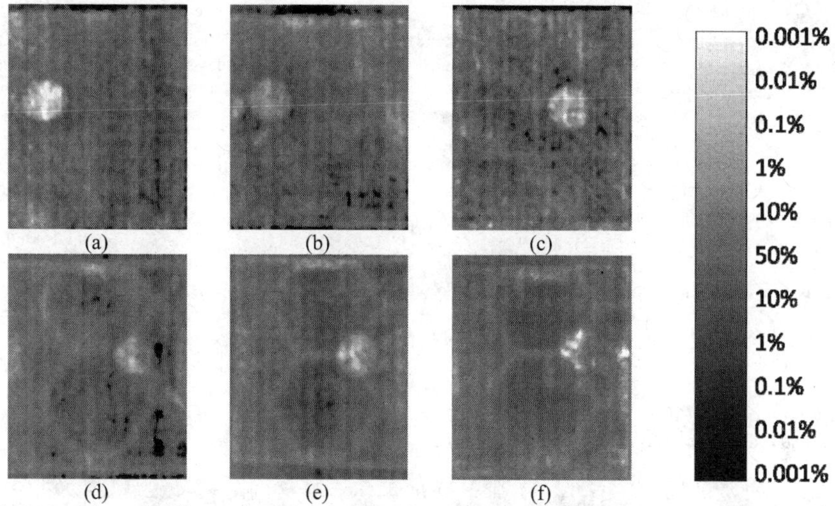

Figure 8. Log probability maps of data for panels N1 – N6 (a – f). The scale shown references the area of the tail of the distribution beyond that gray-level.

Table III. Defect characteristics for the N-series panels. Defect area: 1134 pixels. Shaded defects were definitively, statistically significantly different from the background. SNR calculated using (2).

Panel #	χ^2 Fit R^2 value	Defect Mean	Confidence Interval	Differentiated Area (pixels)	Defect SNR
N1	0.994	152.2	99.99%	5	4.96
N2	0.986	167.0	97.70%	961	2.47
N3	0.993	112.4	99.79%	86	4.03
N4	0.985	107.7	97.80%	922	3.09
N5	0.990	129.4	99.32%	285	3.59
N6	0.968	107.4	97.90%	877	2.19

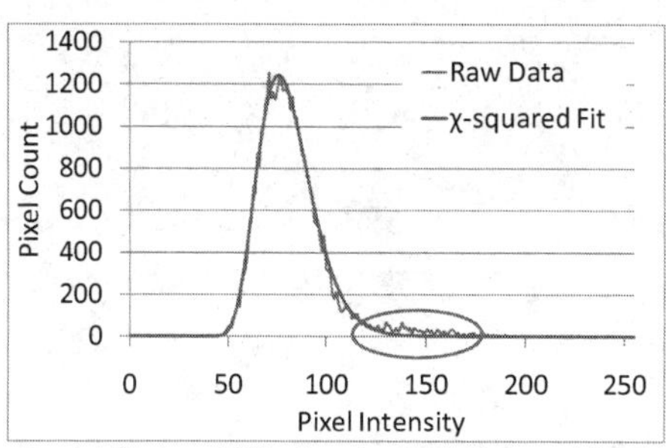

Figure 9. Histogram plot for panel N1, showing both the raw histogram and the χ^2 fit.

Impact or Ceramic-side Inspection

The intentional defects were not visible in the ceramic-side inspections due to the already large mismatch of acoustic impedance between the ceramic layers and the elastomer, hence the lack of a distinct circular region in the centers of Figure 10 and Figure 11. However, these figures do show a significant variation in acoustic amplitude detected (i.e., pixel intensity) so it would seem worthwhile to evaluate the highest and lowest intensity regions in each of these images. Thus, the data presented in Table IV refers to the highest and lowest 0.5" (13 mm) diameter regions within the ROI of each.

As shown in Table IV, some kind of statistically significant defect was found in all of the D-series panels except D3. The high-intensity defects (shown as shaded rows in the table) showed a definitive statistically significant increase from the background (i.e., increased acoustic impedance mismatch) while low-intensity defects (shown as bold rows in the table) showed a definitive statistically significant decrease from the background (i.e., decreased acoustic impedance mismatch). However, because the slope of the χ^2 distribution is much steeper on the low-intensity side, the ability to resolve small differences in acoustic amplitude in this region is very limited, so the relative significance of this kind of defect may be artificially amplified as a result. Furthermore, a decrease in acoustic mismatch would be more likely to enhance armor performance (as it represents a greater cohesion between the layers), and so these "defects" may not be critical at all. Thus, the areas of greatest importance would be represented by the shaded rows.

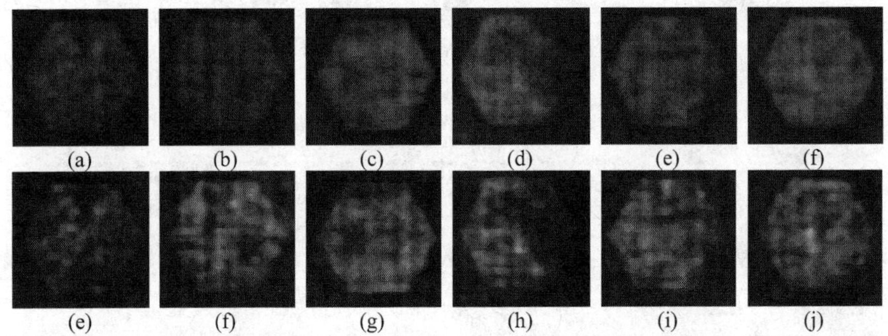

Figure 10. Ceramic-side inspection c-scan data for panels D1 – D6 showing median filtered c-scans derived from Weiner filtering (a – f) and Fourier filtering (e – j) of the a-scans, showing ROI only.

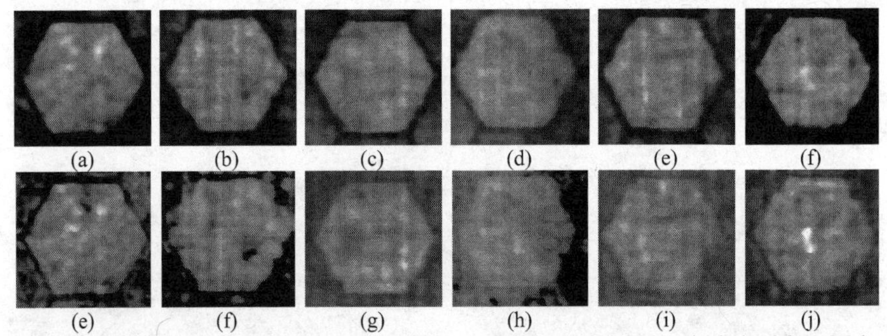

Figure 11. Log probability maps of data from Figure 10 for panels D1 – D6 again derived from Weiner filtered (a – f) and Fourier filtered (e – j) c-scans. Note that only the ROI is shown here.

Table IV. Defect (identified as the 137 pixel areas with the highest and lowest average intensities) characteristics for the D-series panels using both Weiner (Wnr) and Fourier (FFT) filtration of the original PA-UT a-scan data. Shaded rows indicate an increased acoustic impedance mismatch, while bold rows indicate a decreased acoustic impedance mismatch. SNR was calculated using (2).

Panel #	χ^2 Fit R^2 value	Defect Mean	Conf. Interval	Diff. Area	Peak SNR	Valley Mean	Conf. Interval	Diff. Area	Valley SNR
D1-Wnr	0.982	66.0	99.62%	29	2.98	27.949	5.343%	407	1.38
D2-Wnr	0.985	54.1	99.40%	46	2.84	26.453	7.963%	606	1.40
D3-Wnr	0.977	78.1	96.18%	291	2.12	36.774	13.801%	1051	1.14
D4-Wnr	0.945	98.4	97.41%	197	2.53	27.796	1.195%	91	2.05
D5-Wnr	0.971	68.3	92.37%	581	1.54	31.796	3.645%	278	1.76
D6-Wnr	0.966	91.5	98.29%	130	2.29	49.453	6.464%	492	1.33
D1-FFT	0.948	76.4	97.79%	168	2.24	26.445	11.038%	840	1.06
D2-FFT	0.933	116.2	92.05%	605	1.85	25.27	0.157%	12	1.72
D3-FFT	0.942	109.4	96.57%	261	1.90	27.555	3.371%	257	1.81
D4-FFT	0.911	112.9	94.84%	393	2.31	16.292	4.073%	310	1.31
D5-FFT	0.909	109.0	85.32%	1117	1.93	20.212	1.382%	105	1.92
D6-FFT	0.951	128.5	99.78%	17	2.79	34.124	5.321%	405	1.46

Table V. Defect (identified as the 137 pixel areas with the highest and lowest average intensities) characteristics for the G-series panels using both Weiner (Wnr) and Fourier (FFT) filtration of the original PA-UT a-scan data. Shaded rows indicate an increased acoustic impedance mismatch, while bold rows indicate a decreased acoustic impedance mismatch. SNR was calculated using (2).

Panel #	χ^2 Fit R^2 value	Defect Mean	Conf. Interval	Diff. Area	Peak SNR	Valley Mean	Conf. Interval	Diff. Area	Valley SNR
G1-Wnr	0.991	36.7	95.21%	1213	2.20	7.985	2.157%	546	1.68
G2-Wnr	0.986	37.4	94.84%	1305	2.31	11.102	3.536%	895	1.26
G3-Wnr	0.988	37.6	98.14%	471	2.79	8.825	4.401%	1114	1.41
G4-Wnr	0.995	37.8	99.63%	94	3.50	8.912	6.812%	1724	1.22
G5-Wnr	0.939	50.1	93.56%	1629	2.41	7.577	0.605%	153	1.60
G6-Wnr	0.987	53.7	97.94%	521	2.50	10.409	0.550%	139	1.78
G1-FFT	0.981	97.3	95.49%	1141	2.25	29.883	7.963%	2015	1.32
G2-FFT	0.986	106.7	97.76%	568	2.77	22.971	0.060%	15	1.71
G3-FFT	0.986	114.2	99.60%	102	3.17	31.029	2.662%	674	1.63
G4-FFT	0.987	113.7	99.83%	44	3.76	23.526	9.339%	2364	1.09
G5-FFT	0.944	125.6	96.82%	805	2.90	10.839	2.451%	620	1.57
G6-FFT	0.982	115.7	99.34%	166	3.46	11.905	1.142%	289	1.62

The G-series panels also showed statistically significant high-intensity defects in both panels G3 and G4, as given in Table V. The G4 defect was severe enough to be seen in both the Weiner and FFT maps, but the G3 defect was only detected using the FFT method. In addition, a single low-intensity defect was detected in panel G2, but only using the FFT method. Finally, there were two more low-intensity defects detected in panels G4 and G5, but these were much smaller than the 137 pixel critical area used for this evaluation, and so when averaged over a larger area these defects were not statistically significant. Of particular note in this data set was the lack of direct correlation between a traditionally measure SNR (which again assumes a normal distribution) and the χ^2 detection confidence. For example, when examining a 137 pixel region of the G-series panels, a SNR (calculated according to conventional method of (2) above) in excess of 2.55 would determine that the region differed from the random background variation by a statistically significant amount, using a normal approximation to the background variation. Such an approximation applied to the FFT results of Table V would predict that all but panel G1 were defective (i.e., contained a statistically significant high-intensity defect). However, the more detailed χ^2 analysis shows that, although the highest-intensity region in panel G6 showed a SNR of 3.46 while panel G3 showed a SNR of only 3.17, the region in panel G3 was determined to be definitively a defect, while the region of G6 was not, and none of the regions in panels G2 or G5 (both of which also had SNR values above 2.55) were statistically defective either.

CONCLUSION

Phased-array ultrasonic inspection methods have been successfully applied to layered composite ceramic armor structures. We have demonstrated that the distribution of acoustic amplitude comprising the resulting c-scans (aka pixel intensities) follows a χ^2 distribution much more closely than a normal distribution, and so the conventional application of Signal-to-Noise Ratio is not applicable for this data set. Furthermore, owing to the limited nature of this feasibility study, only a portion of each panel was representative of the "normal" construction of the layered armor design, so only a limited region of interest was evaluated for each panel. Thus, a χ^2 distribution was fit to the histogram of pixel intensities for each ROI and used to determine the statistical probability that each pixel or region of pixels could have occurred by random variation of the acoustic amplitude reflected from the interface of interest. Any region whose area exceeded that of the tail of this distribution beyond that region's average intensity was therefore definitively identified as a defect.

Detailed and quantitative statistical analysis of PA-UT data has demonstrated that this method is able to detect known inclusion defects between the support and elastomeric layers. Defects as small as 0.5" (13mm) diameter were detected with an average SNR of 2.65, and 1.5" (38mm) diameter defects were detected with an average SNR of 3.39. In addition, the system has demonstrated the ability to definitively identify as statistically significant all of the 1.5" and some of the 0.5" inclusions as defects. Furthermore, while inclusions between the ceramic and elastomeric layers were not detected (owing to the already extremely large acoustic impedance mismatch between these layers), variation within that interface was observed, and in several cases, statistically significant "natural" defects were detected.

REFERENCES

[1] J.M. Wells, W.H. Green and N. L. Rupert—""On the Visualization of Impact Damage in Armor Ceramics", *Eng. Sci. and Eng. Proc.*, **22** (3), H.T. Lin and M. Singh, eds, pp. 221-230 (2002).

[2] J. S. Steckenrider, W. A. Ellingson, and J. M. Wheeler "Ultrasonic Techniques for Evaluation of SiC Armor Tile," in *Ceram. Eng. and Sci. Proc.*, **26** (7), pp. 215-222 (2005).

[3] R. Brennan, R. Haber, D. Niesz, and J. McCauley "Non-Destructive Evaluation (NDE) of Ceramic Armor Testing," in *Ceram. Eng. and Sci. Proc.*, **26** (7), pp. 231-238 (2005).

[4] G.P. Singh and J. W. Davies, "Multiple Transducer Ultrasonic Techniques: Phased Arrays" In Nondestructive Testing Handbook, 2nd Ed., 7, pp. 284-297 (1991).

[5] J. Scott Steckenrider, William A. Ellingson, Rachel Lipanovich, Jeffrey Wheeler, Chris Deemer, "Evaulation of SiC Armor Tile Using Ultrasonic Techniques," *Ceram. Eng. and Sci. Proc.*, **27** (7), (2006).

[6] D. Lines, J. Skramstad, and R. Smith, " Rapid, Low-Cost, Full-Wave Form Mapping and Analysis with Ultrasonic Arrays", in Proc. 16th World Conference on Nondestructive Testing, September , 2004.

[7] J. Poguet and P. Ciorau, " Reproducibility and Reliability of NDT Phased Array Probes", in Proc. 16th World Conference on Nondestructive Testing, September , 2004.

[8] J.S. Steckenrider, W.A. Ellingson, E.R. Koehl and T.J. Meitzler, "Inspecting Composite Ceramic Armor Using Advanced Signalt Processing Together with Phased-Array Ultrasound," in *Ceram. Eng. Sci. Proc.*, **31** [5] 1-12 (2010).

[9] Russ, John C. *The Image Processing Handbook.* Boca Raton, FL: CRC Press, 2007, p. 26.

[10] Schroeder, Daniel J. *Astronomical Optics.* San Diego, CA: Academic Press, 2000, p. 433.

ULTRASONIC NONDESTRUCTIVE CHARACTERIZATION OF OIL-BASED CLAY

V. DeLucca and R. A. Haber
Department of Materials Science and Engineering, Rutgers University
Piscataway, NJ, USA

ABSTRACT

Oil-based modeling clays are commonly used as a backing material for high strain-rate testing applications. After an initial test, the extent to which the clay properties may have changed has not been extensively studied. The change in properties after testing may affect the outcome of subsequent tests. This research utilizes ultrasonic nondestructive characterization techniques to examine the properties of a clay backing material before testing situations. Sample sets were fabricated to isolate the effect of different variables, including pressure, layering, and foreign material contamination. Controlling these variables leads to a better understanding of how each affects the acoustic properties of the clay. Ultrasound testing includes sonic velocity and frequency-based attenuation coefficient measurements.

INTRODUCTION

When conducting high strain rate tests, oil-based modeling clay is routinely used as a backing material. While performing ballistic testing, the clay acts as a gauge to measure the amount of protection offered by the armor in question [1]. The armor system is determined to be effective for different types of threats based on the depth of the indentation caused by impact or the depth of penetration[2]. While the NIJ sets standards for the conditioning of the clay backing material prior to testing, the uniformity of the backing material before and after each test are not measured [3]. If there are differences in backing material homogeneity from one test to the next, or at different locations in the same test, the results of the testing may not be entirely accurate. The conventional method for determining if the clay backing material is properly conditioned for testing is a drop test where a steel ball is dropped from a predefined height onto the clay and the depth of the indent is compared to a standard depth [3]. This method of examination may slightly change the properties of the backing material thereby affecting the outcome of any subsequent testing. Another issue with the clay backing material is that the same clay is used for many tests over a period of time which may cause degredation of the clay or contamination by foreign materials.

This research was conducted to determine whether ultrasonic nondestructive characterization (NDC) is a viable method for evaluating clay backing materials for high strain rate applications. Ultrasound has been previously proven as a useful characterization technique for a wide variety of materials including ceramics and biological materials without damaging or changing the material [4]. In the case of clay backing materials the goal is to use ultrasonic NDC to determine the properties and conditioning of the clay without altering it in a way that might affect the outcome of subsequent testing. This study used a pulse-echo, immersion based ultrasound scanning system which uses a single transducer to both transmit and receive the ultrasound signal. A-scan point measurements and C-scan property maps were taken according to methodology developed at the Rutgers Center for Ceramic Research [5, 6, 7].

RESULTS AND DISCUSSION

Ultrasonic testing was conducted using 0.5MHz, 1MHz, and 2MHz central frequency immersion-based, pulse-echo, planar transducers. To investigate a wide range of factors that may contribute to backing material inhomogeneity a number of sample sets were made. Each set of samples were made from Roma Plastilina No. 1 modeling clay as specified by NIJ Standard – 0101.06 [3]. The first set of samples were slices of different thicknesses taken from a fresh clay block. The slices varied in thickness from 3mm to 25mm in increments of 3mm to determine the maximum thickness that could be examined while avoiding severe peak attenuation. It was determined that the maximum thickness at which a bottom surface peak could be clearly distinguished was 6mm, while thicker samples attenuated the signal too much to see anything but the top surface reflection. Both a first and second bottom surface reflection were visible in the 3mm sample. Subsequnt clay samples were made to be around this thickness (3 ± 1mm). At the lowest frequencies the peaks became too convoluted and were indistinguishable from each other due to the thinness of the samples used. As a result, time-of-flight and attenuation coefficient could not be accurately measured and only results using the 2MHz transducer will be shown.

A second set of samples were made to examine the effect of small changes in thickness. These samples, reffered to as Set 2, were made by taking five clay balls with masses from 13g-17g and pressing them in a carver press with a 57.15mm die to form flat, uniform disks of varying thickness. The disks were pressed to 15000lb (26.1MPa). The thicknesses of the disks ranged from 2.83mm to 4.05mm. The thicknesses, as well as the mass of the clay before and after pressing are tabluated in Table I. During the pressing process, a small amout of clay was observed flowing out between the die and punches as pressure was applied. This explains the discrepancies between the starting masses and final masses listed in Table I.

To examine the effect of changes in the pressure at which they were pressed (analogous to differences in force applied when packing clay in backing molds) a third set of samples was made by pressing the clay to different pressures. These samples, reffered to as Set 3, were made by taking six 15g clay balls and pressing them in a carver press with a 57.15mm die to form flat, uniform disks. They were pressed from 5000 – 15000lb (8.7 – 26.1MPa) in increments of 2000lb (3.5MPa). These pressures are tabulated in Table I for each sample. As with sample set 2, some clay escaped the die during pressing leading to some variation in the final masses and thicknesses of the samples.

The last set of samples was made to examine the effect of contamination by foreign materials. Such a situation might arise if dust or debris from testing becomes lodged in the clay backing material and is not removed before it is reused. Five of these samples were made by kneading 1 – 5 wt.% A16 alumina powder into the clay. After mixing in the alumina powder the clay was observed to be noticably drier and less pliable. Balls of approximately 12g of each mixture were pressed to 10000lb (17.3MPa) with a 57.15mm die to form flat, uniform disks. These are reffered to as Set 4, and the amount of alumina powder in each sample is tabulated in Table I. As with previous sample sets, some clay escaped the die during pressing, leading to some discrepancy in the final masses and thicknesses of the samples.

One sample was also made to examine the effect of lamination layers in the clay backing material. Such a situation might arise if extra clay was added to the backing material to fill in a hole and was not firmly packed. Two thin disks were made by pressing 7.5g balls of clay to 10000lb (17.3MPa). The two disks were then stacked on top of one another and crimped along the edges so as to leave a thin layer of air between the clay.

Ultrasonic testing was performed on each sample using the 2MHz transducer. For most of the samples point measurements were taken about the center of the sample. Measurements were taken to determine the longitudinal speed of sound through the clay and the frequency dependent acoustic attenuation coefficient. For the layered sample C-scan maps of amplitude and longitudinal velocity were made. Other elastic properties could not be determined by ultrasound in these samples as the shear peak was not resolved. The longitudinal acoustic velocities for each sample set are included in Table I.

Table I. Properties of clay sample sets

Sample #	Starting clay mass (g)	Final clay mass (g)	Thickness (mm)	Pressure (MPa)	Wt. % Al_2O_3	c_L (m/s)
Set 2						
13	13	11.47	2.83	26.1	0	1687
14	14	13.18	3.26	26.1	0	1676
15	15	13.88	3.46	26.1	0	1693
16	16	12.74	3.10	26.1	0	1620
17	17	16.35	4.05	26.1	0	1668
Set 3						
5000	15	14.65	3.65	8.7	0	1604
7000	15	14.25	3.54	12.1	0	1613
9000	15	13.89	3.41	15.6	0	1629
11000	15	14.15	3.49	19.1	0	1670
13000	15	14.42	3.61	22.5	0	1630
15000	15	14.13	3.56	26.1	0	1703
Set 4						
1	12	11.31	2.75	17.3	1	1599
2	12	10.93	2.65	17.3	2	1621
3	12	11.45	2.85	17.3	3	1652
4	12	11.66	2.90	17.3	4	1667
5	12	11.98	2.95	17.3	5	1695

In sample Set 2, testing was performed to examine how the properties of the clay might change with the thickness. As seen in Figure 1a there was no clear relationship between the speed of sound in the clay and its thickness. Figure 1b shows the frequency dependent attenuation coefficient spectra for these samples. Although the attenuation coefficient varies greatly from sample to sample, there is no relationship between the thickness of the sample and its attenuation coefficient relative to the other samples. Attenuation coefficient is a materials property where thickness differences in the same material should not affect this value. The differences in the attenuation coefficient are not due to thickness but differences in the microstructure of the clay disks. The variation in acoustic velocity and attenuation coefficient seen in Figure 1 show the statistical variation in the measurement of these samples caused by variation in the microstructure, likely due to pores or localized regions of high or low density.

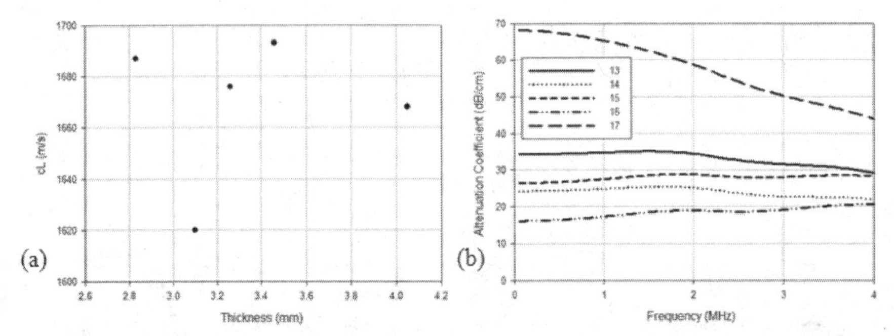

Figure 1. Graphs for sample set 1 showing (a) velocity as a function of thickness and (b) attenuation coefficeint as a function of frequency

In sample Set 3 testing was conducted to examine how the properties of the clay disks change with varying amounts of pressure added. As seen in Figure 2a there is a clear correlation between speed of sound and applied pressure where the longitudinal speed of sound is increasing with increasing pressure. The increase in sound velocity is likely due to an increase in the stiffness of the clay as there was no significant difference observed between the densities of the samples. This could possibly be caused by removal of oil from the clay during pressing; with more oil lost at higher pressures the clay becomes brittle. Figure 2b shows the attenuation coefficeint spectra for these samples. As in Set 1, the attenuation varies greatly between samples with no clear correlation between the applied pressure and the attenuation.

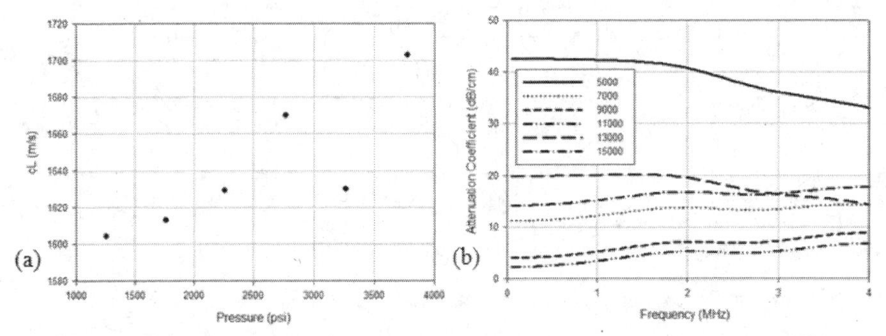

Figure 2. Graphs for sample set 2 showing (a) velocity as a function of pressure and (b) attenuation coefficient as a function of frequency

In sample Set 4 testing was performed to examine how the properties of the clay might change with contamination of a foreign material, simulated by adding alumina powder to the clay before pressing. Figure 3a shows the acoustic velocity of the clay increasing linearly with the concentration of alumina powder added. Figure 3b shows the attenuation coefficeint spectra for these samples. As in the previous sets, there is no clear connection between the amount of alumina powder added and the attenuation coefficient at these frequencies. According to scattering and absorption theory [7] scattering attenuation is negligible at these low frequencies and thermoelastic absorption by alumina grains would be dominant for grain sizes between 2 and 4 microns. As the milled A16 alumina used in these samples was of submicron grain size, this type of absorption would have little effect on the attenuation coefficient in the measured frequency range [8].

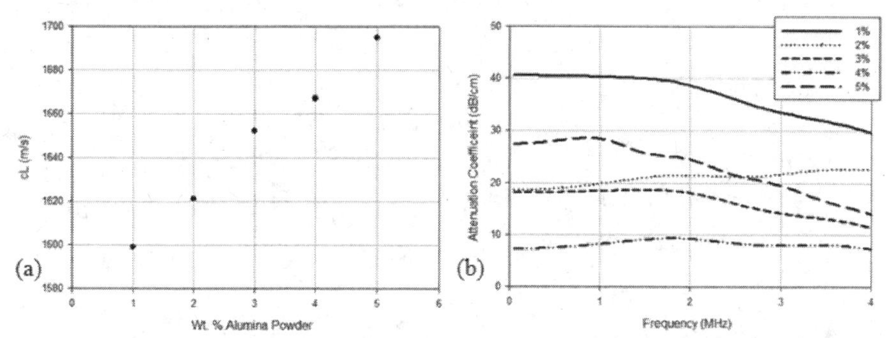

Figure 3. Graphs for sample set 3 showing (a) velocity as a function of wt.% alumina and (b) attenuation coefficient as a function of frequency

For the laminated sample, it was impossible to get attenuation data because the addition of the interface layer attenuated the signal to the point where the second bottom surface was not able to be resolved. Measurements were taken only for the acoustic velocity and amplitudes of the bottom and interface surface reflections. Figure 4 shows an A-scan from the laminated disk sample and a scematic showing where the peaks come from within the sample. The two interface reflections are so close to each other that they become convoluted and combine to form one large peak in the A-scan.

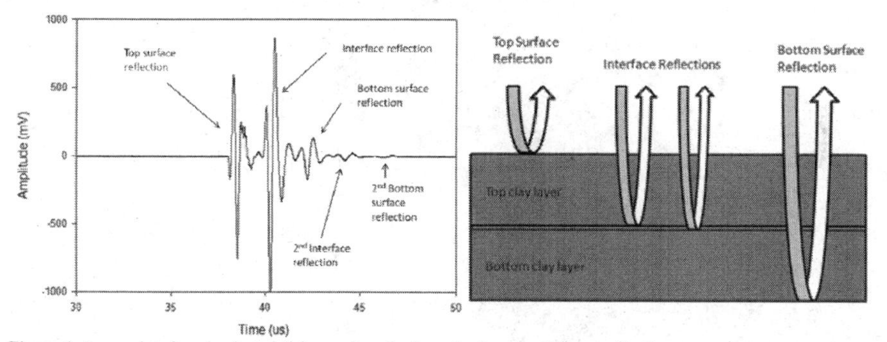

Figure 4. A-scan data from laminated disk sample and schematic showing different reflections

Figure 5 shows C-scan images from the laminated disk sample. The interface surface amplitude C-scan map is fairly homogenous over the entire sample. The bottom surface amplitude c-scan map shows clear regions of relatively high and low amplitude. The large areas of high amplitude on the sides of the image are caused by the glass slides holding the sample off the bottom of the immersion tank. The areas of higher amplitude within the sample area likely have better adhesion and less air at the interface causing signal attenuation than the areas of low amplitude. The longitudinal velocity c-scan map shows that the average speed of sound in the sample is between 1500 and 1600 m/s. This is slightly lower than point measurements taken from the other sample sets and is indicative of the thin layer of air between the clay layers. There is also a distinctive left to right gradient in velocity, which is likely caused by small differences in thickness of the sample from left to right. This gradient may be masking more subtle changes in velocity caused by the interface between the clay layers in the sample.

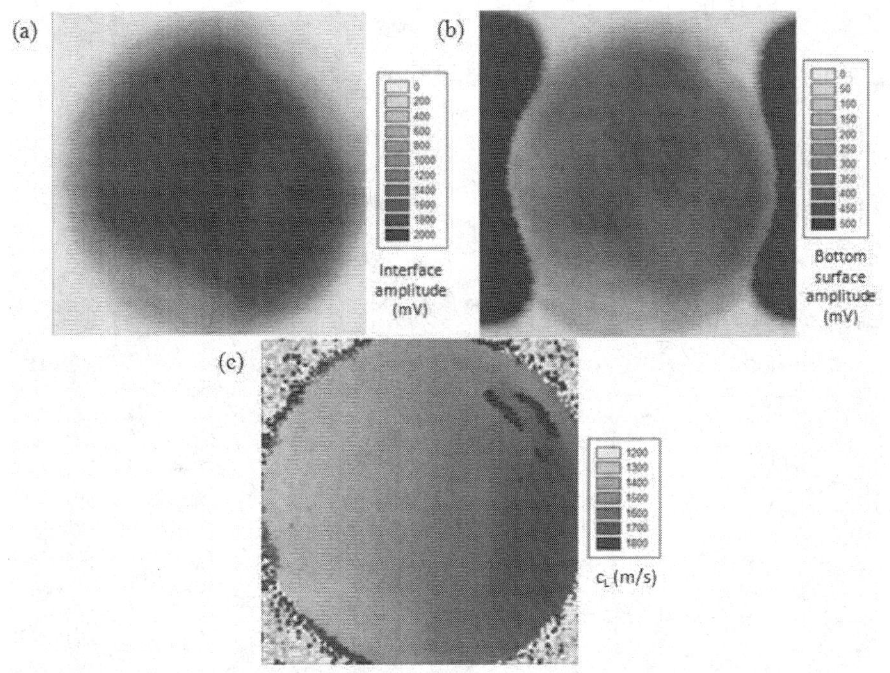

Figure 5. C-scan maps of (a) interface surface amplitude, (b) bottom surface amplitude, and (c) acoustic velocity for the laminated disk sample

CONCLUSION

While the pressure at which the samples were formed and the amount of foreign material added were found to influence the speed of sound in oil-based clay, no such corellations were found in the attenuation coefficient spectra. If future work is to be done in this area, it should include further study of factors that contribute to the attenuation coefficient spectra in this material. The dominant acoustic attenuation mechanisms in the clay system should be investigated to determine why the attenuation coefficeint spectra were so variable between samples that were so similar. All of the clay used in this study was obtained from the same package of clay, so it would also be useful to examine variability between batches of clay. Temperature and aging effects should also be considered as they may also play a role in determining the properties of the clay. It may be possible to examine thicker samples using higher power output transducers or phased arrays. The effectiveness of air-coupled or direct contact transducers on clay should also be investigated as they would be able to examine samples without having to immerse them in a fluid medium.

ACKNOWLEDGEMENTS

The authors would like to thank the NSF IUCRC Ceramic, Composite, and Optical Materials Center and the Army Research Laboratory Materials Center of Excellence for Lightweight Vehicular Armor.

REFERENCES

[1] Prather, R. N., Swann, C. L., and Hawkins, C. E., 1977, "Backface Signatures of Soft Body Armors and the Associated Trauma Effects," Army Armament Research and Development Command Aberdeen Proving Ground Md Chemical Systems Laboratory, Report No. ARCSL-TR-77055.

[2] Committee to Review the Testing of Body Armor Materials for Use by the U.S. Army, "Testing of Body Armor Materials for Use by the U.S. Army – Phase II: Letter Report," National Academy of Sciences, 2010.

[3] National Institute of Standards and Testing, Ballistic Resistance of Body Armor, NIJ Standard 0101.06, NCJ 223054, 2008.

[4] Brennan, R., "Ultrasonic Nondestructive Evaluation of Armor Ceramics", Ph.D. Thesis, Rutgers University (2007).

[5] Bottiglieri, S. and Haber, R. A. (2010) High Frequency Ultrasound of Alumina for High Strain-Rate Applications, in Advances in Ceramic Armor V (eds J. J. Swab, D. Singh and J. Salem), John Wiley & Sons, Inc., Hoboken, NJ, USA.

[6] Bottiglieri, S. and Haber, R. A. (2010) Corrective Techniques for the Ultrasonic Nondestructive Evaluation of Ceramic Materials, in ADVANCES IN CERAMICS ARMOR VI: Ceramic Engineering and Science Proceedings, Volume 31 (eds J. J. Swab, S. Mathur and T. Ohji), John Wiley & Sons, Inc., Hoboken, NJ, USA

[7] Portune, A., "Nondestructive Ultrasonic Characterization of Armor Grade Silicon Carbide" Ph.D. Thesis, Rutgers University (2010).

[8] Zener, C., "Internal Friction in Solids". Proceedings of the Physical Society, 1940. **52**: p. 152-167.

Phenomenology and Mechanics of Ceramics Subjected to Ballistic Impact

2011 OVERVIEW OF THE DEVELOPMENT OF CERAMIC ARMOR TECHNOLOGY: PAST, PRESENT AND THE FUTURE

William A. Gooch Jr.
U.S. Army Research Laboratory
Weapons and Materials Research Directorate
Aberdeen Proving Ground, MD 21005-5066

ABSTRACT

The development and fielding of ceramic armor technology has accelerated with the requirements to provide high ballistic performance at reduced weight for a wide range of military platforms and applications. This overview paper will expand on the technical aspects of the presentations given at the 2001 PACRIM IV Conference in Hawaii and the 2006 30th International Conference on Advanced Ceramics and Composites held in Cocoa Beach, Florida. The intent is to examine the significant developments of ceramic armor technology as seen by the author over the last 40 years with emphasis on the primary evolution and developments that advanced the technology. An understanding of this chronology will set the direction for future developments and applications.

INTRODUCTION

The increasing capability of modern anti-armor threats and the need to field lower-weight combat vehicles, capable of engaging an opponent with little preparation, have intensified the need for highly effective passive armor systems. Ceramic armor technology offers significant advantages for meeting future protection requirements for light vehicles such as the Joint Light Tactical Vehicle (JLTV) to heavier vehicles such as the Ground Combat Vehicle (GCV), personnel protection vests and airborne platforms. The investigation and application of ceramics against small arms threats has a long history, dating back to the early 1960s and the ballistic performance of ceramic armors for personnel protection is very high; the principles governing these defeat mechanisms and the design parameters against such threats are now generally understood. However, achieving similar ceramic performance versus larger caliber, kinetic energy penetrator threats have long presented a difficult challenge.

The application of ceramics for armor continues to be primarily used in lightweight armor systems for protection against small arms and machine gun threats. The design of these systems is typically based upon the mechanical properties of the ceramic to fracture the penetrator and the ability of a rear compliant layer to catch the projectile debris and the damaged ceramic material. For defeat of these low-velocity, short projectiles, the fracture mechanism occurs very early in the process with the majority of the interaction time dedicated to conversion of the kinetic energy of the debris into deformation and delamination of compliant backing. For medium caliber and heavy armor applications, where the dominant threat is modern, high velocity, high-aspect ratio, heavy metal eroding projectiles, the defeat mechanisms are much more complicated and of longer time duration. For the past five decades, a wide variety of research programs, both domestic and foreign, have focused on developing improved ceramic armor systems for the defeat of these threats. This paper presents an overview and chronology of significant developments of ceramics for armor, with emphasis on research conducted on ceramic armors at the primary research facilities of the United States with additional emphasis on research conducted at the U.S. Army Research Laboratory (ARL).

TERMINAL BALLISTIC EFFECTS

A review of the difference in terminal ballistic effects observed during the interaction of different classes and caliber's of kinetic energy (KE) projectiles is important to understand the required defeat mechanisms and armor designs. The delineation between the threat projectiles is primary related to the caliber, velocity and energy available, but is not exact and some projectiles cross over into the two categories discussed below. While the penetrator/target interactions for these two categories involve similar processes, defeat of the higher performance, long rod threats require different emphasis in the armor design parameters to be successful and the progress has been more difficult. The interaction of shaped charge warheads and ceramic materials operate in the penetration regime where mechanical properties, such as strength and hardness are almost negligible; the interactions are approaching hydrodynamic conditions and the comparative densities of penetrator and target drive the resultant ballistic performance.

Small Arms/Heavy Machine Gun Defeat

Historically, ceramic composite armor systems were designed to defeat armor-piercing (AP), kinetic energy projectiles, mainly in the small arms and heavy machine gun category. These AP projectiles are purely inertial rounds, most commonly made of hard steel (HRc 60-64), of moderate density (7.85 g/cm^3) with newer designs utilizing even harder tungsten carbide (WC) cores at higher densities (13.5-15.0 g/cm^3). The hard core is generally encased in a thin jacket of a more ductile metal for interior ballistic or aerodynamic considerations, but penetration performance of the bullet is controlled by the core properties. Such projectiles typically have a length to diameter (L/D) ratio in the range of 3:1 to 5:1 with moderate muzzle velocities of less than 1 km/s. The generally accepted high-end caliber is 14.5-mm, typified by the Soviet KPV family of heavy machine guns. Some saboted, light armor-piercing (SLAP) rounds have velocities up to 1.3 km/s but with reduced core weight. Overall, these projectiles tend to produce a total KE on the order of 10^3 - 10^4 J.

Early research [1–4] discovered that the perforation of ceramic armor systems occurred in three general stages: 1. shattering; 2. erosion; and 3. capture. During the shattering phase, the penetrator fractures and breaks on the surface of the ceramic plate; the high compressive strength of the ceramic overmatches the loading produced by the penetrator impact, and the penetrator material flows and shatters. This initial stage is followed by a period of damage accumulation in the ceramic material initiated by tensile wave reflections, and bending of the ceramic tile and backing plate. During the second stage of ceramic armor penetration, the ceramic material is cracking, but the ceramic material can still contribute to defeat of the penetrator core through erosion mechanisms. In the final capture phase, the ceramic has lost considerable strength, but ceramic and backing combine to reduce the velocity through momentum transfer mechanisms.

The defeat mechanism for hard-core AP projectiles is primarily stages 1 and 3 with projectile fracture upon impact against an armor plate having sufficient hardness and/or high obliquity. The shattering and subsequent dispersion of the fragments result in a dissipation of the kinetic energy of the core over a larger area than if intact, thereby achieving defeat of the round with a reduced amount of armor weight. Monolithic ceramic plates were best suited to produce the shattering phenomena due to their high hardness and low densities, although pelletized ceramic systems have found successful applications. However, ceramic armor requires a backup component to support the ceramic and delay failure during the initial impact/shattering interaction; the backup component then serves to absorb the residual projectile fragments and comminuted ceramic particles (Phase 3). The state of the art in protection against small arms threats is typified in lightweight, two-component ceramic-faced composite armors designed for use in insert plates for personnel body armor, armored helicopter seats and appliques to metal or composite based vehicle structures.

Heavy Metal Long Rod KE Projectile Defeat

The mechanism for defeat of long rod penetrators (LRP) is more complex than for the conventional AP projectiles described above. These penetrators are commonly made of high strength, high density materials, such as tungsten sintered alloy or depleted uranium, having densities near 18 g/cm^3 with moderate hardness, good toughness and ductility; hence, the projectiles are not susceptible to shattering as hard core, relatively brittle, AP projectiles. This category includes armor-piercing, discarding-sabot (APDS) and armor-piercing, discarding-sabot, fin stabilized (APDSFS) projectiles, in calibers from 20-mm to >140-mm. These LRPs are designed with a high L/D ratio (currently fielded examples exceed 30:1) and the high density core material coupled with relatively high muzzle velocity (1.3 - >1.6 km/s), yields KE in excess of 10^6 J, creating a high energy density per unit area of target impacted than with a corresponding hard core AP round. These factors, when combined with the greater projectile length and reduced propensity for fracture, makes the LRP a much more effective penetrator. Even if the frontal portion of the LRP can be effectively damaged, a substantial portion of the rod remains to continue the armor penetration process. Thus, the conditions that allow a simple ceramic composite to function effectively for small arms defeat do not apply when the armor is impacted by a LRP. The primary defeat mechanism is erosion (Phase 2) and the effectiveness is relatively low for simple ceramic armor systems.

Shaped Charge Projectile Defeat and Interaction

While ceramics are mostly utilized in KE defeat, most ballistic applications are also subject to attack by shaped charge munitions. A typical SC is shown in Figure 1 and Figure 2 shows the sequence of flash x-rays illustrating the functioning of the warhead [5-6]. The conical copper liner is embedded in a cylinder of explosive which is detonated at the base of the explosive and the resultant detonation wave collapses the liner on the axis of the charge. This collapse causes a high velocity jet to be ejected forward. Depending on the design, the tip of the jet is traveling about 10 km/s with the tail traveling about 3 km/s. This velocity gradient causes the jet to stretch and elongate, creating very high L/D ratios. As a result of the hypervelocity jet tip, the pressures produced during jet-target impact far exceed the yield strength of most materials and as a first approximation, the strength and viscosities of the jet and target can be neglected, i.e., the hydrodynamic interaction can be considered an incompressible, inviscid fluid flow. One dimensional Bernoulli equations can be employed to estimate the penetration performance of the jet into a ceramic that simplifies to the length of the jet times the square root of the ratio of the density of the jet to the density of the target [7]. Using this simple, well known jet penetration equation, a silicon carbide tile when penetrated by a copper jet will exhibit a mass efficiency of about 1.6 that of steel. This implies that ceramic components in targets have an effectiveness that needs to be considered.

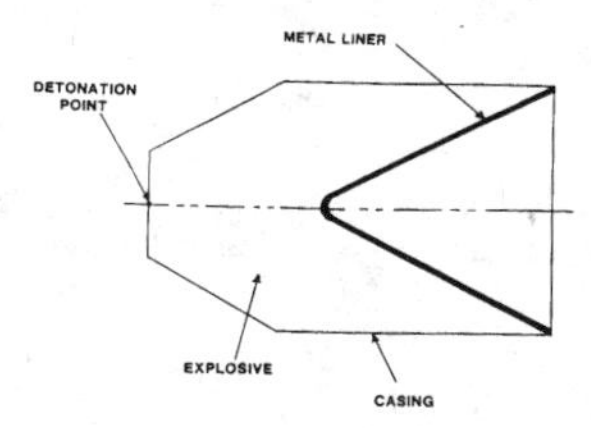

Figure 1. Shaped Charge Warhead

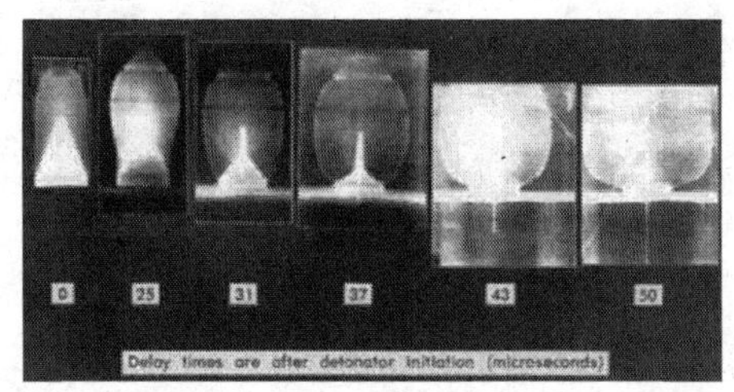

Figure 2. Formation of SC Jet

DEVELOPMENTS IN CERAMIC ARMOR TECHNOLOGY

Introduction

This abbreviated chronology of ceramic armor research and development contains only the highlights of the extensive research work conducted on ceramic technology at facilities of the Department of Defense, Department of Energy, Industry, academic centers in the United States and equivalent centers worldwide. The emphasis is on the development of ceramic armor technology for military application and is not a complete record of work, specifically, indicating major technical events that influenced the state of the art at that time. The chronology is as viewed from the standpoint of the author with an entry into ceramic armor research in 1970 and the interaction with the principle researchers for work prior to and during this period that is available in the open literature. The dates are in relative chronological order, but exact dates vary depending on technology development versus publication. The intent of this chronology is to examine the historical basis and current knowledge of ceramic armor systems to facilitate further increases in ballistic performance for future armored systems.

Patents of Goodyear Aerospace

The development of lower purity aluminum oxide ceramics after World War II facilitated the application of these early ceramics for potential armor designs and the first patented ceramic armors identified were by the Goodyear Aerospace Company between 1963 and 1979. Figure 3 shows a ceramic composite armor design using 85% alumina plates on a woven/roven backing that is not very much different than the designs seen today [8]. Figure 4 dates the first patented design using ceramic balls in polyurethane foam, also with a woven/roven backing that is the predecessor to the large number of ceramic ball/cylinder designs observed today [9]. While earlier research work on employing brittle materials for penetrator interactions is known, these two patents quantify earlier research efforts that led to identifiable armor designs.

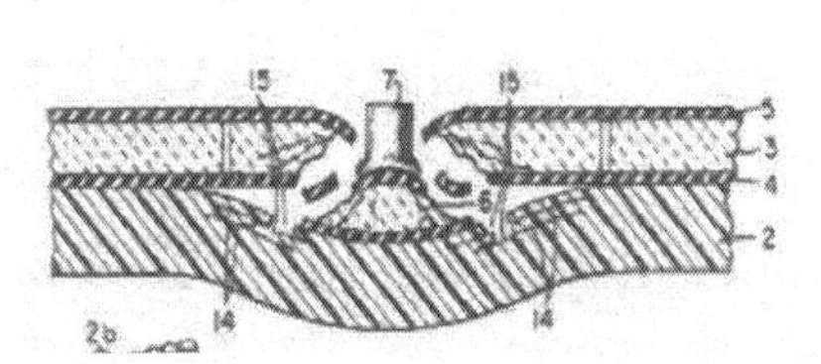

Figure 3. 1963 Goodyear Aerospace Patent 3,509,833

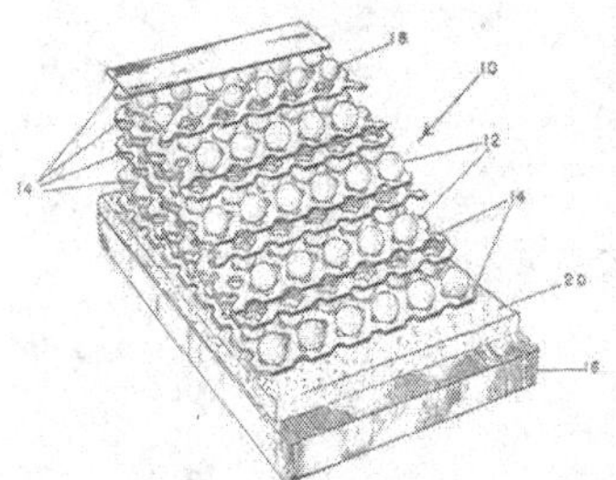

Figure 4. 1967 Goodyear Aerospace Patent 4,179,979

Lawrence Radiation Laboratory Research on Impact on Brittle Materials

The seminal research conducted by Wilkins, Landingham, Honodel, Sawle, and Cline, was funded by the Advanced Research Projects Agency (ARPA) and published in six reports [10-15] between 1967 and 1970. This work provided the first documented technical analysis of the defeat of hard-core projectiles impacting ceramic faced materials and is still considered a fundamental study today and should be reviewed periodically to reconsider the implications of this work. The research documented a large set of ballistic testing that expanded the diagnostic techniques for examining penetrator/target interactions, particularly brittle failure of ceramics. This led to the development of the first theoretical models of dynamic material behavior and development of the 2-D elastic-plastic computer code HEMP that predates today's highly developed finite element codes with advanced ceramic strength and failure models. New ballistic ceramics were examined that laid the foundation for currently used ceramics.

Viet Nam Era Protective Vests

The first fielded protective ceramic composite vests of quantity were in response to the personnel survivability requirements of the Viet Nam war period. Previously, most vests were fiber-based or plastic composites, such as duron or nylon, utilized in aircrew or fixed position ground applications where mobility was not an important factor. The extensive ground combat/counter-insurgency aspects of this era mandated higher mobility and increased personnel protection of soldiers. This set the requirements for lighter and less-intrusive protective vests that are still factors today. In this period, the first ceramic based vests were produced in 1965 by Reflective Laminates of AD85 sintered alumina compound angle plates on roven/woven backings for helicopter crew applications (Figure 5) [16]. The necessity to decrease weight led to the development of hot-pressed boron carbide ceramics by Norton Ceramics on the same backings, basically, a technology still in production today and in use on current protective vests. In 1968, the U.S. Army Natick Laboratory contracted with the Carborundum Company for the development of reaction-bonded boron carbide compound angle plates on woven/roven backings that were the first major U.S. ground force use of ceramic composite technology. With the wind-down of the Viet Nam war and the cancellation of a 30,000 vest procurement in 1970, the cyclic nature of the development and production base of this technology was established and has followed similar changes in demand with combat needs to this date.

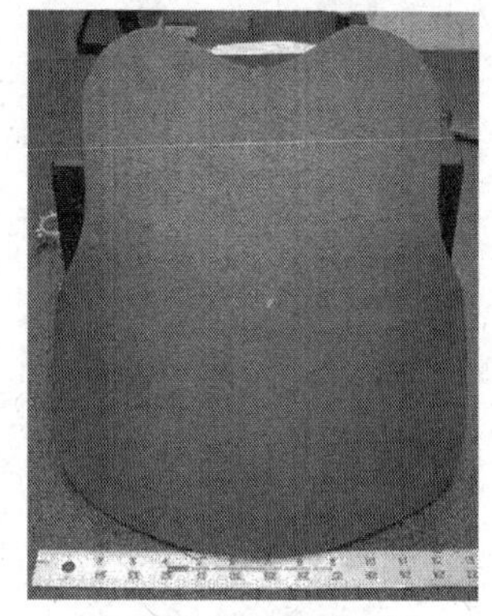

Figure 5. 1965 Reflective Laminates Sintered Alumina

Cobra and Apache Ceramic Helicopter Seats

The increased use of helicopter gunships in close combat situations led to the requirement of increasing the protection of the crew. The first true helicopter ballistic seat for the Cobra Gun Ship utilized Norton hot-pressed boron carbide on Kevlar 29 backing and was fielded in 1969. Ceradyne Incorporated eventually bought out Norton Ceramics and continues to produce similar designs to date. The seat design consisted of large bonded and chamfered ceramic plates that form the seat as seen in Figure 6 [16]. Figure 7 shows the later development of the AH64 Apache Armored Helicopter seat, produced by Simula (now BAE Systems) with hot-pressed boron carbide tiles produced by Cercom Incorporated, now BAE Advanced Materials, on a Kevlar 49 backing [16].

Down Cycle of Ceramic Armor Production

The thirteen year period from 1970-1983 after the Viet Nam War ended was a major down period for armor production, but was a period of increased interest in ceramic armor technology research, particularly under inhouse Army funding at the then US Army Ballistic Research Laboratory (BRL) and the US Army Mechanics and Materials Research Center (AMMRC), both now ARL, and external funding to DoD, DoE and private sectors from the ARPA. The latter research work developed under two programs known as Roof Point and Cater Mill and provided the impetus to renewed interest in ceramic armor technology. Much of the impetus to these programs was the perceived intelligence that the Soviet Union had embarked on developing advanced ceramic armor technologies for their combat systems and the US needed to understand, develop and counter these developments. One short-lived company of interest during this period was Armorflex of Santa Maria, CA that produced a green aluminum oxide resulting from a nickel additive. Two major US technology developments during this period are described below.

Aeronautical Research Associates of Princeton (ARAP) and ABEX/Norton Ceramics

The ARPA program resulted in the development of the concept of precision cast metal surrounds that fitted precision hot-pressed ceramics in a matrix. This was a response to newer penetrators being fielded that primarily utilized erosion defeat mechanisms that required increased fabrication technology. ARAP researchers realized in 1978 that increased efficiency of ceramics might be possible by lengthening the duration of the shattering stage of the penetration process, and/or by increasing the efficiency of the erosion process of the comminuted ceramic material. These researchers found that modest lateral confinement allowed constraint of the broken ceramic pieces, thereby enhancing the erosion phase of the ceramic penetration process. This confinement could be obtained by casting, as seen in the very efficient appliqué armor developed in 1984-86 by ABEX-NORTON who procured this technology from ARAP. Silicon carbide tiles were inserted into very accurately cast aluminum and steel matrices [17] (Figure 8). Significant testing was conducted at ARL that showed the increased efficiencies of these armor designs. As still evident today, the major draw-back was cost due to higher fabrication

Figure 8. Abex-Norton Ceramic/Metal Composite

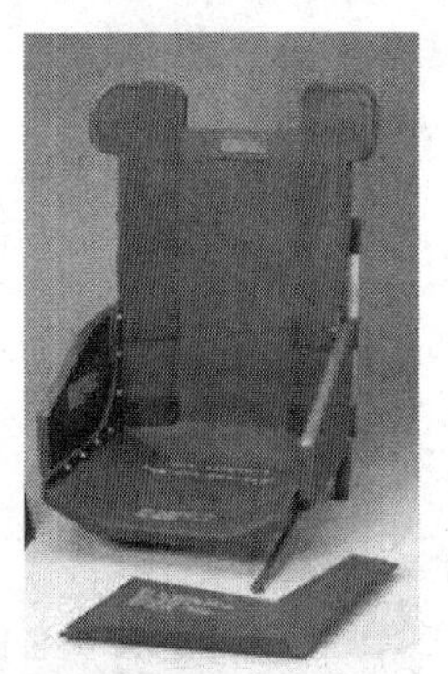

Figure 6. Cobra Gunship Armored Seat

Figure 7. 1983 AH64 Apache Armored Seat

technology needed to produce finished end items.

Observation/Development of Ceramic Dwell Phenomena

The most significant developments during this period, however, were first observed in 1979 by Hauver [18-20] and later by Rapacki [21] who examined test geometries that delayed the generation of damage in the ceramic tile, thereby increasing the duration of the shattering phase of the penetrator defeat process (Figure 9). As penetrator threats increased in length and L/D ratio, Hauver realized that the shattering stage duration was critical to the overall efficiency of the ceramic defeat process. Improved ceramic tile confinement geometries that substantially increased the shattering/erosion phase of the penetrator defeat process, completely eroding the penetrator (dwell) were documented. These experiments employed compressive confinement of the ceramic tile (heat shrink of the metal surround), in combination with techniques to delay tensile wave and bending damage to the ceramic utilizing reverse ballistics and one-megavolt x-rays to view dwell and damage in real time. The ceramic performance was enhanced through control of system geometry to minimize damage and increase the shattering stage of penetrator defeat. However, the overall mass and space efficiencies of these laboratory packages were low, due to the considerable confinement materials employed in the geometry. Orphal expanded these efforts into the hypervelocity regime (1.5-4.6 km/s) for confined SiC [22]. Work in this area continues to today [23-27].

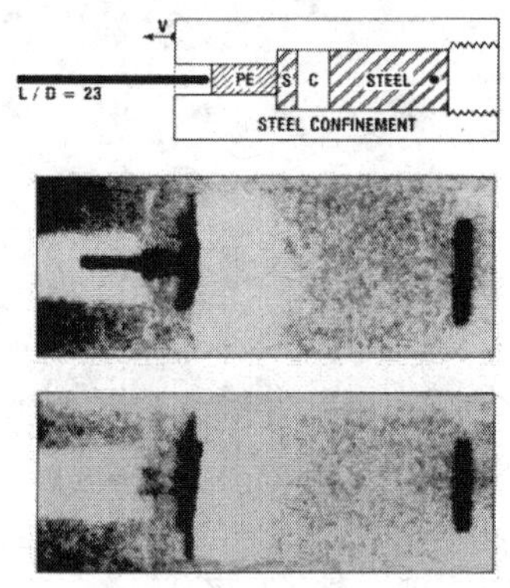

Figure 9. Hauver's Observation of Ceramic Dwell

In later work, the same equipment was used to study the interaction of steel and tungsten carbide projectiles impacting boron carbide that led a better understanding of how the ceramic hard faced component of body armors responded to impact during the initial 50-60 μs [28]. Figure 10 and 11 shows the setup of a typical reverse ballistics test at ARL and Figure 12 shows the resultant timed images of seven tests showing the interaction of a boron carbide tile with the 0.30-cal APM2 projectile. The projectile core can be seen to dwell and erode laterally inside the boron carbide tile. These studies have been used to verify computation codes that utilize brittle failure models. The best image of complete dwell can be seen in the Figure 13 high speed camera images where Southwest Research Institute impacted a 0.30-cal APM2 projectile on a very large tile of aluminum oxide and tile exhibited self-confinement and total dwell was observed with little damage to the tile [29].

Figure 10. ARL 100-mm gas gun with dual 1-MeV x-ray pulsers above impact chamber

Figure 11. Static projectile and x-ray cassettes in front of gun tube

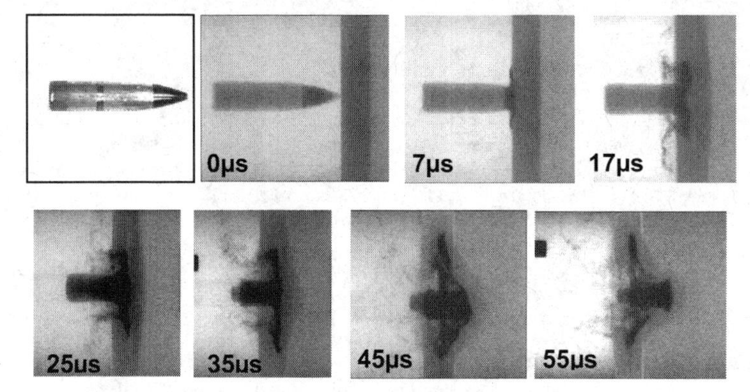

Figure 12. 0.30-cal APM2 Projectile Impacting a Boron Carbide Ceramic Plate During the First 55μs

Four Power Senior National Representatives Target Working Group

A major quadrilateral government program between the U.S., United Kingdom, Germany and France was the Four Power Senior National Representatives Target Working Group that developed common range targets for anti-armor system development. The members were drawn from the research agencies of each country and, for over 25 years from 1979-2004, established a common set of range targets, representing threat systems, that were used to develop the primary anti-armor systems in use in the four countries today. As many of the threat targets were metal, ceramic, glass and composite based, the SNR Target Group developed a wide range of targets, based on a cumulative compilation of over four thousand ballistic tests. Among the ceramic based targets, alumina ceramic tiles as large as 450mmX450mmX100mm and 200mm thick were procured in large numbers for testing of range targets. The knowledge gained and materials developed during this program has migrated into US programs. In the latter years, the program was increased to a five power working group with the addition of Italy.

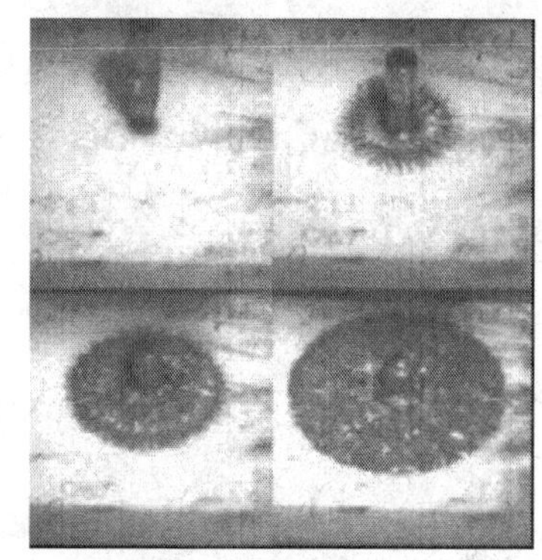

Figure 13. SWRI Image of Dwell of a 0.30-cal APM2 projectile on Large Self-confining Aluminum Oxide Tile

DARPA Armor/Antiarmor Program

In 1985, during the height of the Cold War, the reorganized Defense Advanced Research Projects Agency (DARPA) initiated a major five year armor/antiarmor program to overcome the perceived notion that the US has fallen behind technically in armor protection technology versus the Soviet Union. DARPA injected large funding into many new programs to develop new lightweight and heavy armors from Industry and the DoE, many being ceramic based technologies. Concurrent large funding investments into US Army programs resulted in additional new ceramic products and companies being available today. Significant ceramic developments and companies started include:

- Cercom Incorporated started in 1985 producing hot-pressed ceramics including boron carbide, silicon carbide, titanium diboride, Ebona-A alumina and tungsten carbide by pressure assisted densification.
- Lanxide Corporation started in 1986 producing DIMOX AS and AT by the Directed Metal Oxidation process
- Dow Chemical developed aluminum nitride as an armor material
- Coors Ceramics developed AD995 CAP3 high purity alumina and AD90 large tiles for testing
- AMMRC and ARL funding of large procurement contracts of aluminum oxide, boron carbide, silicon carbide, titanium diboride ceramics in a wide range of ceramic tile sizes. Large dimension tiles of SiC and Ebona-A were produced at Cercom (Figure 14) [16].
- ARL testing of large full scale glass/ceramic composite targets against KE and Shaped charges in the US (Figure 15) and with the German Company Ingenieurbüro Deisenroth (IBD) of Lohmar, Germany (Figure 16) [30].
- Development of 1/4/3 ceramic screening tests at the BRL [31] and the Depth of Penetration (DOP) ceramic screening tests at AMMRC [32].

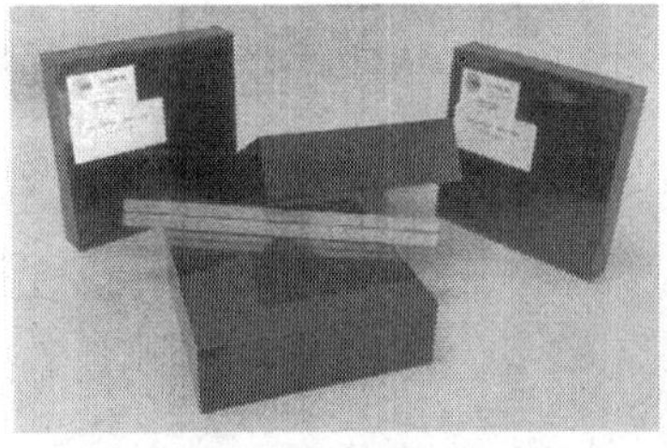

SiC-B Tiles 8"x8"x2" TiB$_2$ Tiles 18"x6"x4" Ebon-A Tiles 18"x18"x10"

Figure 14. Large Ceramic Tiles produced by Cercom Incorporated (BAE Advanced Materials)

Figure 15. Full Scale Ceramic/Glass Target Developed with Coors Ceramics

Figure 16. Full Scale Ceramic Target Developed by IBD and Tested at ARL

Tandem Ceramic Armor

In 1990, research at ARL lead to the demonstration of a set of medium caliber and full-scale armor targets that incorporated existing ceramic defeat knowledge into an armor technology known as tandem ceramic armor (TCA) [33]. TCA determines the optimum performance of a specific cross-sectional ceramic armor design and then repeats the designs in multiple, shock-isolated sections; the performance is thus additive (Figure 17). Laboratory targets, utilizing conventional laminated ceramic-metal technology, demonstrated system designs that produced the state of the art for KE performance. A limiting factor, however, was the space requirements that grew as the penetrator performance increased and cross-sectional thickness also increased.

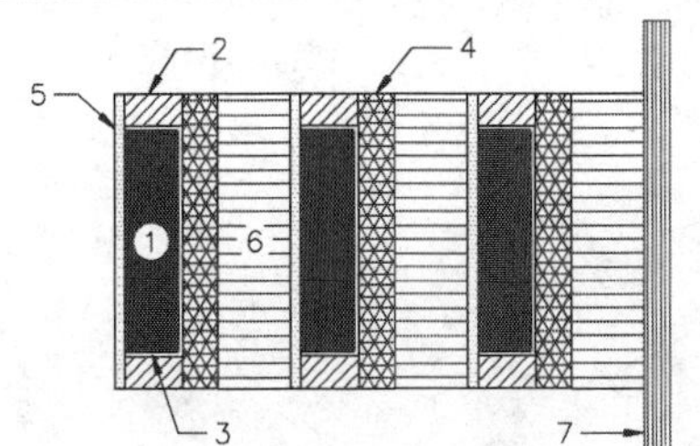

TANDEM ARMOR SYSTEM
1. CERAMIC TILE
2. CONFINEMENT FRAME
3. POLYMERIC ADHESIVE
4. SUPPORTING PLATE (METAL/COMPOSITE)
5. THIN GRP SECTION (OPTIONAL)
6. HONEYCOMB/ISOLATION MATERIAL
7. VEHICLE HULL

Figure 17. Tandem Ceramic Armor Concept

German Ernst-Mach-Institute Edge on Impact Tests

The requirement to understand brittle failure in ceramics led to a number of improved diagnostic capabilities to observe the dynamic processes. The high-speed photographic technique of the Edge-on Impact (EOI) test was developed in the early 1990's at the Ernst-Mach-Institute (EMI) of Freiburg, GE in order to visualize dynamic fracture in brittle materials. In a typical EOI test, the projectile impacts one edge of a ceramic tile and the fracture propagation is observed during the first 20 μs after impact by means of a Cranz-Schardin high-speed camera with the setup shown in Figure 18. EOI tests allow a characterization of different ceramics by the macroscopic fracture patterns, single crack velocities and crack front velocities (damage velocities); a typical 20 image photograph is shown in Figure 19. ARL and EMI continue examining both opaque and transparent ceramics [34-36].

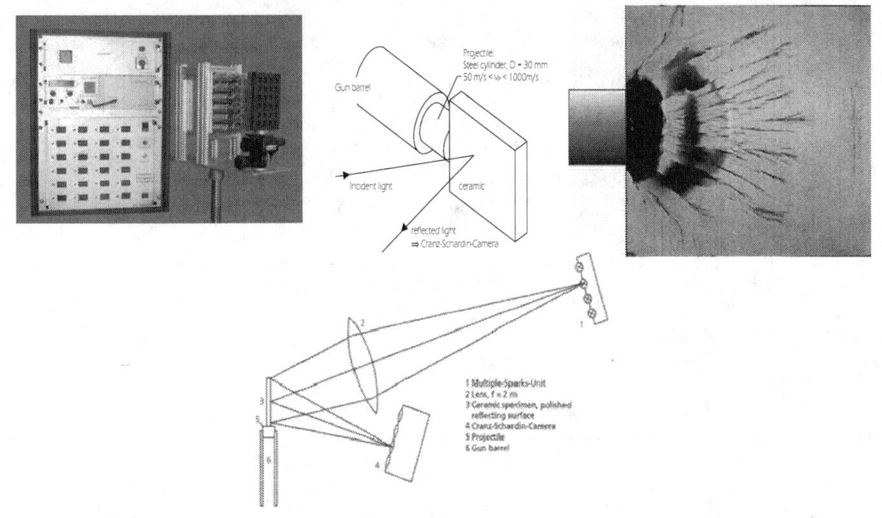

Figure 18. Ernst Mach Institute Edge on Impact Experiments on Glasses and Ceramics

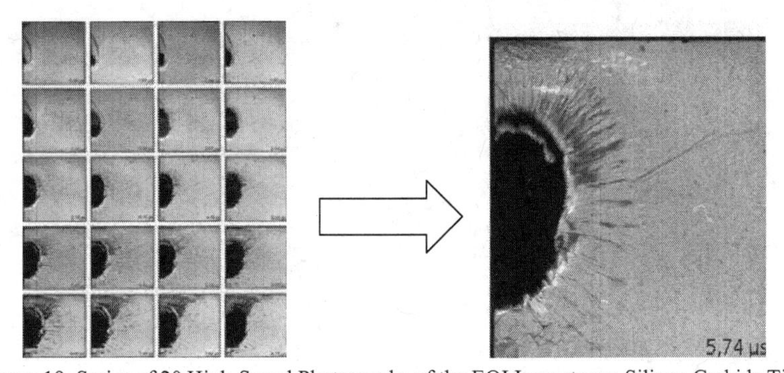

Figure 19. Series of 20 High-Speed Photographs of the EOI Impact on a Silicon Carbide Tile

Hot-Isostatically Pressed Encapsulated Ceramics

ARL continued to generate increased efficiency in ceramic armors by enhancing both the erosion and "dwell" mechanisms of ceramic armor for penetrator defeat. The development of hot-isostatic-press (HIP) processing of ceramics with metal surrounds in the 1990's (Figure 20) has demonstrated dwell on the ceramic front surface of laboratory scale threats at efficient armor system areal densities [37-38]. This HIP processing forms a macro-composite through the generation of residual compressive stresses (mismatch of thermal expansion coefficients of the ceramic tiles and metal confining plates) in the ceramic tile during cool down of the HIP assembly from the pressing temperature. The macro-composite is then able to withstand the large ballistic bending loads during round impact, so that the ceramic tile resists fracture and retains a high compressive strength. The macro-composite formed by HIP processing also keeps the broken ceramic pieces confined during the second erosive phase of the ceramic armor defeat process, should it occur, thus maintaining a high erosive efficiency. The main issues include the parasitic weight incurred by the confinement and the higher fabrication costs to make finished products.

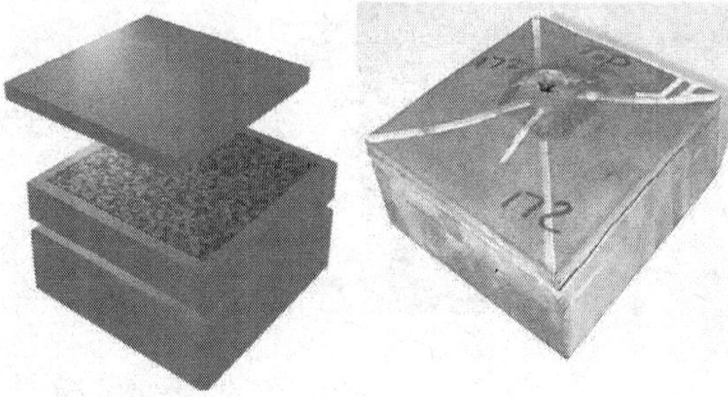

Figure 20. Hot-Isostatic Pressed Metal Encapsulated Ceramic

Dwell Transition Studies at the Swedish Defense Research Agency FOI.

The observation of the dwell phenomena in ceramics and the desired efforts to delay the onset of ceramic failure led to the concept of a critical impact velocity at which point the surface load generated by the projectile impact transitions from interface defeat to normal penetration (Figure 21); the gray area indicates the transition interval. This critical velocity is directly related to the maximum strength of the ceramic. The experimental technique, developed by Lundberg at the Swedish Defense Research Agency FOI is shown in Figure 22 where the confined ceramic target is fired in reverse ballistics into a static penetrator [39-42]. The ceramic test cylinder is confined in a shrunk-fit steel

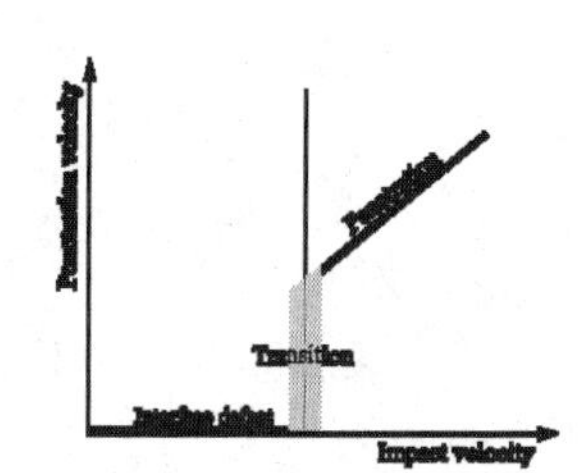

Figure 21. Penetration Velocity versus Impact Velocity.

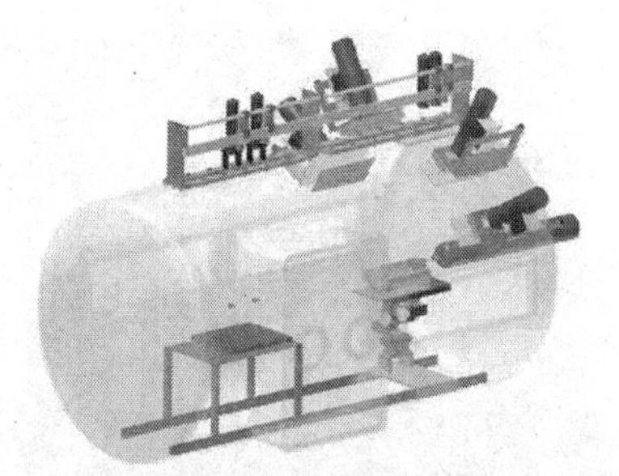

Figure 22. Experimental Reverse Ballistics Test Array and Ceramic Projectile

confinement sleeve and is fit into a ballistic sabot with a shock attenuating copper tip. Four X-ray flashes (450 kV) were used to capture the interaction between the target and the projectile as the flash X-rays make it possible to see through the target and the confinement material, offering the possibility to study the penetration process at different times. Figure 23 shows two x-ray images of dwell transition in a silicon carbide target. The difference in velocity between the two impacts was 35 m/s and shows the changeover to penetration at the interface. The critical transition velocity has become an important penetrator/target parameter to know when designing ceramic armors.

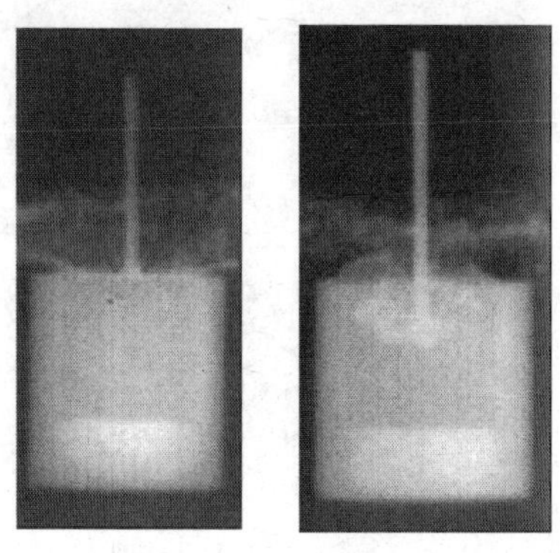

Figure 23. Transition from Interface Defeat to Penetration in a SiC Ceramic.

Pelletized Ceramic Armor Systems.

The early patents of Goodyear Aerospace in 1963 defined two ceramic designs, the first being homogenous ceramic tiles that were adhered to a composite backing and the second being non-homogeneous encapsulated ceramic balls in a matrix with a composite backing. Much of the ceramic armor technology to date centered on the first design where the primary defeat mechanism was projectile breakup and capture for small arms or enhancing erosion through delayed penetration techniques such as dwell for longer rod penetrators. The designs were generally arrays of fitted flat tiles or a single compound angle tile on various backings or encapsulation. In the 1990's, the re-emergence of ceramic armors based on pelletized or discrete shaped ceramic components embedded in a compliant matrix were seen in use in a wide range of applications. The defeat mechanism has been enhanced by the non-homogeneous cross-section where a myrid of ceramic shapes and head designs not only breakup the projectile, but induce additional rotational effects that enhance the performance. The primary threats are small arms up to 14.5mm and fragments from artillery or improvised explosive devices. Performance against LRP drops off as penetrator length and interaction time increases. The defeat mechanisms of these pelletized armors are, in fact, related to effects induced by the non-homogeneous cross-sections in perforated or cast P900 armors or expanded metal armors, defined as tipping plate metal armors [43]. While the technical literature identifies many variations and producers of pelletized armor design, two companies at the forefront in transitioning this ceramic technology into fielded applications.

LIBA™ (Light Improved Ballistic Armor) is a composite ballistic armor panel developed by Mofet Etzion Ltd. of Kibbutz Kfar Etzion, Israel [44]. Applications have found wide use in Europe and this technology was licensed to General Dynamics in the US as SURMAX™. The ceramic design is based on a wide range of ceramic cylinder shapes and balls with patents on specific head shapes to induce additional rotation (Figure 24); the primary ceramic is sintered aluminum oxide that provides good cost effectiveness for the performance. Figure 25 illustrates the fabrication advantages of pelletized systems with ease of encapsulation, form fitting and excellent multi-hit capability.

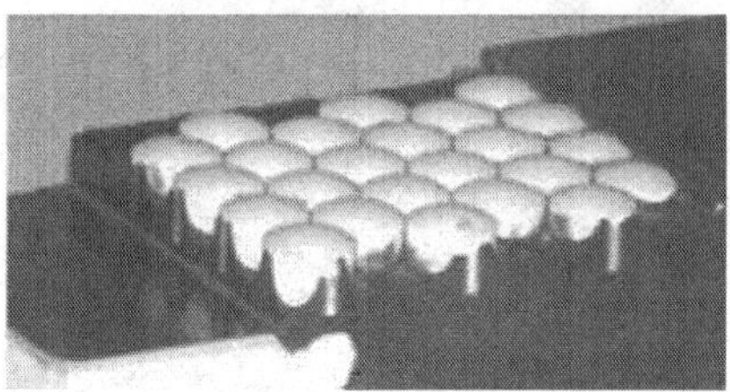

Figure 24. Mofet Etzion Ceramic Cylinders with Patented Head Shapes for LIBA™ Armor

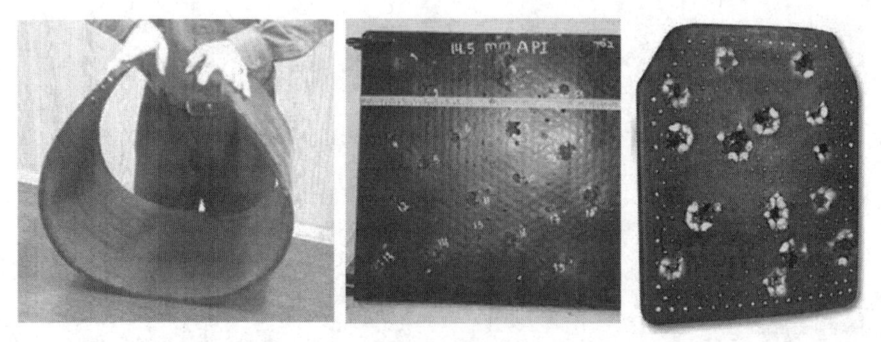

Figure 25. Fabrication and Multi-hit Advantages of LIBA™ Pelletized Ceramic Systems

SMART™ (Super Multi-Hit Armor Technology) Armor is produced by Plasan Sasa Israel at Kibbutz Sasa in northern Israel and Plasan Sasa North America of Bennington, VT and is widely used in a number of armored vehicles worldwide [45]. The armor integrates external armor modules to the base armor equipped with internal high-performance protective spall liners. The external module consists of ceramic cylinders in a polymer matrix and metal core that provides the required projectile breakup (Figure 26). The metal core contributes to increased multi-hit capability, ease of vehicle mounting and increased resistance to shock. Figure 27 illustrates the excellent inherent multi-hit capability of pelletized systems as compared to tile designs due to damage limitation of the discrete ceramic cylinders. Pelletized systems also allow repair of ceramic composite by mechanically replacing the cylinders and repairing the damaged areas as seen in truncated process shown in Figure 28 [45].

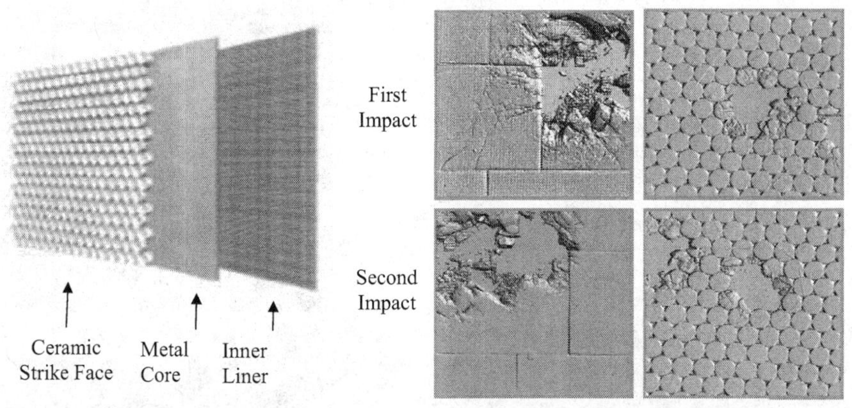

Figure 26. SMART™ Armor Produced by Plasan Sasa of Israel

Figure 27. Multi-hit Capability of Ceramic Tiles versus Cvlinders

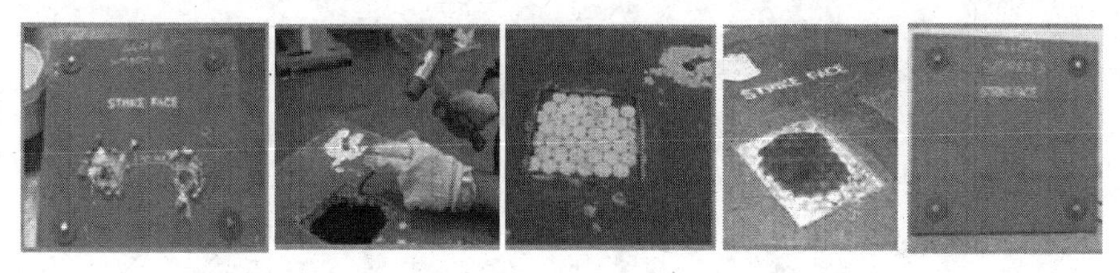

Figure 28. Repair of SMART™ Pelletized Ceramic Composite Armor System

The cost effectiveness of these pelletized armors is directly related to the use of lower cost aluminum oxide ceramic components that can be sintered in a myriad of shapes. A number of producers provide the components in different alumina content, but two major suppliers produce the bulk of these aluminum oxide shapes for worldwide applications, Coorstek of Golden, CO and Industrie Bitossi SpA of Vinci, Italy. As an example, Figure 29 shows the large array of cylinders, hexagonal tiles and balls in various sizes and thicknesses that are produced by Industrie Bitossi SpA, including the various head shapes that can be used to induce rotation or breakup of the penetrator; the cylinders are produced from 98% alumina under the Corbit™ 98 designation [46].

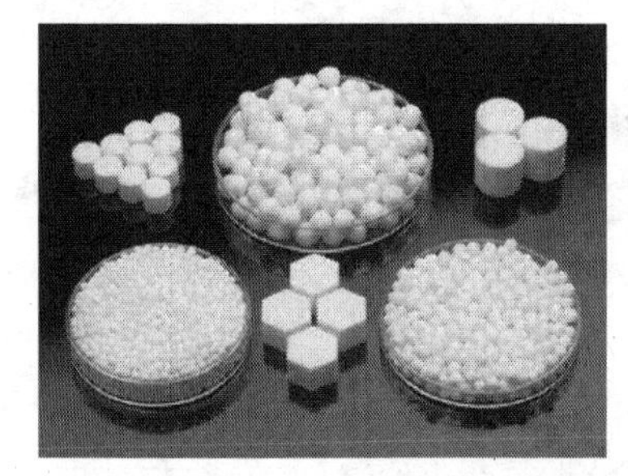

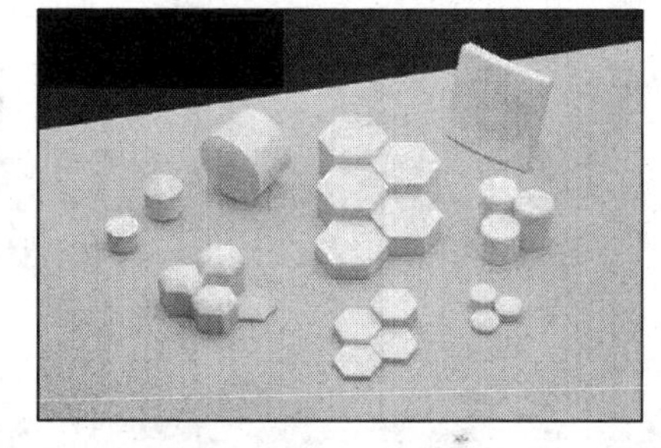

Figure 29. Ceramic Components for Use in Pelletized Armor Systems from Industrie Bitossi SpA

The use of closely fit tiles in a repetitive array facilitates the use of robotic technology to reduce manufacturing costs and handling of large numbers of components. Figure 30 shows the automated lay-up from Plasan Sasa of a metal door panel with aluminum oxide cylinders that is typical of what can be automated to reduce costs of production [45]. Figure 31 shows a similar arrangement on an uparmored door by Mofet Etzion [44].

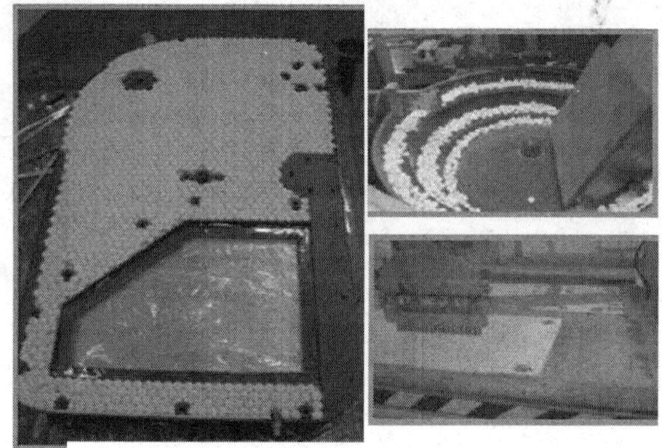

Figure 30. Robotic Fabrication of SMART™ Armor Cylinder Technology on an Armored Door

Figure 31. LIBA™ Door Armoring with Robotic Aluminum Oxide Cylinder Emplacement

High Density Ceramics

While the preponderance of ceramic armors take advantage of the inherent lower densities of the ceramic component in the design, usually between 2.5-4.0 g/cm^3, specific applications can actually take advantage of high density ceramics where space efficiency and ceramic thickness becomes a design issue. For protection against high performing long rod penetrators where steel thickness or low density ceramic tile thickness exceed the space bounds, the lower density ceramics can actually become too thick for the application in areas such as roofs or hatches where vision sight-lines become critical. In 1994, ARL and Cercom Incorporated (now BAE Advanced Materials), developed ballistic grade high-purity, hot-pressed tungsten carbide and ditungsten carbide ceramics for ballistic applications. At densities of 15.7 and 17.2 g/cm^3, these ceramic tiles were one-fifth the thickness of similarly processed silicon carbide and inherently achieve mass efficiencies of 2 and space efficiencies of 4 compared to steel [47]. This class of ceramics has very specific applications, but can be used to provide protection where other ceramic armors cannot be fit.

Functionally Graded Metal/Ceramics

In 1993, ARL and Cercom Incorporated, now BAE Advanced Materials developed a process to hot-press large near net-shape Functionally Graded Materials (FGM) tiles in a single stage utilizing titanium and titanium/titanium diboride (TiB_2) powder mixtures, forming a titanium monoboride (TiB) hard face/titanium metal substrate that grades through intermediate layers (Figure 32) [48]. The TiB ceramic was formed through a reaction sintering process between the TiB_2 and titanium powders during the hot-press phase and was a new ballistic ceramic available to the armor community. In this development, the TiB is densified as a cermet (ceramic in a metal matrix) to aid in fabrication. A major development in the process was overcoming the inherent thermoelastic properties of the constituent layers and the resultant stresses that arise from the differences in thermal expansion coefficients and elastic moduli of the layers. Analytical and finite element modeling techniques were used to determine the residual stresses and modify the processing parameters. The resultant tiles produced are among the largest functionally gradient materials produced by a practical process and represent advancement in this technology area.

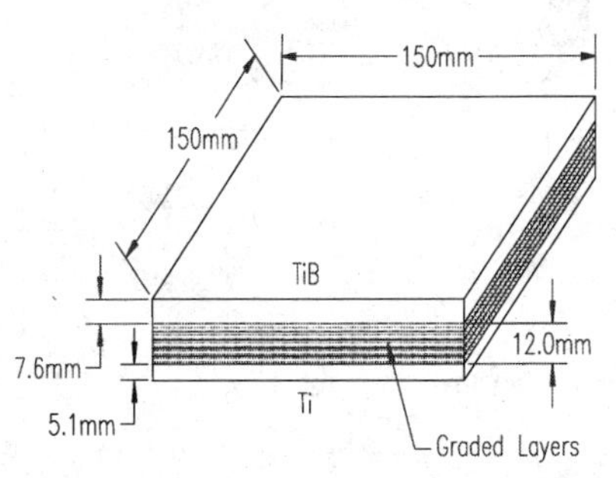

Figure 32. Typical TiB Cermet/Titanium FGM

Numerical Modeling of Brittle Failure of Ceramics

The concurrent development of advanced diagnostic capabilities and ever-increasing computational speed of modern parallel processing computers has culminated in the ability to obtain good agreement between experiment and simulation of ceramic armors [49-52]. This has resulted from an expanding set of failure models for brittle materials as well as the documentation of the required dynamic properties that comprise the input data for simulation. Today, simulations are being used to establish the designs for experimental confirmation and will expedite the development of final designs and a large computational capability exists to assist developers.

CONCLUSIONS

This paper has presented an abbreviated overview and chronology of significant U.S. developments in ceramic armor technology over the last 40 years as observed by the author. This overview paper has expanded the technical aspects of the presentations given at the 2001 PACRIM IV Conference in Hawaii and the 2006 30th International Conference on Advanced Ceramics and Composites at Cocoa Beach, Florida. The applications of ceramics for armor are growing rapidly as the need for lighter and more agile combat vehicles increases. Ceramic armor technology offers the best potential for meeting future protection requirements at the lightest areal weights, but the increased cost and complexity of the target design needs to be offset by the increased utility of the overall design.

REFERENCES

[1] C. Donaldson, "The Development of a Theory for the Design of Lightweight Armor", Aeronautical Research Associates of Princeton, Inc., Technical Report AFFDL-TR-77-114.

[2] A. Florence, "Interaction of Projectiles and Composite Armor", Stanford Research Institute, AMMRC-CR-69-15, August 1969.

[3] A. Prior, "The Penetration of Composite Armor by Small Arms Ammunition", *Proceedings of the International Ballistic Symposium*, 1986.

[4] A. Wong and I. Berman, "Lightweight Ceramic Armor - A Review", Army Materials and Mechanics Research Center, Report AMMRC-MS-71-1, 1971.

[5] W. Gooch, M. Burkins, W. Walters, A. Kozhushko and A. Sinani, "Target Strength Effect on Penetration by Shaped Charge Jets", *Hypervelocity Impact Society Symposium*, Galveston, Texas, 6-10 November 2000

[6] W. Gooch, "An Overview of Protection Technology for Ground and Space Applications", *Second Australian Congress on Applied Mechanics*, Canberra, Australia, 10-12 February 1999

[7] W. Walters and J. Zukas, Fundamentals of Shaped Charges, Wiley, New York, CMC Press, MD 1998.

[8] US Patent 3,509,833, Filed March 28, 1963, Issued May 5, 1970, Goodyear Aerospace

[9] US Patent 4,179,979, Filed May 10, 1967, Issued December 25, 1979, Goodyear Aerospace

[10] M. Wilkins, C. Honodel and D. Sawle, "An Approach to the Study of Light Armor", Lawrence Radiation Laboratory, UCRL-50284, June 13, 1967.

[11] M. Wilkins, "Second Progress Report of Light Armor", Lawrence Radiation Laboratory, UCRL-50349, Nov 1967.

[12] M. Wilkins, "Third Progress Report of Light Armor Program", Lawrence Radiation Laboratory, UCRL-50460, July 9, 1968.

[13] M. Wilkins, C. Cline and C. Honodel, "Fourth Progress Report of Light Armor Program", Lawrence Radiation Laboratory, UCRL-50694, 1968

[14] M. Wilkins, R. Landingham and C. Honodel, "Fifth Progress Report of Light Armor Program", Lawrence Radiation Laboratory, UCRL-50980, 1971.

[15] R. Landingham and A.W. Casey, "Final Report on Light Armor Materials Program", Lawrence Radiation Laboratory, UCRL-57269, 1972.
[16] Photos Courtesy BAE Advanced Materials, Vista, CA
[17] Contract, "Demonstration of Cast, Composite Ceramic Armor (C^3A), BRL Contract DAAA-15-86-C-0014, 1990.
[18] G. Hauver, P. Netherwood, R. Benck, W. Gooch, W. Perciballi and M. Burkins, "Variations of Target Resistance During Long-rod Penetration into Ceramics", *13th Int. Ballistics Symposium, Stockholm*, Sweden, 1992.
[19] G. Hauver, P. Netherwood, R. Benck and L. Kecskes,"Ballistic Performance of Ceramics", *U.S. Army Symposium on Mechanics*, Plymouth, MA, 17-19 August 1993.
[20] G. Hauver, P. Netherwood, R. Benck and L. Kecskes, "Enhanced Ballistic Performance of Ceramics", *19th Army Science Conference*, Orlando, FL, 20-24 June 1994.
[21] G. Hauver, E. Rapacki, P. Netherwood and R. Benck, "Interface Defeat of Long Rod Projectiles by Ceramic Armor", ARL Tech Report ARL-TR-3590, September 2005
[22] D. Orphal and R. Franzen, "Penetration of Confined Silicon Carbide Targets by Tungsten Long Rods at Impact Velocities from 1.5 to 4.6 km/s," *Int. J. Impact Engng.*, **19**(1): 1-13 (1997).
[23] T. Holmquist, C. Anderson and T. Behner, "Design, Analysis and Testing of an Unconfined Ceramic Target to Induce Dwell," *Proc. 22th Int. Symp. Ballistics*, **2**: 860-868, DES*tech* Publications, Inc., Lancaster, PA (2005).
[24] T. Behner, C. E. Anderson, Jr., T. J. Holmquist, D. L. Orphal, M. Wickert, and D. W. Templeton, "Penetration Dynamics and Interface Defeat Capability of Silicon Carbide against Long Rod Impact," *Int. J. Impact Engng.*, accepted for publication (2010).
[25] S. Chocron, C. Anderson, T. Behner and V. Hohler, "Lateral Confinement Effects in Long-rod Penetration of Ceramics at Hypervelocity", *Int. J. Impact Engng.*, 33(1-12): 169-179, 2006.
[26] C. Anderson, T. Behner, D. Orphal, A. Nicholls, T. Holmquist, and M. Wickert, "Long-rod Penetration into Intact and Pre-damaged SiC Ceramic," *Proc. 24th Int. Symp. on Ballistics*, **2**: 822-829, DES*tech* Publications, Inc., Lancaster, PA (2008).
[27] J. C. LaSalvia, E. J. Horwath, E. J. Rapacki, C. J. Shih and M. A. Meyers. "Microstructural and Micromechanical Aspects of Ceramic/Long-rod Projectiles Interactions: Dwell/Penetration Transitions. In *Fundamental Issues and Applications of Shock-Wave and High-Strain-Rate Phenomena*, 437-446, edited by K. P. Staudhammer, L. E. Murr and M. A. Meyers, Elsevier Science, New York, (2001).
[28] W. Gooch, M. Burkins, P. Kingman, G. Hauver, P. Netherwood, and R. Benck, "Dynamic X-ray Imaging of 7.62mm APM2 Projectiles Penetrating Boron Carbide", *18th International Symposium on Ballistics*, San Antonio, TX, 15-19 November 1999.
[29] T. Holmquist, C. Anderson, T. Behner, and D. Orphal, "Mechanics of Dwell and Post-Dwell Penetration," Advances in Applied Ceramics, 109(9): 467-479 (2010).
[30] W. Gooch, "Overview of the Development of Ceramic Armor Technology - Past, Present and the Future", *30th International Conference on Advanced Ceramics*, Coco Beach, FL, 22-26 January 2006.
[31] M. Burkins and W. Gooch, "U.S. Ceramic Ballistic Test Methodology and Data", TTCP-WTP1 Meeting, Maribyrnong, Australia, 10 May 1995.
[32] P. Woosley, "Ceramic Materials Screening by Residual Penetration Ballistic Testing", *13th International Symposium on Ballistics*, June 1992.
[33] W. Gooch, "An Overview of Ceramic Armor Applications", *International Conference on Advanced Ceramics and Glasses, PAC RIM IV*, Maui, Hawaii, 4-8 November 2001.

[34] E. Strassburger, H. Senf, "Experimental Investigations of Wave and Fracture Phenomena in Impacted Ceramics and Glasses", U.S. Army Research Laboratory contractor report ARL-CR-214, 1995.

[35] H. Senf, S. Winkler, "Experimental Investigation of Wave and Fracture Phenomena in Impacted Ceramics: Sapphire", U.S. Army Research Laboratory contractor report ARL-CR-310, 1997

[36] E. Strassburger, "Visualization of Impact Damage in Ceramics using the Edge-on Impact Technique", *International Journal of Applied Ceramic Technology*, Vol. 1, No. 3, pp. 235-242, 2004.

[37] E. Horwath and W. Bruchey, "The Ballistic Behavior of HIP Encapsulated Ceramic Tiles", *8th Annual TARDEC Ground Vehicle Survivability Symposium*, Monterey, CA, March 1997.

[38] W. Bruchey and E. Horwath, "System Considerations Concerning the Development of High Efficiency Ceramic Armors", *17th Int. Sym. on Ballistics*, Midrand, South Africa, March 1998.

[39] P. Lundberg, R. Renström and B. Lundberg, "Impact of Metallic Projectiles on Ceramic Targets: Transition between Interface Defeat and Penetration", *Int. J. Impact Engng*, 24, 259-275 (2000).

[40] P. Lundberg and B. Lundberg, "Transition between Interface Defeat and Penetration for Tungsten Projectiles and Four Silicon Carbide Materials", *Int. J. Impact Engng*, 31, 781-792 (2005).

[41] R. Renström, P. Lundberg and B. Lundberg, "Stationary Contact between a Cylindrical Metallic Projectile and a Flat Target under Conditions of Dwell", *Int. J. Impact Engng*, 30, 1265-1282 (2004).

[42] O. Andersson, P. Lundberg, and R. Renstrom, "Influence of Confinement on the Transition Velocity of Silicon Carbide," *Proc. 23rd Int. Symp. on Ballistics*, **2**, 1273-1280, Taragona, Spain (2007).

[43] W. Gooch, M. Burkins, L. Mills, J. Ogilvy, and A. Ricchiazzi, *Cast Single Plate P900 Armor*, U.S. Patent 5007326, April 16, 1991

[44] Photos and data courtesy of Mofet Etzion website www.mofet-etzion.co.il

[45] Photos and data courtesy of Plasan Sasa NA

[46] Photos and data courtesy of Industrie Bitossi SpA

[47] W. Gooch and M. Burkins, "Ballistic Development of U.S. High Density Tungsten Carbide Ceramics", *Dymat 2000*, Krakow, Poland, 23-29 September 2000.

[48] W. Gooch, M. Burkins and R. Palicka, "Development And Ballistic Testing Of A Functionally Gradient Ceramic/Metal Applique", *NATO Applied Vehicle Technology Panel*, Loen, Norway, 7-11 May 2001.

[49] C. Anderson and W. Gooch, "Numerical Simulations of Dynamic X-Ray Imaging Experiments of 7.62-mm APM2 Projectiles Penetrating B_4C", *19th International Ballistics Symposium*, 7-11 May 2001, Interlaken, Switzerland.

[50] C. Anderson, M. Burkins, J. Walker, and W. Gooch, "Time-Resolved Penetration of B4C Tiles by the APM2 Bullet", CMES, Vol. 8, No. 2, pp 91-104, 2005

[51] J. Walker, "Analytic Model for Penetration of Thick Ceramic Targets," *Ceramic Armor Materials by Design* (J. W. McCauley, *et al*, Eds.), *Ceramic Transactions*, 134: 337-348 (2002).

[52] J. Walker, "Analytically Modeling Hypervelocity Penetration of Thick Ceramic Targets," *Int. J. Impact Engng.*, 29(1-12): 747-755 (2003).

IMPACT STRENGTH OF GLASS FOR ARMOR APPLICATIONS

Xu Nie, Weinong W. Chen
School of Aeronautics and Astronautics
Purdue University
West Lafayette, IN, USA

ABSTRACT

This paper reviews three recent high rate characterization techniques modified from Kolsky bars for brittle materials. In addition to axial compression, a compression/shear technique uses inclined specimen geometry to introduce extra shear stresses in the glass specimen to examine the effects of variation in stress state on the dynamic axial failure strength. Dynamic bending strength is determined by extending the loading rate range of the standard quasi-static four-point bending and ring-on-ring techniques into dynamic region. This is achieved by adapting the corresponding specimen configuration into the test section of a compression Kolsky bar (SHPB). The results for a borosilicate glass obtained from these techniques are discussed.

INTRODUCTION

Silicate glasses have been used as transparent armors for decades. Although several other transparent ceramic armors (e.g. AlON, spinel) have demonstrated superior performances against ballistic impact, the silicate glass is considered to offer an excellent balance among manufacturing costs, areal density and penetrations resistance. A typical case showing the fracture and failure process of a thin-layered transparent armor subjected to impact loading is summarized in Fig. 1. Upon the impact of a projectile, armor material on the impact side is subjected to combined high strain rate compressive and shear loading which lead to localized fracture and pulverization. The back side of the plate may fail due to spalling and bending-induced tension.

The effective use of glasses in optimized structures requires thorough understanding and quantitative modeling of the mechanical response and failure behavior of glasses at high rates of deformation. Recent research efforts have been invested in exploring failure wave propagation through glass [1-5] and glass response to particle and cylindrical projectile impact [6-9]. In pressure/shear plate impact experiments on soda-lime glass, it was shown the shock-induced microcracking considerably reduced the shear resistance of the glass [10, 11]. Despite these research efforts, the fracture and failure behavior of glasses under high-rate complex stress states have not been well studied. Recently, lateral confinement was applied on borosilicate glass samples in order to achieve a dynamic multi-axial stress state [12].

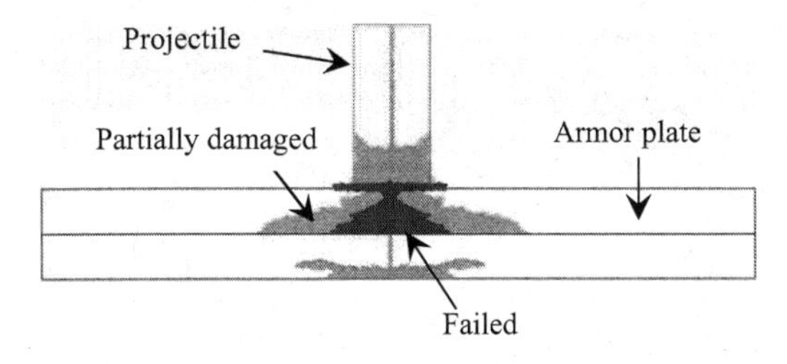

Figure 1: Penetration process of a single projectile on thin-layered armor

Flexural loading is typically seen by thin transparent armors under impact. Under such loading conditions, glass strength is typically limited by surface defects introduced by after-manufacturing handling, but rarely by bulk defects [13]. Therefore, for the glasses in armor applications, it is important to understand the loading rate and surface condition effects on the tensile/flexural strength. Under quasi-static loading conditions, intensive research efforts have been invested in exploring the influence of various surface treatments on the flexural strength of glass and ceramic materials [14-20]. It has been identified that surface conditions play a significant role in both the fracture strength and crack propagation of glasses [17-20]. Compared to ceramics, glass materials are more susceptible to surface flaws due to the lack of bulk defects such as large grains, grain boundaries and secondary phase inclusions [13]. Although the theoretical strength of glass is considered to be in the order of 10 GPa at room temperature [21], very little work has been documented to achieving this strength mostly due to the damage to the glass samples and moisture effect in the testing environment. Different methods to improve the surface quality of oxide glasses have been reviewed in [22]. As a conclusion, surface etching is considered to be one of the most effective and simple methods to remove or reduce surface defects. The principle for surface etching is to either remove surface cracks completely or blunt the crack tip significantly through material removal. Bulk glass strengths exceeding 1 GPa have been produced by this method [23, 24]. Recent research results on ballistic response of thin glass plates have revealed significantly enhanced impact resistance in HF-acid etched glass target [25]. No fracture was initiated on the back side of the glass plate target, which was in tension, at impact speeds of up to 700 m/s. Although the effects of surface modification on impact resistance are clearly shown, the surface morphologies of treated and untreated glass and their influences on the material strength were not reported. It is of significant interest to establish the relationship between dynamic flexural strength and surface conditions in order to optimize the ballistic resistance in light transparent armors.

The penetration of thin glass armor is a rather complicated process that involves different loading rates, stress states, temperatures, surface flaws and materials at different damage levels. The combination of all these variables poses significant challenges in both material characterization and numerical simulations. However, effective design and modeling of any armor structure would first require the input of fundamental materials parameters. This paper reviews three recent experimental methods to (1) dynamically load the sample under different stress states and surface conditions, and (2) recover the deformed/fractured sample after a single, well defined loading pulse for the analysis of failure mode. The details of experiment design, results and discussions are presented in the following sections.

EXPERIMENTAL TECHNIQUES

Kolsky pressure bar (or split Hopkinson pressure bar, SHPB) is a well-established apparatus commonly utilized in the high-strain-rate testing of materials to provide a complete family of dynamic stress-strain curves as a function of strain rates [26]. Originally developed by Kolsky [27], this technique was initially used for the characterization of the dynamic flow behavior of ductile materials at strain rates up to 10^4 s^{-1}. When this device is used to determine the dynamic properties of other materials, modifications are needed to ensure that the specimen deforms under desired testing conditions. For example, when the specimens are brittle materials, e.g., ceramics and concretes, a nearly linear loading pulse becomes necessary in order to deform the brittle specimens at a nearly constant strain rate [28-30]. A set of modified Kolsky bar was used to generate nearly linear ramp pulses to load the glass specimen, see Fig. 2. A pulse shaper, which was made of a thin metal disc, was placed between the striker and incident bar. By adjusting the dimensions of this annealed copper pulse shaper, ramp loading pulses with different slopes may be generated. This modification is necessary for obtaining a constant loading rate and for preventing the specimen from failing before dynamic stress equilibrium and constant strain rate are achieved.

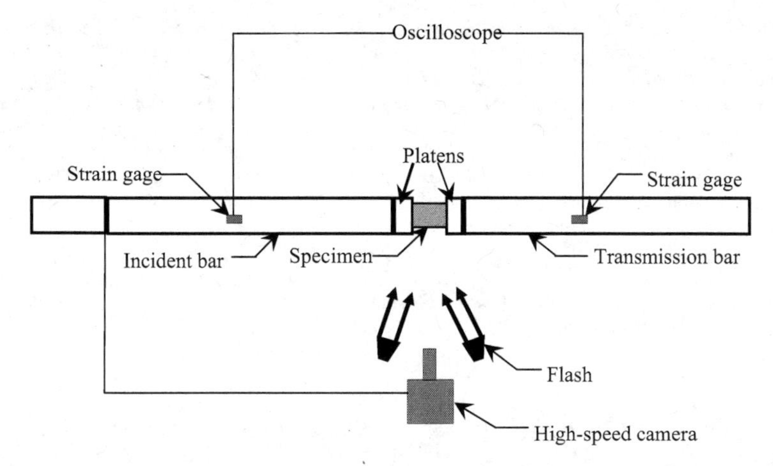

Figure 2: Modified Kolsky bar, with pulse shapers, platens and a high-speed camera.

The borosilicate glass investigated in this study has a high compressive strength. Upon impact, the debris from the broken specimen may indent the metal bar ends. Since it is very time-consuming to re-surface the damaged bar ends after a few experiments, disposable tool steel platens were placed between the glass specimen and the maraging steel bar end faces. These platens were also highly polished to minimize end friction on the specimen ends. To record the high-speed deformation/failure processes in the glass specimens, a Cordin 550 high-speed digital camera was used in the experiments, synchronized with the loading stress pulses. Two high-intensity strobe lights were placed along side of the camera as light source. Thirty two frames were captured for each experiment at a frame rate of 200,000 frames per second.

The high rate tensile strength of glass is characterized using a recent modification on Kolsky bars which utilizes four-point bending configuration to investigate the dynamic fracture toughness of brittle materials [31]. This experimental method was designed based on the ASTM Standard C 1421-01b [32]. To load the specimen at high constant rates while maintaining equilibrated loading across the gage section, the pulse shaping technique was employed to generate controlled loading profiles. The measured flexural strength values are used to make inferences to the uniaxial tensile strength of glass. A set of 4 point bending fixtures were designed according to ASTM C1161-02c [33]. The spacing between the two loading pins and that between two support pins are 10 mm and 20 mm respectively. Pin rollers were made of hardened M-2 steel with a hardness of RC 60. Those rollers were fixed to the aluminum back fixture. The low rate (0.7 MPa/s, 50 MPa/s, and 2500 MPa/s) experiments were performed on a close-loop controlled servohydraulic testing machine (MTS 810).

The surface tensile strength of glass is also investigated through another type of flexural testing method—equibiaxial ring-on-ring (ROR) flexural technique. The implementation of ROR technique is mainly motivated by the desire of eliminating edge failures that are frequently encountered in four-point bending experiments. By this means the high rate surface tensile strength of glass material can be interpreted as a function of surface defects without the interference of any geometrical stress concentration at the edges. This technique was developed according to ASTM C1499 [34], and incorporated to a modified Kolsky bar. A pair of concentric steel rings was attached to concentric aluminum substrates on the Kolsky bar so that the system alignment is secured. The rings were hardened to HRC 60 and then polished to ensure smooth contact with glass samples. The diameters of those concentric rings are 12.5 mm and 25 mm, respectively, while the ring radius is 2.5 mm. The incident and transmission bars of the Kolsky bar setup are made of 6061-T6 aluminum alloy with a common diameter of 31.75 mm. An image of this testing configuration is shown in Fig. 3. In addition to the ring fixture, a pair of universal joints which are of the same diameter of the bars was also placed between the gage section and the transmission bar. Universal joints were adopted in Kolsky bar system to eliminate possible misalignment in the gage section [35]. This modification is very important in brittle-material testing because these materials are susceptible to failure initiation from concentrated stresses. The joints used in this research are composed of a convex plane and a concave plane which are of equal curvature and facing each other. During specimen-installation procedure, this pair of surfaces is the last to engage, eliminating misalignment and ensuring an even contact between the loading rings and the specimen surface.

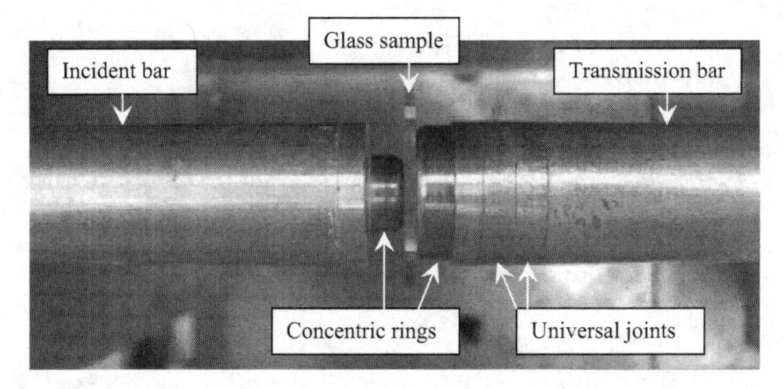

Figure 3: Dynamic ring-on-ring test section configuration.

IMPACT STRENGTH OF GLASS

The glass investigated in this research is Borofloat® 33 manufactured by Schott Glass and supplied by Army Research Laboratory (ARL). The raw material is processed by the float technique and provided in the form of 300 mm x 300 mm x 12.7 mm square plates without coating and thermal strengthening. To avoid the disturbance from possible chemical interference between the glass plate and the liquid tin during manufacturing, all the specimens for mechanical testing were cut from the air side. As a typical borosilicate glass, Borofloat® 33 has an amorphous network structure modified from the silica network. Besides B_2O_3, which is the main additive for this material, there are two other additional network modifiers: Na_2O and Al_2O_3. The compositions of this material are shown in Table 1. Other important physical and mechanical properties of interest are: density ρ = 2.21 g/cm^3, Young's modulus E = 61 GPa, Poisson's ratio ν = 0.19, longitudinal wave speed C_L = 5508 m/s, shear wave speed C_S = 3417 m/s.

Table 1: Main chemical compositions of the borosilicate glass.

Composition	SiO_2	B_2O_3	Na_2O	Al_2O_3
Percentage (%)	80.5	12.7	3.5	2.5

Dynamic Compression/shear Strength

Specimens were cut from a glass plate, and then ground to specified dimensions. The specimens were 9 9 mm in cross section and 12.5 mm in length. The specimen shape is shown in Fig. 4. Four cuboid specimen geometries, with the tilting angle of 0°, 3°, 5° and 7°, respectively, were used in the dynamic compression experiments. All the specimens were polished to reduce the influence of surface flaws on measured glass strength. Furthermore, polished steel platens were placed in between the glass specimen and Hopkinson bar ends. The purposes of using these steel platens were to prevent the bars from being indented by the high strength glass fragments and to minimize friction on the specimen-bar interface.

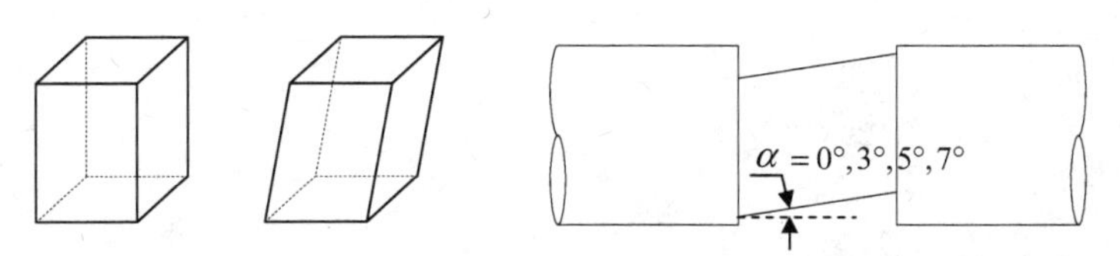

Figure 4: Specimen geometry (a) 0 specimen, (b) Specimen with a tilting angle, (c) Side view of the specimen

The purpose of introducing inclining angles to the specimens is to facilitate various degrees of shear stresses in the specimens while they are loaded in compression. Tilted specimens were initially used for the investigation of dynamic shear banding in metallic materials [36]. We used similar tilted specimens in this study to investigate the effects of the introduction of shear stress on the damage and failure behavior of the glass specimens. Typical results and related high-speed images for 7° samples are shown in Fig. 5. As is evident in the figure, damage/cracking in the tilted specimens always initiate from the specimen corners with stress-concentration. The subsequent crack propagation into the specimen is roughly along the specimen axis, rather than along the Kolsky bar compressive loading direction. The stress-time history curves of all the angled specimens are also marked with the instants when the corresponding high-speed images were taken. A circled area in the figure indicates the initiation of damage at the corners. Time interval between points 1 and 2 is 5 μs, whereas the interval thereafter is 10 μs. The damage initiation starts earlier with increasing tilting angle due to the stress concentration and the increased amount of shear stress at the corners. For an angled specimen, the stress distribution on the specimen/bar interfaces is no longer uniform. The stress in stress-time history curves shown in Fig. 5 is the average stress history captured by the strain gages mounted on the transmission bar. The local stress in the vicinity of the obtuse corner is considerably larger than the average stress in the specimen. Thus, damage always initiates from the obtuse angle area, which is verified in our observations as presented in Fig. 5. Furthermore, the observed average stress at failure decreases with increasing specimen angles (from 1302 MPa for 0° specimen to 560 MPa for 7°

specimen). While the material fails in the stress concentration area, the other regions of the specimen are still in a relatively low stress state. Therefore, the damage initiates in the tilted specimens at lower average stress levels but the catastrophic failure of the entire specimen comes much later. It took only ~20 μs for a 0° specimen to experience the entire failure process. However, it took as long as 35 μs for a 7° specimen to go from damage initiation to catastrophic collapse.

The results shown in Fig. 5 illustrate that the onset of damage in the glass specimen can be accurately determined by the high-speed camera. This capability of damage initiation determination in angled specimens facilitates the determination of dynamic strength of the borosilicate glass under compression/shear loading. Stress concentration is inevitably introduced by the geometry of the tilted specimens. To identify the failure stress that corresponds to damage initiation in the specimen, local stress state at the corners, rather than average stress, needs to be used. However, the Kolsky bar measurements, shown in Fig. 5, provide only the averaged stress transmitted from the specimen. Finite element analysis with a commercial code ABAQUS was used together with the Kolsky bar stress history and the high-speed images to determine the stress state in the specimen when damage initiates. Table 2 summarizes

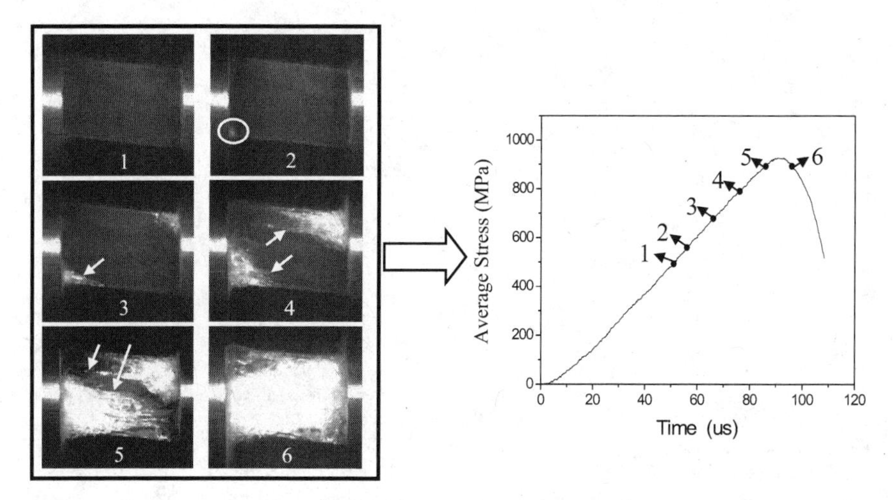

Figure 5: High speed images and stress histories showing the failure processes 7° glass specimen. White arrows show the propagation and coalescence of macroscopic cracks.

the compressive and shear stress components in the obtuse corner for the angled specimens obtained through the hybrid numerical/experimental analysis, along with that of the 0° specimen. The 0° sample experienced a relatively homogeneous deformation process so the failure stress is determined directly by experiments. The results shown in the table clearly show the trend that the dynamic compressive strength of the borosilicate glass decreases with increasing shear stress. It is noted that the compressive strength was taken at the onset of damage in the specimen, not the stress at which the sample collapsed catastrophically. To further illustrate the trend on the strength dependence on shear stress, the biaxial stress components tabulated in Table 2 are used and converted into equivalent stress, the results of which are displayed in Fig. 6. Figure 6 again shows the trend that the failure strength, in terms of equivalent stress, of the borosilicate glass is sensitive to the imposed shear component.

Table 2: Biaxial failure stress of all degree glass specimens.

Specimen angle	0°	3°	5°	7°
Compressive stress (MPa)	1302	1266	1233	1185
Shear stress (MPa)	0	66	107	145

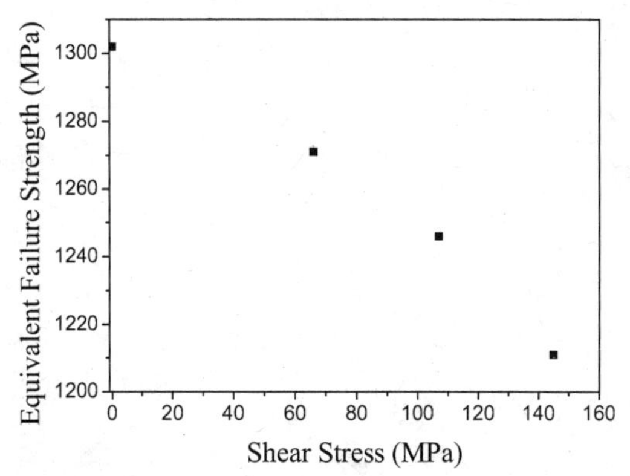

Figure 6: Variation of dynamic equivalent failure strength of borosilicate glass with applied shear stress.

Four-point Bending and Ring-on-ring (ROR) Flexural Strength

For four-point bending experiments, the samples were first machined into 2 mm×3 mm×30 mm glass prisms, and then polished to a surface finish of 80/50 scratch/dig. All the four edges of the as-received samples were chamfered before surface treatments were taken. The chamfer angle was 45°, with a dimension of approximately 0.12 mm. The polished and chamfered samples were then divided into three groups. The first group was ground with 220-grit sandpaper, and the second group was ground with 1500-grit sandpaper. Both grinding treatments were done only on the tensile surface and were perpendicular to the tensile axis. The specimens from the third group were immersed in 5% HF acid for 15 minutes. The principal of HF acid etching is that HF in the aqueous solution attacks SiO_2 in the glass to form water-soluble reaction products. By this way glass material is striped off from the surface layer by layer so the pre-existed surface flaws were eliminated. The ROR samples are in the form of cylindrical disks with a dimension of 2 mm in thickness and 45 mm in diameter. The surface treatments for ROR samples are similar to that for the four-point bending samples, except that 1500-grit sandpaper grinding was replaced by fine polishing, and the 200-grit sandpaper was replaced by a coarser 180-grit sandpaper.

The variation in four-point bending strength as a function of loading rate for all surface conditions is shown in Fig. 7. It is clear that the flexural strength of borosilicate glass increases with increasing loading rates in the loading-rate range achieved in this study, regardless of surface conditions. Below the rate of 2,500 MPa/s, the flexural strength is roughly a linear function of logarithm loading rate, whereas not much rate sensitivity is observed above 2,500 MPa/s. For the group #1 and #2 samples, particle size difference in sandpapers only imposes relatively small variations in the flexural strength. The average strength of group #2 samples is 60%-90% (depending on loading rates) higher than that of group #1 samples. This strength improvement is achieved by an approximately 80% reduction in surface roughness. For group #3 samples which were etched by HF acid, the average flexural strength is 700%-1,500% higher than that of group #1 samples over the same loading-rate range. The average flexural strength of etched samples achieved at the highest loading rate is 1.1 GPa. The surface roughness values of the etched samples are actually higher than those of fine ground samples (group#2). This observation indicates that surface roughness value alone is not sufficient to relate the flexural strength to the surface quality of glass materials.

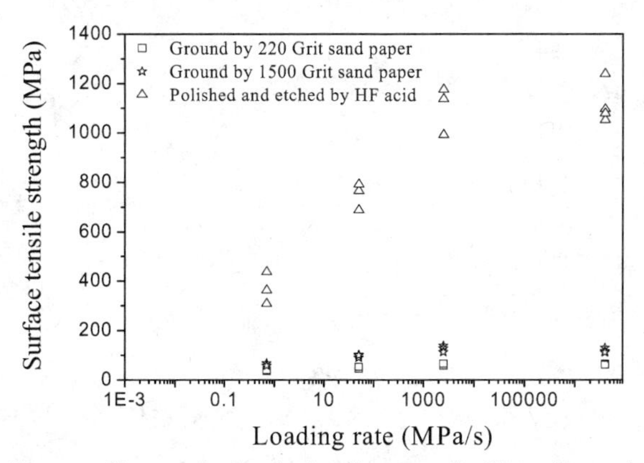

Figure 7: Loading-rate effects on the 4- point bending strength of borosilicate glass with different surface conditions.

Scanning Electron Microscopy (SEM) images on fracture surfaces are therefore taken to further investigate the failure mechanism of different groups of specimens. Failure origins in sandpaper ground glass samples are observed to locate at the sub-surface microcrack front, as is evident in Fig. 8 (a) and (b), which are most likely being induced by indentation of abrasive particles. So the flexural strength of sandpaper ground samples is governed by the sharp sub-surface flaws. The observed differences in strength at different grinding conditions could also be explained by the differences in the flaw size. Fig. 8 (c) is the fracture surface of HF acid etched sample. The acid-etching induced surface flaw is much smaller in size compared to those semi-elliptical cracks with no visible sub-surface cracks. These surface pits are blunt in nature so the stress concentration is much less than that around a sharp sub-surface crack front, which results in the high strength of etched samples.

A further look on the loading rate dependence on flexural strength suggests that for sandpaper ground samples, the flexural strength increased by about 90% when the loading rate changed from 0.7MPa/s to 4 x 10^6MPa/s, whereas for HF acid etched samples such a strength increase is about 200%. Since the only differences between those 2 groups of samples are the critical flaw types, the difference in the rate dependence of strength is considered to be the difference in the shapes of defects. Under a sharp indentation pit, cracks are formed and connected to the pit before global mechanical loading. On the other hand, under a pit blunted by HF acid, new cracks will have to form under the loading. Higher loading rate will cause multiple cracks to form simultaneously, which may account for the difference in rate dependency.

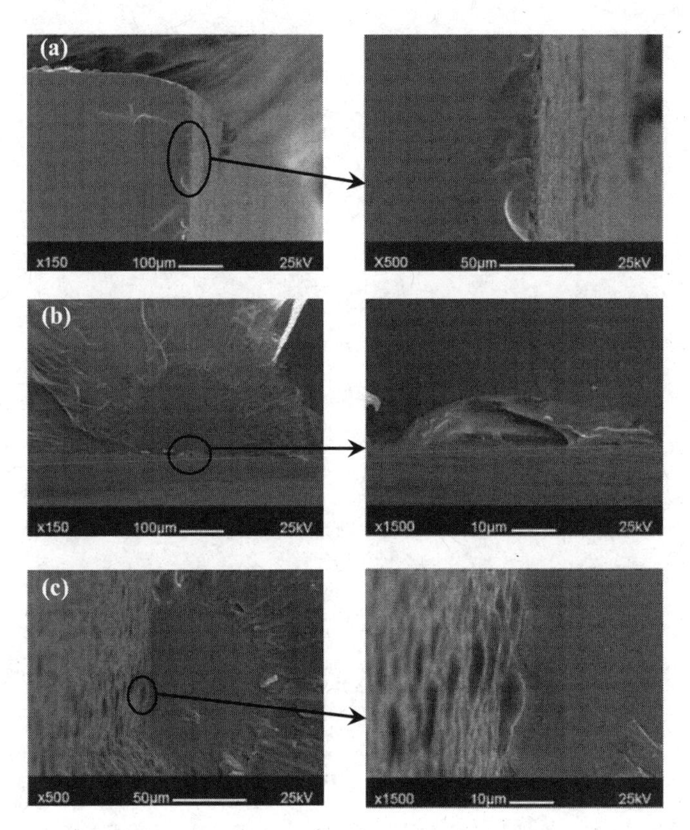

Figure 8: Fracture surface and failure origins of different groups of glass samples. Flaw size is measured in terms of length and depth. (a) ground by 220 grit sandpapers (approximate flaw size $125\mu m \times 29\mu m$); (b) ground by 1500 grit sandpapers (approximate flaw size $57\mu m \times 11\mu m$); (c) polished and etched by HF acid (approximate flaw size $12\mu m \times 4\mu m$).

The calculated ROR equibiaxial flexural strength values for borosilicate glass samples at different loading rates and surface conditions are summarized in Fig. 9. The scatters (error bars) in the plot represent the whole range of experimental data for each testing condition, and the symbol in between is the arithmetic average of strength. The results indicate that the surface modifications significantly affect the flexural strength of the glass material. The sandpaper grinding degrades the strength by 60-70% from the as-polished surface condition. However, HF acid etching on as-polished specimens promotes the surface tensile strength by 200-400%, depending on the applied loading rates. The experimental results also indicate that the loading rate has remarkable effects on the flexural strength. Under all surface conditions tested, the strength universally increases with loading rates. But the rate of strength increase levels out at the loading rate of ~3,500 MPa/s, which is similar to that in the 4-point bending experiments. The fractography for different surface conditions also resembles those discovered in the 4-point bending experiments.

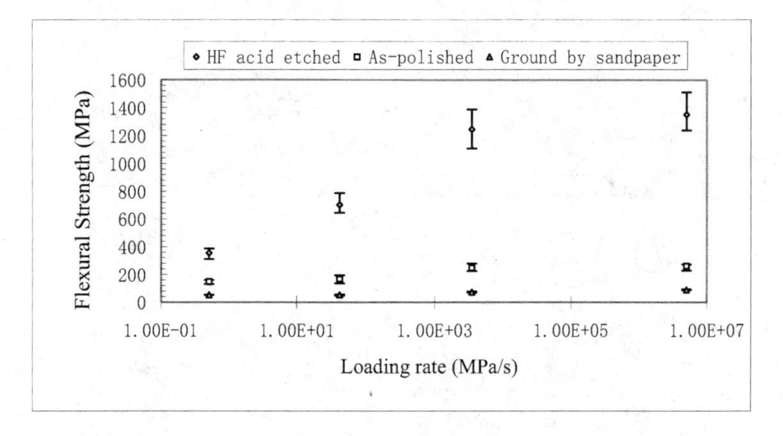

Figure 9: Equibiaxial ROR flexural strength of borosilicate glass as a function of loading rates and surface conditions.

SUMMARY

Borosilicate glass specimens with different tilting angles were dynamically loaded with a modified Kolsky bar. The tilted specimen geometry introduces a shear component to axial compression, creating a principal stress state of axial compression and lateral tension. This facilitates the investigation of the effects of added shear on the dynamic failure process of the glass material at high rates of deformation. Linear ramp loading was generated through pulse shaping on the Kolsky bar to load the linear material at a constant rate. A high-speed digital camera was used to record the damage initiation and failure process in the glass specimens. The images were synchronized with the loading history. Numerical stress analysis was used to determine the local stress state at the onset of damage at the obtuse corners in the specimens. The local stress, in the form of both axial compressive component and the equivalent stress at the onset of damage was found to decrease with increasing shear stress. Therefore, the failure strength of this borosilicate glass is sensitive to the introduction of shear stress. Cracks in specimens with or without tilting angles all propagate along the specimen axial direction, instead of the global loading direction, which are driven by the axial compression/lateral tension stress state.

Quasi-static and dynamic flexural experiments on a borosilicate glass at different surface conditions were conducted in terms of both 4-point bending and ROR equibiaxial bending. The dynamic experiments were carried out on a modified Kolsky bar setup. A pulse-shaping technique was employed to subject the specimen at constant loading rates. The surface conditions of the tensile surfaces of the samples were varied to study their effects on the flexural strength of the glass. Sample flexural strength of over 1 GPa at high loading rates was measured on specimens where the tensile surfaces were chemically etched. The flexural strength of the borosilicate glass increases with increasing loading rate for all the surface conditions studied. For the ground samples, the flexural strength decreases with increasing surface roughness, while the etched samples possess high strength although at high surface roughness. SEM image analysis on fracture surfaces show that small, blunt surface pits are the failure initiation sites for acid etched samples, while large, sharp sub-surface cracks are identified as fracture origins for the sandpaper ground samples.

REFERENCES

[1] A. Ginzburg and Z. Rosenberg, "Using reverberation techniques to study the properties of shock loaded Soda-Lime glass", Shock Compression of Condensed Matter,AIP, 1997, 529-53

[2] N. S. Brar, "Failure waves in glass and ceramics under shock compression", Shock Compression of Condensed Matter, AIP, 1999, 601-606

[3] J. Millett and N. Bourne, "The shear strength of a shocked borosilicate glass with an internal interface", Scripta Materialia, 2000, 681-685

[4] G. R. Willmott and D. D. Radford, "Taylor impact of glass rods", Journal of Applied Physics, 2005, 97, 093522

[5] H. D. Espinosa, Yueping Xu and N. S. Brar, "Micromechanics of failure waves in glass: I, experiments", Journal of the American Ceramic Society, 1997, 80, 2061-2073

[6] N. H. Murray, N. K. Bourne, J. E. Field and Z. Rosenberg, "Symmetrical taylor impact of glass bars", Shock Compression of Condensed Matter, 1997, 533-536

[7] C. G. Knight, M. V. Swain and M. M. Chaudhri, "Impact of small steel spheres on glass surfaces", Journal of Materials Science, 12, 1573-1585, 1977

[8] M. M. Chaudhri and C. R. Kurkjian, "Impact of small steel spheres on the surfaces of "Normal" and "Anomalous" glasses", Journal of American Ceramic Society, 1986, 69 [5] 404-410

[9] J. E. Field, Q. Sun and D. Townsend, "Ballistic impact of materials", Inst. Phys. Conf. Ser., 102(7) 1989

[10] N. S. Brar, S. J. Bless and Z. Rosenberg, "Impact-induced failure waves in glass bars and plates", Applied Physics Letters, 1991, Vol. 59, Issue 26, 3396-3398

[11] S. Sundaram and R. J. Clifton, "Flow Behavior of Soda-Lime glass at high pressures and high shear rates", Shock Compression of Condensed Matter, 1997, 517-520

[12] K. A. Dannemann, A. E. Nicholls, C. E. Anderson Jr., I. S. Chocron and J. D. Walker, "Compression Testing and Response of Borosilicate Glass: Intact and Damaged", Proceedings of Advanced Ceramics and Composites Conference, January 23-27, 2006 Cocoa Beach, FL

[13] M. Hara, "Some aspects of strength characteristics of glass", Glastechnische Berichte, 61, 191-196, 1988

[14] B. P. Bandyopadhyay, "The effects of grinding parameters on the strength and surface finish of two silicon nitride ceramics", Journal of Materials Processing Technology, 53, 533-543, 1995

[15] D. M. Liu, C. T. Fu, L. J. Lin, "Influence of machining on the strength of $SiC-Al_2O_3-Y_2O_3$ ceramic", Ceramics International, 22, 267-270, 1996

[16] M. Guazzato, M. Albakry, L. Quach, M. V. Swain, "Influence of grinding, sandblasting, polishing and heat treatment on the flexural strength of a glass-infiltrated alumina-reinforced dental ceramic", Biomaterials, 25, 2153-2160, 2004
[17] J. E. Ritter, Jr., C. L. Sherburne, "Dynamic and static fatigue of silicate glasses", Journal of the American Ceramic Society, 54, 12, 601-605, 1971
[18] J. E. Ritter, M.R. Lin, "Effect of polymer coatings on the strength and fatigue behavior of indented soda-lime glass", Glass Technology, 32, 2, 51-54, 1991
[19] John. J. Mecholsky, Jr., S. W. Freiman, Roy W. Rice, "Effect of grinding on flaw geometry and fracture of glass", Journal of the American Ceramic Society, 60, 3-4, 114-117
[20] G. Scott Glaesemann, Karl Jakus, John E. Ritter, Jr., "Strength variability of indented soda-lime glass", Journal of the American Ceramic Society, 70, 6, 441-444, 1987
[21] C. Gurney, "Source of weakness in glass", Proceedings of the Royal Society of London, 282[1388], 24-33, 1964
[22] I. W. Donald, "Methods for improving the mechanical properties of oxide glasses", Journal of Materials Science, 24, 4177-4208, 1989
[23] C. Symmers, J. B. Ward and B. Sugarman, "Studies of the mechanical strength of glass", Physics and Chemistry of Glasses, Vol. 3, 76-83, 1962
[24] C. K. Saha and A. R. Cooper, "Effect of etched depth on glass strength", Journal of the American Ceramic Society, 67[8], C158-C160, 1984
[25] A. S. Vlasov, E. L. Zilberbrand, A. A. Kozhushko, A. I. Kozachuk, and A. B. Sinani, "Behavior of strengthened glass under high-velocity impact", Strength of Materials, 34 [3], 266-268, 2002
[26] G. T. Gray, "Classic Split Hopkinson Pressure Bar Technique,"ASM Handbook, Vol. 8, Mechanical Testing and Evaluation, ASM International, Materials Park, OH (2000)
[27] H. Kolsky, "An investigation of the mechanical properties of materials at very high rates of loading", Proceedings of the Royal Society of London, 1949, B62, 676-700
[28] C. A. Ross, D. M. Jerome, J. W. Tedesco and M. L. Hughes, "Moisture and Strain Rate Effects on Concrete Strength", ACI Mater. J., 93, 293-300 (1996).
[29] W. Chen and G. Ravichandran, "Dynamic Compressive Behavior of a Glass Ceramic under Lateral Confinement", J. Mech. Phy. Solids, 45 [8] 1303-1328 (1997)
[30] S. Sarva and S. Nemat-Nasser, "Dynamic Compression Strength of Silicon Carbide Under Uniaxial Compression", Mat. Sci. & Engrg. A, A317, 140-144 (2001)
[31] T. Weerasooriya, P. Moy, D. Casem, M. Cheng and W. Chen, "A four-point bend technique to determine dynamic fracture toughness of ceramics", Journal of the American Ceramic Society, 89[3], 990-995, 2006
[32] ASTM C 1421-01b, "Standard test methods for determination of fracture toughness of advanced ceramics at ambient temperatures", Annual Book of ASTM Standards, ASTM, West Conshohocken, PA, 2001
[33] ASTM C 1161-02c, "Standard test method for flexural strength of advanced ceramics at ambient temperature", Annual Book of ASTM Standards, ASTM, West Conshohocken, PA, 2003
[34] ASTM C1499 – 05, "Standard Test Method for Monotonic Equibiaxial Flexural Strength of Advanced Ceramics at Ambient Temperature", Annual book of ASTM standards, ASTM, West Conshohocken, PA, 2003
[35] W. Chen, H. Luo, "Dynamic Compressive Responses of Intact and Damaged Ceramics from a Single Split Hopkinson Pressure Bar Experiment", Experimental Mechanics, 44(3), 295-299, 2004
[36] L. W. Meyer, E. Staskewitsch and A. Burblies, "Adiabatic shear failure under biaxial dynamic compression/shear loading", Mechanics of Materials, 17, 203-214, 1994.

MEASUREMENT OF DEFORMATION IN ALUMINA SAMPLES INDENTED AT HIGH STRAIN RATES

C E J Dancer[a*], H M Curtis[a], S M Bennett[a], N Petrinic[b] and R I Todd[a]
[a] Department of Materials, University of Oxford, Parks Road, Oxford, OX1 3PH, UK
[b] Department of Engineering, University of Oxford, Parks Road, Oxford, OX1 3PJ, UK

*Corresponding author: claire.dancer@materials.ox.ac.uk

ABSTRACT

In this paper we describe the use of Cr^{3+} fluorescence mapping for the study of residual stress and plastic deformation induced in alumina targets by dynamic impact with sharpened tungsten carbide projectiles. A range of alumina samples in the form of small discs with varying glass content, purity and grain size were impacted by small projectiles accelerated to around 100 m/s using a gas gun. Post-test characterisation of the targets was carried out using 3D optical microscopy, scanning electron microscopy, and by Cr^{3+} fluorescence mapping of the residual stress and plastic deformation in the impacted region. The response of the alumina was found to vary according to its glass content and grain size, and the materials were ranked according to their response under impact.

INTRODUCTION

The widespread use of alumina as an armour material[1] is due to its low density, relatively inexpensive manufacturing routes and its durability which enables alumina to defeat a large number of threats[2]. Alumina can defeat impacting projectiles by a number of mechanisms, including the shattering or blunting of the incoming projectile, dissipation of projectile energy by cracking or plastic deformation, and erosion of the projectile, leading to it breaking up or even its complete destruction[3].

Alumina has been extensively studied by a variety of high strain rate tests to assess its performance as an armour material[4-6]. However the highly destructive nature of tests carried out at ballistic velocities generally allows only limited post-deformation microstructural information to be gathered. More extensive microstructural characterisation can be more easily carried out following less destructive quasi-static indentation testing. However, the failure mechanisms in alumina are different for quasi-static and dynamic tests due to the ability of ceramics to plastically deform at high strain rates. In dynamic impact failure is caused by coalescence of many micro-fractures which are activated at high strain rates[7] to accommodate the strains generated by plastic deformation[8]. The extent of plastic deformation depends on the pre-existing flaw population (e.g. pores and glassy grain boundaries) in the alumina specimen[9]. In addition, the velocity of impact and the microstructure of the specimen affect the response (elastic, inelastic or mixed) of the alumina to dynamic impact[8]. The different failure mechanisms in dynamic testing mean that results from quasi-static indentation are difficult to correlate with performance under dynamic impact.

However, by carefully controlling the degree of damage sustained by the alumina tiles, characterisation methods normally used to measure residual stress around hardness indentations, for example, can be adapted for use on dynamically tested alumina specimens. For quasi-static hardness indentations, the position and width of Cr^{3+} luminescence peaks in alumina have been shown to be related to the residual stress and plastic deformation in the ceramic respectively[10]. By collecting spectra at points across the sample surface this technique has previously been used to study the deformation caused by quasi-static indentations[11,12]. In this paper we demonstrate the use of this technique on a number of different alumina samples which have been tested under dynamic conditions to assess both surface and sub-surface damage.

EXPERIMENTAL

Alumina discs were produced from as-purchased commercially produced powders, either TM-DAR alumina powder (Tai-Mei, 99.99% purity) or AES11c alumina powder (Sumitomo, 99.9% purity). Both powders were pressed into discs under 100 MPa uniaxial pressure. AES11c samples were also cold-isostatically pressed at 200 GPa. All discs were sintered at 1550°C for 2 hours in air. Following sintering the discs were all approximately 25 mm in diameter and 5 mm in thickness. These materials are denoted A (TM-DAR) and B (AES11c). Samples were also produced from TM-DAR powder with added MgO to reduce the grain size. 0.25 wt.% MgO as a sintering aid and 2.1 wt% of organic dispersant (Dispex A40, Allied Colloids) were added to the alumina powder (TM-DAR) and mixed by attrition milling in water (Szegvari 01HD attrition mill, yttria-stabilized zirconia tank and blade and 3 mm diameter zirconia balls) for 2 h. The powder was then freeze-dried and calcined for 1 h at 600 °C. This powder was then pressed into discs as previously and sintered under vacuum for 1 h at 1500°C (Sample C). In addition samples of three commercial alumina materials (D, E and F) were obtained. Sample types D and E contained approximately 5% and 2% glassy phase respectively. Sample type F was a commercially produced alumina composite with a two-phase microstructure. The grain size of each material was measured from SEM images (JEOL 6300 and Carl Zeiss EVO LS15) of a polished and thermally etched cross-section of each sample using the linear intercept method. The density was measured by the Archimedes method using distilled water.

Before impact testing, the surfaces of the discs were lapped and polished to a 3 μm finish using diamond pastes. This was required to minimise noise in the optical fluorescence spectra obtained post-impact testing.

High strain rate impact was carried out by firing sharpened projectiles at polished ceramic discs using a gas gun (Figure 1). Tungsten carbide (Tungstcarb products, grade K20) rods were sharpened using a diamond engraving tool to produce sharp projectiles similar to the cores of armour piercing bullets. Tip sharpness was checked using 3D optical microscopy (Alicona InfiniteFocus) and projectiles were rejected if their tip diameter was greater than 200 μm. All alumina discs were glued to a thick piece of alumina to avoid failure by bending at the back surface and mounted to a maraging steel bar of similar diameter to the samples using a small amount of vacuum grease. The sharpened tungsten carbide projectile was attached to a nylon sabot and fired towards the samples over a distance of approximately 2 m using a gas gun. The sabot and the projectile were separated before impact by means of a splitter. The velocity was measured using light gates and was adjusted by control of the gas pressure to approximately 100 m/s. Tests were filmed using a high speed camera (Phantom 7, Vision Research, USA). At least 2 samples were impacted of each alumina type.

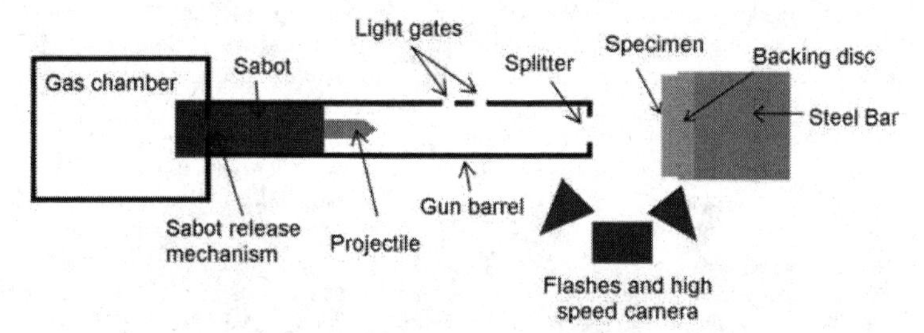

Figure 1: Setup of dynamic impact testing apparatus

Following dynamic testing, the impacted surfaces of the specimens were examined by 3D optical microscopy (Alicona InfiniteFocus), and scanning electron microscopy (JSM6300, tungsten filament and Carl Zeiss EVO LS15, LaB_6 filament). Cr^{3+} fluorescence peak positions were mapped as a function of position on the face of the specimen using a 1000 series Renishaw Raman microscope with a 50mW He-Ne laser and an automated X-Y stage (Prior, UK). To determine the unstrained reference peak positions, spectra were taken at four positions on the sample at the maximum possible distance from the centre and averaged. All images were taken in non-confocal settings using a 50x objective lens, numerical aperture of 0.75 and working distance of 370 μm. Spectra were acquired from the surface of the specimen including the crater region at a spacing of 50 μm in x- and y-directions and were fitted using GRAMS/32 software (Galactic Industries Corporation, UK). Samples were then cross-sectioned through the indent region, polished and examined by fluorescence microscopy and SEM.

RESULTS AND DISCUSSION

MACROSCOPIC OBSERVATIONS OF DAMAGE

Most of the alumina discs remained intact after impact testing. Impact of the projectile was confirmed from high speed video images. In addition an indentation mark in the centre of the disc was always visible following successful tests. A minority of the discs were completely fractured into several large pieces. Radial cracks originating from the impact site were visible on the surfaces of several discs. These observations are analysed systematically in Table I, which gives a qualitative ranking of the performance of each alumina sample. The composition of sample F is commercially sensitive so its glass content is not given.

Table I: Ranked performance of alumina specimens under impact by tungsten carbide materials

Rank	Sample Type	Number of radial cracks per sample	Number of completely fractured discs	Mean Grain Size (μm)	Glass Content (%)
1	B	0	0	2.0	0
2	F	~4	0	1.7	-
=3	C	~5	0	1.4	0
=3	A	~5	0	6.8	0
5	E	~5	1	4.7	2
6	D	~7	1	8.3	5

From Table I it is evident that the best performance was for those materials with the finest grain sizes. Of the pure alumina specimens, materials with high glass content suffered more damage than purely crystalline materials. Sample F, which is an alumina composite material, had similar performance to pure alumina specimens with similar grain size. This indicates that, in general, fine-grained crystalline materials sustain the least damage in this test. It is interesting to note from the SEM images of the three materials with the smallest average grain sizes, that those materials (B and F) which perform best contain a range of grain sizes, while Sample C has a more uniform grain size throughout (Figure 2).

The projectile velocity varied between 76-121 m/s with an average of 101 m/s due to the method of accelerating the projectile. No systematic variation of performance with velocity was observed.

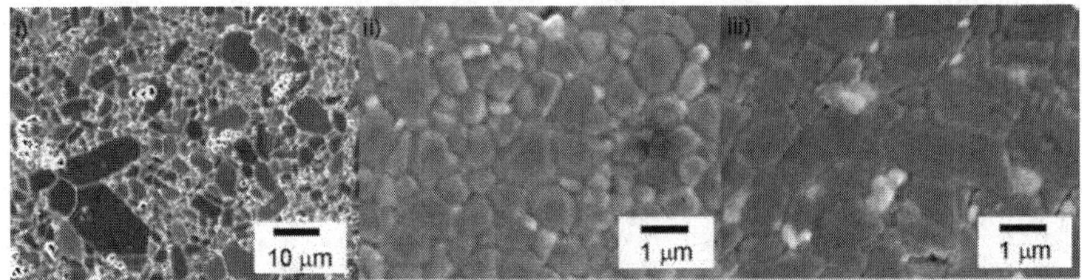

Figure 2: SEM images of samples of type i) B, ii) C and iii) F, illustrating the differences in their grain size distributions.

MICROSCOPIC OBSERVATIONS OF DAMAGE

A 3D optical microscope (Alicona Infinite Focus) was used to assess the size and depth of the crater formed by the impact of the projectile on each sample. This microscope takes images of the surface at a range of working distances and combines them to form a 3D image. Figure 3 shows images of selected samples as 2D projections. A profile of the height change across the crater region is shown for each image. By measuring the maximum height change from these profiles for a number of lines across the crater, an average crater depth could be calculated. The diameter of the crater was measured from the images, estimating the crater edge position from the depth profiles and visual inspection of the optical micrographs.

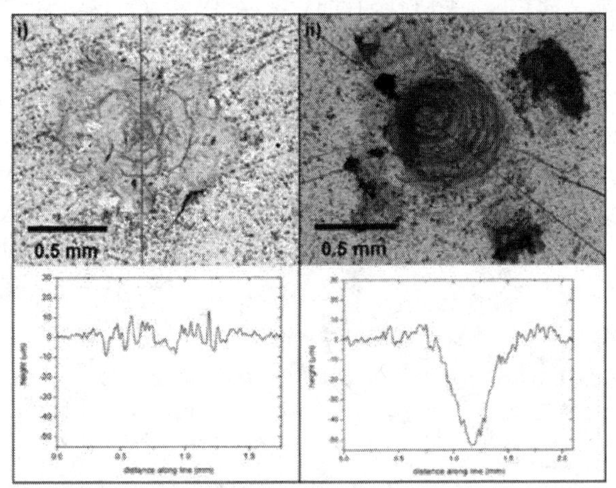

Figure 3: 2D projections of 3D optical micrographs for selected samples characteristic of each observed behaviour type, with the profile measured along the line running across each image. i) Sample type C, showing a relatively flat and shallow indentation characteristic of sample types A-C and F. ii) Sample type D with a deep indentation characteristic of the more glassy materials D and E.

The crater shape and size was found to vary according to the material composition. The pure alumina samples (A, B and C) with zero glass content had average crater depth of around 10 µm and average diameter of approximately 1.3 mm. The alumina composite sample (F) had a similar shallow crater type. The samples with significant glass content (D and E) had deeper and more variable crater depths of 15-40 μm and diameter of approximately 1.6 mm. These craters are much deeper than those for the pure alumina materials, and it is possible that the crater is not solely due to plastic deformation and compaction, but that material has additionally been lost during impact.

SEM images of the indented surface of the samples revealed that the region in and around the indentation contains significant micro-cracking (Figure 4). These micro-cracks were observed in all specimens. The micro-cracks are observed to follow some grain boundaries, though not all, and have no preferential direction, suggesting a biaxial stress state existed in the surface at some time during the dynamic impact testing. The region containing micro-cracks had a diameter of 0.75-1.5 mm, and the edge of the micro-cracked zone corresponds to the start of radial cracks where such cracks were observed.

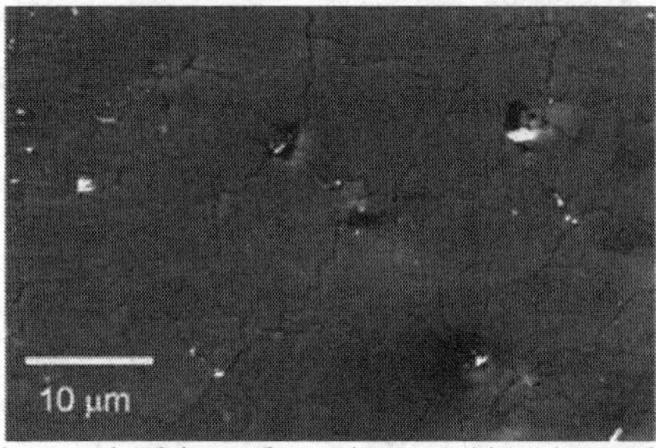

Figure 4: Typical SEM micrograph of the surface micro-cracking observed in the impacted region of all samples. This micrograph was taken from a sample of type A.

Following characterisation of the surface, samples were cross-sectioned through the indented region using a diamond saw and polished to allow the examination of the region below the indentation (Figure 5). The sub-surface region immediately below the impact site closely resembled the plastic zone observed below Vickers indentations[13]. The size of this plastic zone was larger for the glassy materials D and E than for the other samples. In all cases the micro-cracked region on the surface (described above and shown in Figure 4) had a larger diameter than the sub-surface plastic zone indicating that the micro-cracked region on the surface is not merely the top surface of the sub-surface plastic zone but rather is due to a different stress state in the bulk to the surface.

In addition to the sub-surface plastic zone, cracks running in various directions were observed in this cross-section. Selected images are shown in Figure 5 to illustrate the different crack behaviours. Cracks were observed running perpendicular to the surface of the sample (Figure 5i), and in other samples at angles of up to 78° to the surface (Figure 5ii). In some samples, the crack direction changes from approximately perpendicular to the surface to a lateral direction (Figure 5iii), indicating a change occurred in the stress state during the impact testing. It is interesting to note that the performance with the best resistance to impact has a noticeably different cross-section. The damaged region is spread more widely across the cross-section, appearing to consist of two separate regions with associated cracks underneath (Figure 5iv), such that the overall damage to any one region is lessened, which seems likely to contribute to the ability of this alumina type to withstand impact effectively with minimal damage.

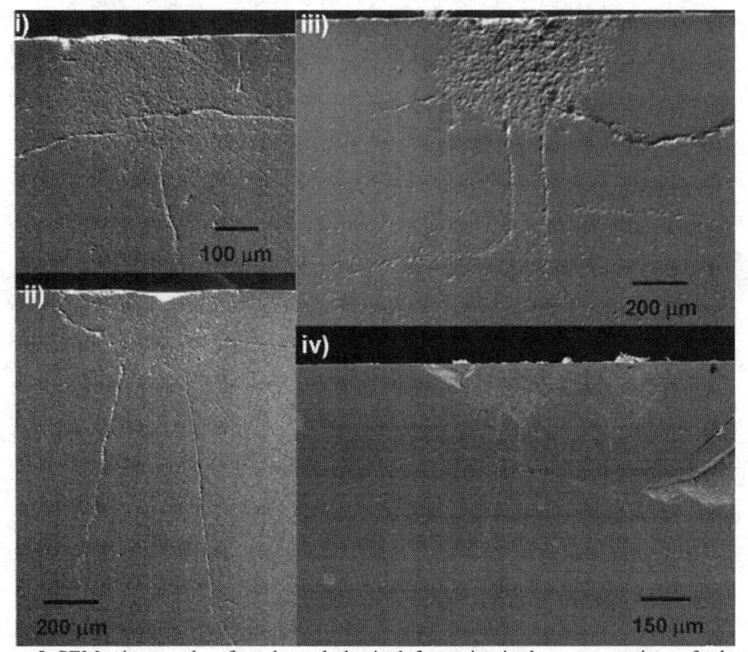

Figure 5: SEM micrographs of cracks and plastic deformation in the cross-sections of selected impacted samples. Samples were chosen to illustrate particular characteristic crack patterns. i) Perpendicular and lateral cracks in Sample type F, ii) Non-perpendicular cracks in Sample type D , iii) Perpendicular cracks changing to lateral cracks in sample type A, iv) Complex widespread crack pattern with two origins in sample type B

CR^{3+} FLUORESCENCE MAPPING

Cr^{3+} fluorescence maps of the impact sites on intact samples have a number of features in common. We first examine the maps of peak position shift, which indicates the degree of residual stress in the material (Figure 6). Immediately around the impact site, a region of negative peak position shifts corresponding to compressive residual stress is observed. This gradually relaxes to the neutral peak position of the bulk of the material. Small tensile (positive position shift) regions can be observed in many samples at a certain distance away from the impact site. By comparison with optical and SEM micrographs we can see that these tensile regions correspond to the tips of radial cracks. The persistence of the crack tip stress field after the test shows that the crack continues to be loaded by the residual stresses resulting from inelastic deformation of the specimen.

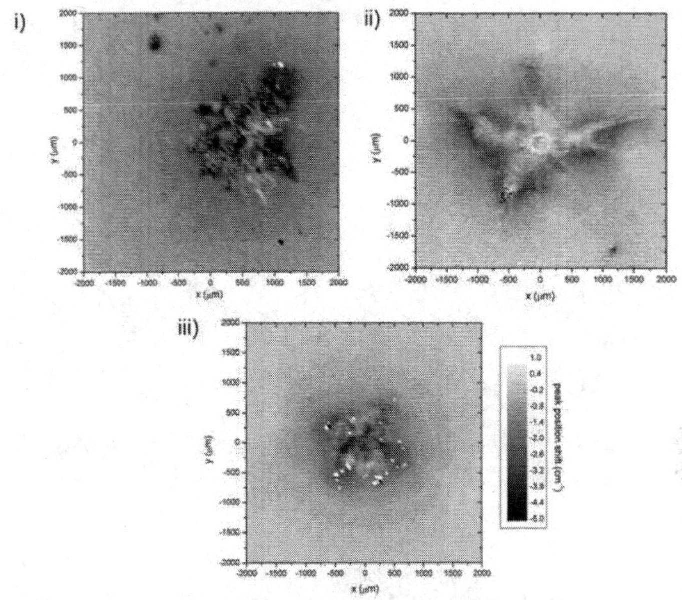

Figure 6: Cr^{3+} fluorescence peak position shift maps of the impacted surfaces of selected samples, chosen to illustrate each type of residual stress distribution observed.
i) sample type B, ii) sample type D, iii) sample type F.

The peak position shifts (Δv_{R1}) measured in these samples are in the range -5 to +1 cm^{-1}, corresponding to a hydrostatic stress σ_H range of -659 to +132 MPa as calculated from:

$$\Delta v_{R1} = \Pi_H \sigma_H \qquad (1)$$

where Π_H is the hydrostatic loading pressure coefficient, taken as 7.59 cm^{-1} GPa^{-1} for a polycrystalline alumina with grain size smaller than the sampling volume, as is the case here[14]. The zero position for the peak position shift calculation was assessed individually for each specimen by taking the average of the peak position measured at 4 points around the edge of each sample. The size of the region of compressive residual stress was larger for the glassy materials D and E than for the crystalline materials produced from TM-DAR alumina (A and C) or the alumina composite (F). However in the samples produced from AES11c (B) the compressive region was much larger, so clearly mean grain size and glass content are not the only factors which affect the size of the compressive zone.

The maps of the fluorescence peak width change images (Figure 7) show fewer differences between samples than the peak position shift maps. In general the size of the affected region is smaller. This is expected because the width change records the size of the plastically deformed zone, rather than the larger zone affected by the residual stress given by the fluorescence peak position maps which includes also the region deformed only elastically. This agrees with observations described earlier regarding the relative sizes of the micro-cracked surface zone versus the sub-surface plastically deformed zone. Specimen type B shows a larger plastic zone than the other specimens examined, with a larger total area and more numerous regions of significant broadening. This agrees with the observations from microscopy that the damaged region is spread more widely in this specimen than the others, which is likely to be related to its superior resistance to impact. Samples A, C and F have similarly sized plastic zones and similar degrees of peak broadening. Sample type D which has the highest glass content shows a very intense peak broadened region at the centre of the impact region. Sample E is similar to D, but with lower values of peak broadening and a smaller intense region.

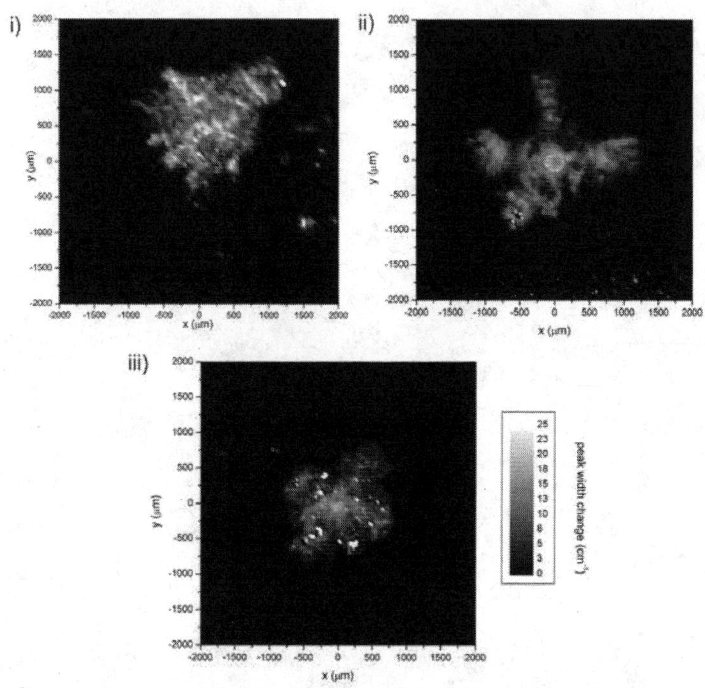

Figure 7: Cr^{3+} fluorescence peak width change maps of the impacted surfaces of selected samples, i) A ii) D and iii) B, chosen to illustrate each type of residual stress distribution observed. This data is taken from the same discs as shown in Figure 6. i) sample type B, ii) sample type D, iii) sample type F.

In addition to Cr^{3+} measurements on the impacted surfaces, the polished cross-sections were examined to give further insight into the likely stress-states in the samples during impact testing. Figures 8 and 9 show the results of these measurements for selected samples. The sectioning process will cause significant stress redistribution in the cross-section; therefore it is not appropriate to directly compare the quantitative values of the residual stress between the maps for the impacted surface (Figure 6) and those for the cross-section. However a number of features are of interest. The largest degree of tensile stress is observed in the region directly below the impact point. The value of the peak position shift decreases away from the indentation, and eventually become compressive (negative peak position shift) (Figure 8). The size of the region where peak broadening is observed (Figure 9) corresponds exactly to that of the observed sub-surface plastic zone observed by SEM (Figure 5). As for the surface, greater peak broadening is observed for the two glassy materials (D and E) than for the crystalline materials, indicating that these materials underwent a significant degree of plastic deformation.

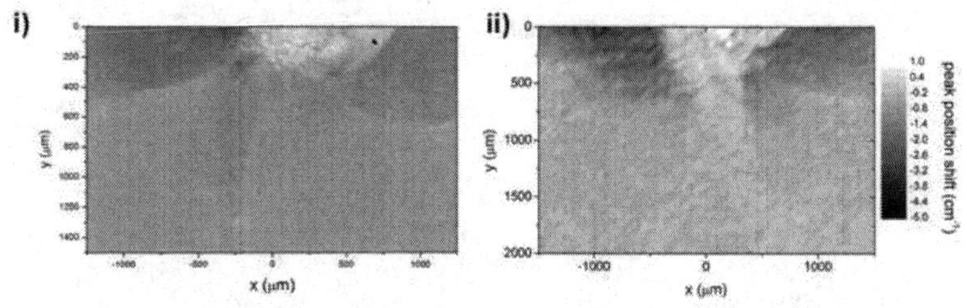

Figure 8: Cr^{3+} fluorescence peak position shift maps of the cross-section under the impact site of selected samples, chosen to illustrate each type of residual stress distribution observed. i) sample type B, ii) sample type D.

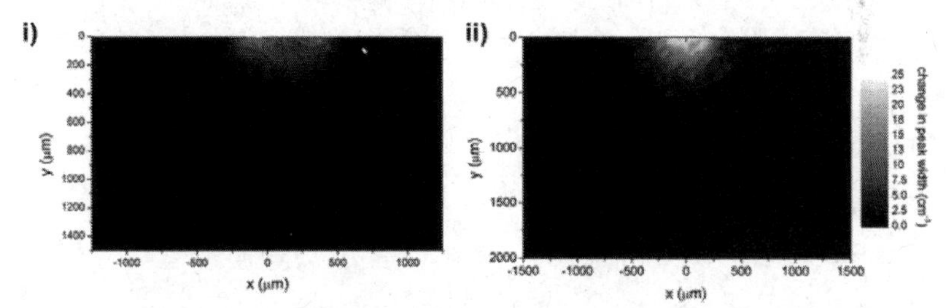

Figure 9: Cr^{3+} fluorescence peak width change maps of the cross-section under the impact site of selected samples. i) sample type B, ii) sample type D.

The sub-surface pullout zone observed in the polished cross sections of indents (Figure 5) is due to the presence of plasticity-induced micro-cracks, by comparison with work on Vickers hardness indentations[11]. For the latter case, this intense micro-cracking is nucleated only within the plastic zone, beneath the indenter[13]. A similar situation is observed here for the region beneath the WC projectile, as minimal peak broadening is observed outside the pullout zone diameter (as measured from SEM images). The width of the pullout zone is somewhere between the tip radius (estimated as ~20-150 μm) and the maximum width of the projectile body (3 mm). This is probably because the projectile is blunted during contact with the ceramic. The pullout/plastic zone shapes observed by SEM (Figure 5) and Cr^{3+} fluorescence microscopy (Figure 9) do not agree with the hemispherical form known for sharp indentation[15] or the spherical form attributed to blunt indentation[16], rather being somewhere between the two. The lower values of fluorescence peak broadening recorded in these dynamic indentations (Figures 7 and 9) compared to quasi-static Vickers indentations[11] are probably because of the indenter shape discrepancy, as the degree of plasticity observed in Vickers (sharp pyramidal) indentations is greater than in Hertzian (spherical or flat-ended cylindrical) indentations.

CONCLUSIONS

We have studied the response of a range of alumina materials to dynamic impact at sub-ballistic speeds by a novel impact test using sharpened tungsten carbide projectiles. By characterising the response of the alumina specimens, a ranked order of performance has been developed and compared to material properties including glass content and grain size. The best performance was achieved by a purely crystalline alumina with relatively small average grain size, but a relatively large grain size distribution rather than a uniform grain size.

The residual stresses and plastic deformation in discs which remained intact after dynamic testing were measured spatially over the surface by Cr^{3+} fluorescence mapping of the indented region. These were compared to micrographs of the damage which revealed micro-cracking in the surface over a wide region and, in the cross-section, median or slightly angled cracks often turning to a lateral direction. The highest degree of plasticity without fracture was observed for samples with the highest glass content (5%). These findings will be useful for the design of materials with optimal properties for use in armour applications.

ACKNOWLEDGEMENTS

This research was carried out as part of the Understanding and Improving Ceramic Armour (UNICAM) Project. We gratefully acknowledge funding from EPSRC and the Ministry of Defence, UK. We also thank Laurie Walton, Richard Duffin, Stuart Carter and Dr Kalin Dragnevski for their assistance with dynamic impact testing experiments.

REFERENCES

[1] Viechnicki D J, Slavin M J and Kliman M I (1991) Ceramic Bulletin 70 pp.1035-1039
[2] Sujirote K, Dateraksa K and Chollacoop N (2007) Am. Ceram. Soc. Bull. v.86 pp.20-25
[3] Chen W W, Rajendran A M, Song B and Nie X (2007) J. Am. Ceram. Soc. v.90 pp.1005-1018
[4] Woodward R L, Gooch Jr W A, O'Donnell R G, Perciballi W J, Baxter B J and Pattie S D (1994) International Journal of Impact Engineering 15 pp.605-618
[5] O'Donnell R G (2001) J Mater. Sci. v.10 p.685-688
[6] Luo H and Chen W (2004) Int. J. Appl. Ceram. Technol. 1 pp.254-260
[7] Muson D E and Lawrence (1979) J. Appl. Phys. 50 pp.6272-6282
[8] Bourne N K, Millett J C F, Chen M, McCauley J W and Dandekar D P (2007) J. Appl. Phys. v.102 p.073514
[9] Lankford J, Predebon W W, Staehler J M, Subhash G, Pletka B J and Anderson C E (1998) Mechanics of Materials 29 p.205-218
[10] Molis S E and Clarke D R (1990) J. Am. Ceram. Soc. v.73 pp.3189-94
[11] Guo S and Todd R I (2010) submitted to Acta Materialia
[12] Wu H Z, S G Roberts and B Derby (2008) Acta Materialia v.56 pp.140-149
[13] Limpichaipanit A and Todd R I (2009) J. Eur. Ceram. Soc. v.29 pp.2841-2848
[14] He J and Clarke D R (1995) J. Am. Ceram. Soc. v.78 pp.1347-1353
[15] Yoffe E H Philos. Mag. (1982) v.40 pp.617-628
[16] Iyer K A International Journal of Fracture (2007) v.146 pp.1-18

MESOSCALE MODELING OF DYNAMIC FAILURE OF CERAMIC POLYCRYSTALS

J.D. Clayton
U.S. Army Research Laboratory
RDRL-WMP-B
Aberdeen Proving Ground, MD 21005-5066
USA
john.d.clayton1@us.army.mil

R.H. Kraft
U.S. Army Research Laboratory
RDRL-WMP-B
Aberdeen Proving Ground, MD 21005-5066
USA
reuben.kraft@us.army.mil

ABSTRACT

Mesoscale models are used to study dynamic deformation and failure in silicon carbide (SiC) and aluminum oxynitride (AlON) polycrystals. Elastic and anisotropic elastic-plastic crystal models represent mechanical behavior of SiC and AlON and grains, respectively. Cohesive zone models represent intergranular fracture. Failure data that can be used to inform macroscopic continuum models of ceramic behavior are collected and analyzed. Studied are effects of grain morphology, specimen size, and applied stress state on behavior of polycrystalline aggregates loaded dynamically at applied strain rates on the order of 10^5/s. Results for SiC demonstrate shear-induced dilatation, increasing shear strength with increasing confinement or pressure, increasing strength with decreasing specimen size (in terms of number of grains), and decreasing strength variability with decreasing size. Results for AlON demonstrate increased initiation of slip activity—particularly in the vicinity of constrained grain boundaries—with confinement.

1. INTRODUCTION

Ballistic performance of a ceramic depends on a number of factors associated with the material response to high loading rates and high pressures that arise during impact [1-3]. The performance difference of two monolithic ceramic materials of comparable mass density necessarily originates from microstructure: crystal structure and composition, grain morphology, grain boundaries, and defects such as pores, inclusions, and secondary phases. Crystal structure and composition affect bulk mechanical properties such as elastic stiffness, hardness, cleavage strength, dislocation slip resistance, twinning resistance, and possible phase transformations. Grain boundaries and defects can affect failure properties such as fracture toughness and spall strength.

Mesoscale modeling, wherein geometries of individual grains are resolved explicitly, offers insight into effects of microstructure on dynamic performance of polycrystalline solids. Previous modeling efforts on ceramics have often considered only two spatial dimensions [4-6]. The present representation is three-dimensional, extending work of Kraft et al. [7] and Gazonas et al. [8]. In addition to AlON addressed previously [7, 8], the present work considers SiC (specifically SiC-N, primarily consisting of the 6H polytype). AlON and SiC exhibit some noteworthy physical differences. SiC (α-phase, 6H) belongs to a hexagonal crystal system, while AlON (γ-phase) is cubic. Polycrystalline SiC is opaque, while polycrystalline AlON is transparent. SiC has a lower mass density and higher fracture strength and toughness than AlON, with a typical grain size smaller than that of AlON by a factor of ~40. Upon optical examination, AlON [9] appears to have fewer processing defects (e.g., voids and inclusions) than SiC-N [10]. Both can exhibit limited plasticity under loading involving confining pressure: SiC by propagation of partial dislocations (and associated stacking faults) on {0001} basal planes [11, 12], and AlON by glide of partial dislocations on {111} octahedral planes [13, 14].

In addition to providing insight of effects of microstructure on performance, mesoscale modeling can yield information to motivate or parameterize macroscopic models that do not explicitly resolve features of the microstructure. Brittle materials are known to exhibit a size effect, wherein smaller samples are often stronger, typically assumed a result of a lower probability of containment of a critical flaw. Numerical failure modeling of brittle materials is also prone to mesh sensitivity. One representative macroscopic model that addresses issues of failure statistics and size effects is Kayenta [15, 16]. Statistics gathered from mesoscale computations can provide input to Kayenta failure surfaces. Mesoscale simulations can address small specimen sizes comparable to finite element sizes used in macroscopic representations, and can consider homogeneous boundary conditions not easily applied in standard characterization tests for failure of brittle materials. Limitations of the present approach are that polycrystalline samples comparable in size to those tested experimentally cannot be simulated directly due to computing constraints (particularly for SiC with its small grain size) so that validation of mesoscopic computations becomes difficult. Furthermore, a number of microscopic properties are not known precisely (e.g., grain boundary fracture properties), and their effects on macroscopic response must be estimated through extensive parameter studies [4-6, 17].

This paper is organized as follows. Section 2 summarizes constitutive models for bulk behavior of single crystals of SiC and AlON. Section 3 summarizes fracture models for interfaces. Section 4 discusses microstructures represented in mesoscale finite element simulations of ceramic polycrystals. Sections 5 and 6 report on results for SiC and AlON, respectively. Section 7 concludes the paper.

2. CERAMIC SINGLE CRYSTALS: CONSTITUTIVE MODELING

Principles of continuum mechanics are used to represent mechanical behavior of individual grains within a ceramic polycrystal. Notational conventions of nonlinear continuum mechanics [18] are used, with all vectors and tensors referred to a single fixed Cartesian coordinate system. Thermal effects are not considered. Local balances of mass, linear momentum, angular momentum, and energy are

$$\rho_0 = \rho J\,,\ \nabla\boldsymbol{\cdot}\boldsymbol{P} = \rho_0\ddot{\boldsymbol{x}}\,,\ \boldsymbol{P}\boldsymbol{F}^T = \boldsymbol{F}\boldsymbol{P}^T\,,\ \dot{W} = \boldsymbol{P}:\dot{\boldsymbol{F}}\,, \tag{1}$$

where ρ_0 is the reference mass density, $\boldsymbol{F} = \nabla\boldsymbol{x} = \boldsymbol{1} + \nabla\boldsymbol{u}$ is the deformation gradient, ∇ is the referential gradient operator, $\boldsymbol{1}$ is the unit tensor, $\boldsymbol{x}(\boldsymbol{X},t) = \boldsymbol{X} + \boldsymbol{u}$ are spatial coordinates with $\boldsymbol{u}$ the displacement, $J = \det\boldsymbol{F}$, $\boldsymbol{P}$ is the first Piola-Kirchhoff stress, and W is the energy density. The deformation gradient and plastic velocity gradient can be written, respectively, as

$$\boldsymbol{F} = \boldsymbol{F}^E\boldsymbol{F}^P\,,\ \boldsymbol{F}^P\dot{\boldsymbol{F}}^{P-1} = \textstyle\sum_i \dot{\gamma}^i\boldsymbol{s}^i\otimes\boldsymbol{m}^i\,, \tag{2}$$

where $\boldsymbol{F}^E$ accounts for elasticity and rigid rotation, $\boldsymbol{F}^P$ accounts for plastic slip, and $\dot{\gamma}^i$, $\boldsymbol{s}^i$, and $\boldsymbol{m}^i$ are the slip rate, slip direction, and slip plane normal for slip system i. Relations in (2) can be extended to account for deformation twins [14, 19, 20] that may be of importance in AlON [9]; discussion of twinning models is omitted here for brevity. Internal energy density is

$$W = \hat{W}(\boldsymbol{E}^E) + f(\xi)\,,\ \boldsymbol{E}^E = (1/2)(\boldsymbol{F}^{ET}\boldsymbol{F}^E - \boldsymbol{1})\,, \tag{3}$$

where $\hat{W}$ depends on elastic strain $\boldsymbol{E}^E$ via second-order elastic constants and possibly higher-order elastic coefficients or pressure-sensitive second-order coefficients. Stress $\boldsymbol{P}$ and Cauchy pressure p are

$$\boldsymbol{P} = \boldsymbol{F}^E(\partial W/\partial\boldsymbol{E}^E)\boldsymbol{F}^{P-T}\,,\ p = -[1/(3J)]\mathrm{tr}(\boldsymbol{F}\boldsymbol{P}^T)\,. \tag{4}$$

In (3), function f accounts for contributions from defects such as dislocations, vacancies, stacking faults, and twin boundaries, and ξ is a generic internal state variable representing such defects. In the absence of slip or defects, and when pressure dependence of elastic coefficients is omitted, (3)-(4) become the usual relations of a Kirchhoff-St. Venant hyperelastic solid with second-order moduli $\boldsymbol{C}$:

$$W = (1/2)\boldsymbol{E}:\boldsymbol{C}:\boldsymbol{E},\ \boldsymbol{E} = (1/2)(\boldsymbol{F}^T\boldsymbol{F} - \boldsymbol{1}),\ \boldsymbol{P} = \partial W/\partial \boldsymbol{F} = \boldsymbol{F}(\partial W/\partial \boldsymbol{E}) = \boldsymbol{F}(\boldsymbol{C}:\boldsymbol{E}). \quad (5)$$

In the geometrically linear approximation, (5) is replaced with the familiar linear elasticity relations

$$W = (1/2)\boldsymbol{\varepsilon}:\boldsymbol{C}:\boldsymbol{\varepsilon},\ \boldsymbol{\varepsilon} = (1/2)[\nabla\boldsymbol{u} + (\nabla\boldsymbol{u})^T],\ \boldsymbol{P} = \boldsymbol{P}^T = \partial W/\partial\boldsymbol{\varepsilon} = \boldsymbol{C}:\boldsymbol{\varepsilon} = \boldsymbol{C}:\nabla\boldsymbol{u}. \quad (6)$$

Kinetic equations for slip and possible strain hardening are of the functional form

$$\dot{\gamma}^i = \dot{\gamma}^i(\tau^i/\tau_c),\ \tau_c/G = g(\xi),\ \dot{\xi} = \dot{\xi}(\dot{\gamma}^i), \quad (7)$$

with τ^i the resolved Kirchhoff stress on shear system i, τ_c the shear strength that is assumed equal for all slip systems, G the shear modulus that can depend on pressure, and g a scalar function of dislocation density. More detailed forms of (7) are available elsewhere [14, 19, 20].

Mechanical properties are listed in Table 1 for the two ceramics studied in the present work. Supporting references with extensive lists of properties are available for SiC [12] and AlON [14]. Cubic elastic constants for AlON are available only from first-principles calculations [7, 8, 21] and are subject to uncertainty, since values quoted have not been validated through experiments. Properties of SiC are representative of SiC-N manufactured by BAE [3]. Properties of AlON are representative of a standard composition having 35.7 mol % AlN [22]. Single crystals are elastically anisotropic; listed bulk and shear moduli and their pressure derivatives at the reference state are representative values for fully dense polycrystals. The initial dynamic slip resistance τ_c, in each material approximately 2% of effective shear modulus G, reflects the experimentally measured polycrystalline shear strength under impact conditions at or above the HEL (strain rate ~10^5/s, pressure >10 GPa) [11, 14, 23].

Table 1: Bulk mechanical properties of SiC and AlON

Property	SiC	AlON
Structure	6H	spinel
Phase	α	γ
Crystal system	hexagonal	cubic
Mass density ρ_0	3227 kg/m^3	3714 kg/m^3
Elastic constant C_{11}	501 GPa	377 GPa
Elastic constant C_{12}	112 GPa	133 GPa
Elastic constant C_{44}	161 GPa	125 GPa
Elastic constant C_{13}	52 GPa	$(= C_{12})$
Elastic constant C_{33}	549 GPa	$(= C_{11})$
Bulk modulus B	222 GPa	214 GPa
Shear modulus G	194 GPa	124 GPa
$B' = dB/dp$	3.10	4.20
$G' = dG/dp$	0.90	0.95
Primary slip plane	{0001}	{111}
Strength τ_c/G	0.022	0.019
Typical grain size	5 μm	200 μm

3. GRAIN BOUNDARY INTERFACES: FRACTURE MODELING

A cohesive zone model is used to represent intergranular (i.e. grain boundary) fracture. First consider mode I fracture. Crack opening initiates when resolved normal traction σ on an interface exceeds fracture strength σ_c. A simple cohesive law of the following form [7, 8] then relates crack opening displacement δ and normal traction σ:

$$\sigma = \sigma_c(1-\delta/\delta_c)\ [\text{for } 0<\delta/\delta_c<1];\ \sigma = 0\ [\text{for } \delta/\delta_c \geq 1]. \tag{8}$$

In the context of linear elastic fracture mechanics, fracture toughness K_c, surface energy Γ_c, strain energy release rate G_c, strength σ_c, critical separation δ_c, and cohesive zone length l_c are related by [17, 24]

$$K_c^2(1-\nu^2)/E = 2\Gamma_c = G_c = (1/2)\sigma_c\delta_c,\ l_c = \pi E\Gamma_c/[\sigma_c^2(1-\nu^2)], \tag{9}$$

where elastic modulus $E = 9BG/(3B+G)$ and Poisson's ratio $\nu = (3B-2G)/(6B+2G)$. Properties are listed in Table 2 for SiC [3, 16] and AlON [25], estimated from flexure strength and static fracture toughness measurements. For comparison, spall strengths of SiC [26] and AlON [27] are also listed in Table 2; spall strength values are comparable to flexure strengths, though spall strengths are known vary considerably with impact pressure. Relations analogous to (8) and (9) are used to describe the relationship between shear traction and tangential crack opening (i.e., mode II/III fracture), with equivalent properties. In the finite element implementation of the fracture model, after complete interfacial separation ($\delta/\delta_c > 1$) occurs, interactions between grains are addressed via a multi-body contact algorithm in the SIERRA software framework [28] that prohibits interpenetration of material. Post-fracture contact is assumed frictionless.

More sophisticated cohesive laws with strength or toughness depending on mode mixity, loading rate, and/or temperature are available, and have been examined in detail elsewhere [17, 24, 29]. Transgranular (i.e., cleavage) fracture within individual grains is not modeled in the present work. The assumption that dynamic fracture occurs predominantly at grain boundaries is appropriate for compositions of SiC (e.g., SiC-N) of present interest that fail statically and dynamically in an intergranular fashion [10, 30] and was followed in a previous mesoscale model of spall fracture in SiC [6]. The assumption that dynamic fracture in AlON occurs predominantly at grain boundaries was used by Kraft et al. [7, 8], though recovered samples from dynamic compression experiments suggest cleavage occurs on {111} planes [9, 13]. A two-dimensional numerical model of transgranular fracture in alumina has been implemented elsewhere [4]; a three-dimensional capability is unavailable in the finite element software [28] used in the present work for materials of current interest. All grain boundaries within a sample of a given material are assumed to have the same properties [6-8]. Fracture strength and/or toughness distributions are expected to exist in real materials (resulting from lattice misorientation distributions and grain boundary defects) and can affect computed macroscopic mechanical properties and failure statistics [4, 5, 17]. As-processed polycrystalline SiC and AlON contain defects such as limited porosity, second-phase inclusions, and small micro-cracks [10, 22]; explicit incorporation of such defects in mesoscale models will be investigated in future work. The present study focuses on effects of grain morphology, bulk material behavior (e.g., crystal structure, elasticity, and plasticity), applied stress state, and specimen size (i.e., number of grains).

Table 2: Fracture properties of SiC and AlON

Property	SiC	AlON
Fracture strength σ_c	570 MPa	306 MPa
Fracture toughness K_c	5.1 MPa m$^{1/2}$	2.5 MPa m$^{1/2}$
Surface energy Γ_c	28.1 J/m^2	9.4 J/m^2
Critical opening δ_c	0.197 μm	0.122 μm
Cohesive length l_c	125.8 μm	104.8 μm
Spall strength σ_s	0.54 GPa -1.3 GPa	0.14 GPa -1.7 GPa

4. MICROSTRUCTURAL REPRESENTATIONS

The procedure for generating grain geometries, employing methods and software developed by Rollett and colleagues [31] is discussed in detail elsewhere [7]. Two finite element meshes, each of a different microstructure, are considered here, as shown in Figure 1. Pertinent properties of each microstructure are listed in Table 3. Cumulative grain size distributions are shown in Figure 2; microstructure I displays a smaller variation in grain size than microstructure II, with more grains of size closer in size to the average grain size. Microstructure I features larger, more equiaxed grains, of more uniform size. Microstructure II features smaller grains, with more jagged grain boundaries. Microstructure I is deemed qualitatively representative of AlON (large equiaxed grains, smooth clean grain boundaries), while microstructure II is more representative of SiC-N (smaller grains, less uniformity, rough boundaries associated with grain boundary segregants). Mesh refinement is sufficient to fully resolve cohesive lengths and grain boundary surface morphology.

Figure 1: Finite element meshes (left) microstructure I, 50 grains (right) microstructure II, 126 grains [microstructures generated in collaboration with A.D. Rollett, Carnegie Mellon University]

Table 3: Finite element meshes of ceramic microstructures

Microstructure	No. grains	No. elements	Volume [mm^3]	Grain boundaries
I	50	1593788	8.0	smooth
II	126	1133743	1.0	rough

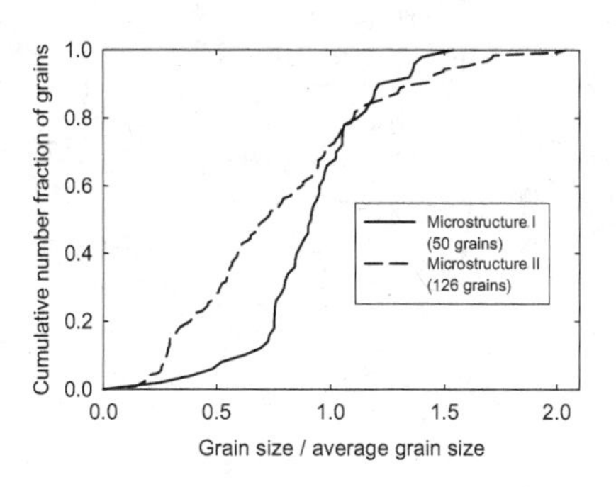

Figure 2: Normalized cumulative grain size distributions

5. SILICON CARBIDE: ELASTICITY AND DYNAMIC FRACTURE

A number of simulations using both microstructures (i.e., meshes) listed in Table 3 were conducted using properties of SiC listed in Tables 1 and 2. Simulations of SiC reported here prescribe linear elastic bulk behavior described by (6), with isotropic properties. Boundary conditions are assigned as follows. The microstructure (cube) is fixed along one surface (i.e., one face of the cube), and shear displacement is applied to the opposite face. Lateral sides remain traction-free. One of three additional conditions is simultaneously applied to the sheared face: (i) the face is left free to expand in the normal direction (referred to as free or unconfined shear), (ii) the face is fixed in the normal direction (referred to as fixed or confined shear), or (iii) the face is displaced in the normal direction causing simultaneous compression and shear (referred to as compression + shear). In all cases, the applied shear deformation rate $\dot{\gamma} = 10^5/\text{s}$. For compression + shear loading, the applied compressive strain rate is also 10^5/s. Microstructures are assigned an initial velocity gradient that matches the applied boundary conditions so as to minimize elastic shocks that would arise if nodal velocities were increased from zero in a stepwise manner; however, release waves do originate from traction-free lateral bounaries. Simulations are listed in Table 4; both microstructures are considered. Different simulations for the same microstructure and boundary condition are delineated by shear loading in different directions. For example, forward and reverse loading on one of two orthogonal directions on each of three orthogonal faces of the cube provides up to twelve simulation cases.

Table 4: Numerical simulations of SiC microstructures

Simulation #	Microstructure	Boundary condition
1-4	I	free (unconfined) shear
5-8	I	fixed (confined) shear
9-12	I	compression + shear
13-24	II	free (unconfined) shear
25-36	II	fixed (confined) shear
37-48	II	compression + shear

Representative results from six simulations (one representative of each row in Table 4) are shown in Figure 3. Contours of the shear stress component work conjugate to the loading mode are shown for deformed meshes at an applied shear γ of 5%. Figure legends are truncated to best display the stress distribution; local maxima and minima at stress concentrations may exceed upper and lower bounds of the legends. Displacements are magnified to highlight trends in dynamic fracture behavior. The following results are noteworthy. Stress magnitude tends to increase with increasing confinement, as stress states progress in severity from (a) to (c) to (e) and (b) to (d) to (f). Microstructure I (50 grains) tends to support larger shear stresses, while fractures seem more pronounced for microstructure II (126 grains). For microstructure II, some grains are ejected from the specimen as fragmentation proceeds. For free shear boundary conditions (Figure 3(a) and (b)), shear stresses have relaxed except near a few critical locations at contact surfaces, and significant bending and dilatation are evident along the top (free surface), the latter phenomenon reminiscent of shear-induced porosity described by Shockey et al. [2] and Curran et al. [32].

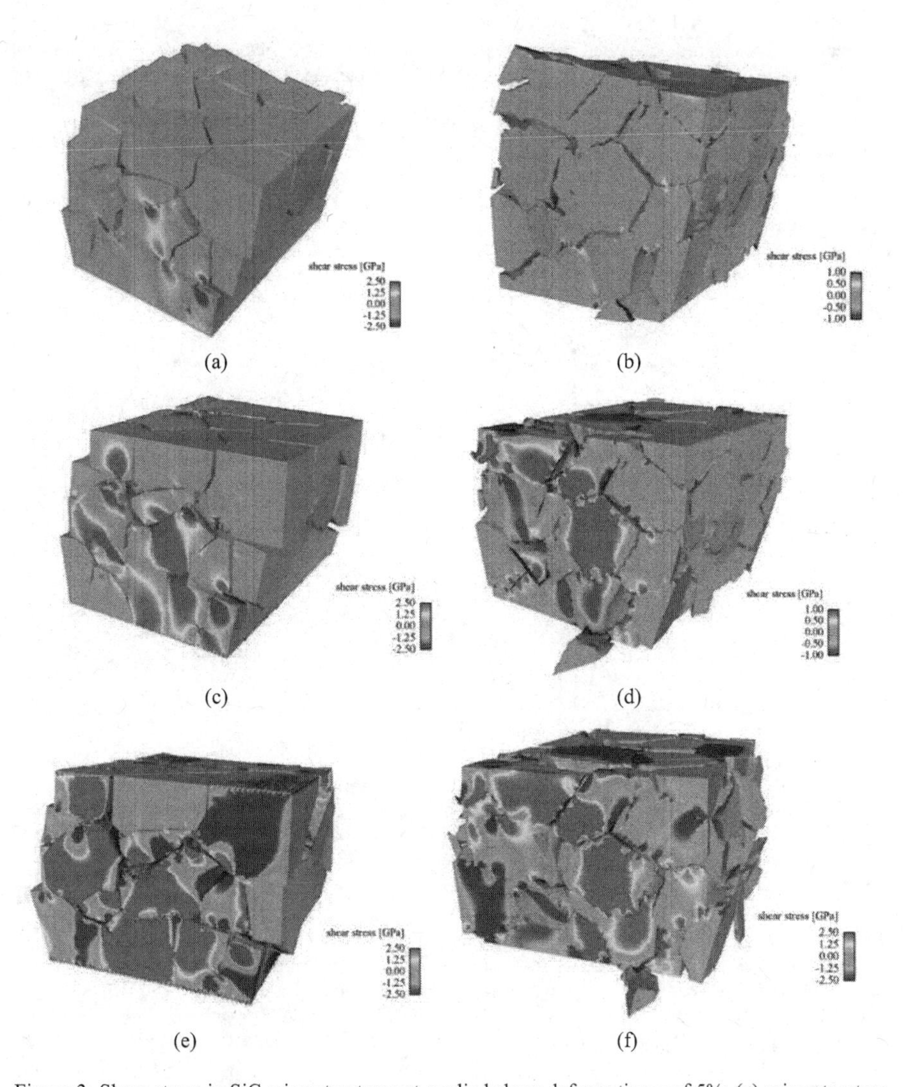

Figure 3: Shear stress in SiC microstructures at applied shear deformation γ of 5%: (a) microstructure I, free shear (b) microstructure II, free shear (c) microstructure I, fixed shear (d) microstructure II, fixed shear (e) microstructure I, compression + shear (f) microstructure II, compression + shear [displacement magnified 5× for (a)-(d); magnified 2× for (e)-(f)]

Figure 4 shows average applied shear stress τ and normal stress Σ, obtained from respective nodal reaction forces tangential and normal to the sheared face of the specimen. Results shown correspond to the six cases in Figure 3. Shear strength and normal stress both increase with confinement and further increase with applied compression. Normal stress is negligible for cases corresponding to free shear boundary conditions. Oscillations in shear stress arise as various fracture surfaces are activated, grains slide relative to one another, and sliding is impeded by constraints of neighboring grains or boundary conditions (for cases involving confinement). It is suggested that strength and pressure are greater in microstructure I because fewer, larger grains are favorably oriented for intergranular fracture, and hence fewer fracture sites are available to accommodate the imposed shear deformation. For the same reason, increased dilatation (i.e., bulking) would be expected for samples with fewer, larger grains since crack paths to achieve percolation across whole aggregates would be more tortuous.

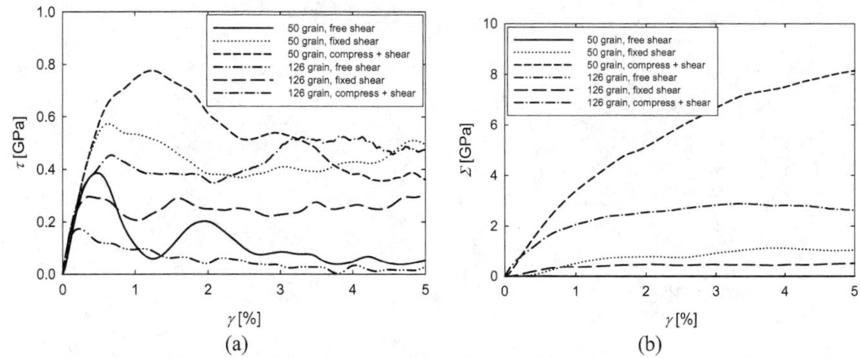

Figure 4: Average shear stress (a) and average normal stress (b) versus applied shear deformation for simulations of SiC microstructures shown in Figure 3

Figure 5(a) shows peak average shear strength $\sqrt{J_2}$ of each aggregate versus average pressure p. Results shown are for all simulations in Table 4 for which stable solutions were obtained. The peak shear strength was obtained at an applied strain level corresponding to the first local maxima in the simulation's average τ-γ curve (e.g., Figure 4(a)), i.e., when $\partial\tau/\partial\gamma = 0$. Strength is computed via

$$J_2 = (\Sigma^2 + 3\tau^2)/3. \tag{10}$$

For both microstructures, shear strength appears to increase linearly with pressure. Slopes of each linear fit are nearly identical, but the strength intercept at zero pressure is lower for microstructure II. Extrapolation of each curve to null shear provides average hydrostatic tensile strengths of 0.23 GPa and 0.10 GPa for microstructures I and II, respectively. Average strengths can be lower than prescribed cohesive strengths because local fractures initiate and propagate early in the simulations as a result of mixed-mode loading (e.g., combined bending and shear) and inertial effects (e.g., release wave interactions) arising at high loading rates. Though the present results are limited in scope, a decrease in strength with increase in size (measured by number of grains) is apparent from Fig. 5(a). This trend of decreasing strength with increasing size has been reported for static flexure (ring crack) and Hertzian indentation experiments on SiC [33]. The present results suggest that qualitatively similar trends may apply for smaller aggregates of material deformed dynamically at much higher rates.

Figure 5(b) shows peak average strength versus applied shear strain for all cases with free shear boundary conditions (i.e., no confining pressure). Stress relaxation at lateral boundaries and the cohesive zone-contact model increase the apparent compliance of the aggregate relative to that of an elastic element of material deformed in simple shear (dotted line). Microstructure I (50 grains) exhibits a larger average strength and lower variation in strength (in terms of standard deviation) than microstructure II (126 grains). Static experiments [33] have similarly demonstrated higher characteristic strength and lower variation (i.e., higher Weibull modulus) as the sample size is decreased. Possible correlation between experimentally measured static strength projected to a low failure probability and ballistic performance has been noted [34]; however, correlation of performance with grain size or hardness was not verified in those experiments. The present results suggest that a microstructure containing fewer, larger grains would provide superior dynamic shear strength, with less variation in failure strength, than would a microstructure with smaller, less equiaxed grains, were all other properties (e.g., elastic and interfacial behavior) held fixed. The former class of microstructure would have fewer grain boundary planes favorably oriented for fracture initiation and crack extension and percolation. However, the present computations omit the possibility of grain cleavage. A high purity SiC microstructure with larger grains would be more prone to transgranular fracture than would a microstructure with smaller grains and grain boundary impurities (e.g., SiC-N) that promote crack deflection and toughening [30, 35].

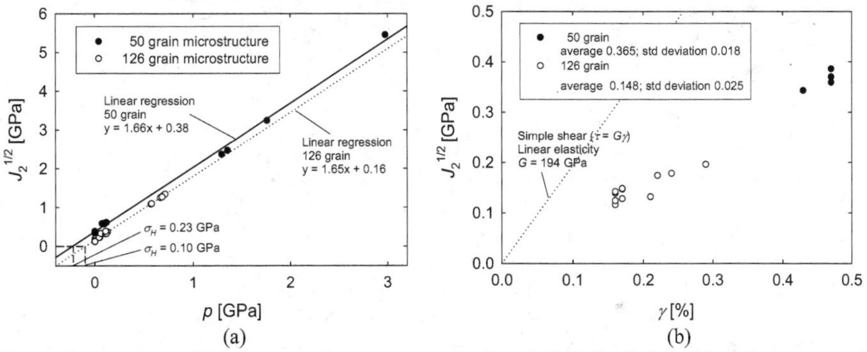

Figure 5: Average peak shear strength versus pressure for all SiC simulations (a) and average peak shear strength versus applied deformation for SiC simulations subjected to free (unconfined) shear (b)

6. ALON: ELASTICITY, PLASTICITY, AND DYNAMIC FRACTURE

A geometrically nonlinear anisotropic elastic-plastic model [14] is used to represent behavior of AlON single crystals, corresponding to (2)-(4) and (7) and Tables 1 and 2. The isothermal version of the model used here considers only the primary slip mode: <110>{111} dislocation glide, with up to twelve active, signed slip systems. Numerical implementation follows Dingreville et al. [36], modified to account for pressure-dependent cubic elastic coefficients. Results of two simulations are reported, as listed in Table 5. Boundary and initial conditions correspond to those discussed already for SiC polycrystals in Section 5. Stability and computational cost restrict the magnitude of deformation achieved in each simulation to ~0.5%, which correlates with initiation of plastic yield.

Table 5: Numerical simulations of AlON microstructures

Simulation #	Microstructure	Boundary condition
1	I	free (unconfined) shear
2	I	fixed (confined) shear

Figure 6 shows that increased plastic slip activity—particularly in the vicinity of constrained grain boundaries—occurs in conjunction with confinement associated with fixed shear boundary conditions. Slip initiates due to stress concentrations from intergranular incompatibility (e.g., elastic anisotropy) and contact interactions. Figure 7 shows average shear and normal stresses computed from nodal reaction forces along the sheared (upper) face. Normal stress is negligible for free shear as the upper face is free to expand, but increases in response to bending, nonlinear elasticity (e.g., a Poynting effect [18]), and dilatation resisted by the fixed boundary. Shear stress and pressure for fixed shear loading begin to increase with loading, relative to the free boundary case, as early as 0.2 % shear deformation.

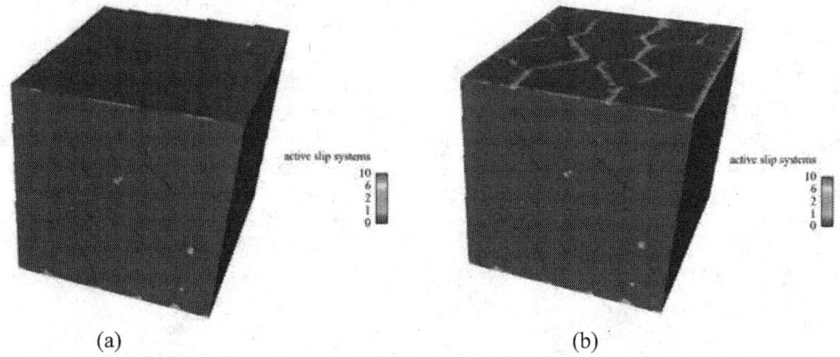

Figure 6: Number of active slip systems in AlON polycrystal deformed under (a) free shear and (b) fixed (confined) shear [shear deformation γ of 0.45%; displacement magnified 20×]

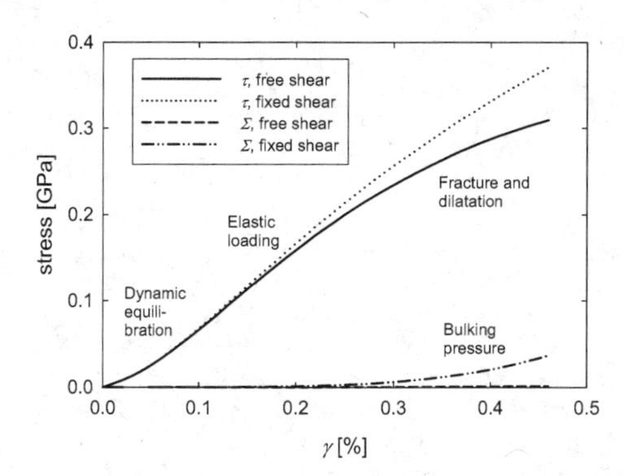

Figure 7: Average shear stress τ and normal stress Σ versus shear deformation for AlON polycrystal

7. CONCLUSIONS

Mesoscale simulations have been conducted for SiC and AlON polycrystals subjected to high strain rate shear loading (10^5/s) with varying confinement or superposed compression. Results of numerous simulations of SiC, idealized with linear elastic bulk behavior and uniform grain boundary fracture properties, suggest the following trends: (i) shear strength increases linearly with confining pressure and (ii) average shear strength is greater and statistical variation in strength lesser for microstructures with fewer, larger, more uniformly sized grains relative to those with more, smaller, less uniformly sized grains. Pressure-dependent strength is correlated with shear-induced dilatancy, and such pressure dependence is significant despite the assumption of frictionless post-fracture grain boundary sliding. Variations in failure result only from grain geometry since constitutive properties are uniform. Results of dynamic simulations on AlON demonstrate increased plastic slip initiation with increasing confinement, particularly in the vicinity of grain boundaries along confined surfaces.

REFERENCES

[1] J. Sternberg, 1989. Material properties determining the resistance of ceramics to high velocity penetration. *Journal of Applied Physics* 65: 3417-3424.

[2] D.A. Shockey, A.H. Marchand, S.R. Skaggs, G.E. Cort, M.W. Burkett, R. Parker, 1990. Failure phenomenology of confined ceramic targets and impacting rods. *International Journal of Impact Engineering* 9: 263-275.

[3] J.C. LaSalvia, J. Campbell, J.J. Swab, J.W. McCauley, 2010. Beyond hardness: ceramics and ceramic-based composites for protection. *JOM* 62, 16-23.

[4] R.H. Kraft, J.F. Molinari, 2008. A statistical investigation of the effects of grain boundary properties on transgranular fracture. *Acta Materialia* 56: 4739-4749.

[5] R.H. Kraft, J.F. Molinari, K.T. Ramesh, D.H. Warner, 2008. Computational micromechanics of dynamic compressive loading of a brittle polycrystalline material using a distribution of grain boundary properties. *Journal of the Mechanics and Physics of Solids* 56: 2618-2641.

[6] J.W. Foulk, T.J. Vogler, 2010. A grain-scale study of spall in brittle materials. *International Journal of Fracture* 163: 225-242.

[7] R.H. Kraft, I. Batyrev, S. Lee, A.D. Rollett, B. Rice, 2010. Multiscale modeling of armor ceramics. In: J.J. Swab (ed.), *Ceramic Engineering and Science Proceedings: Advances in Ceramic Armor VI*, John Wiley and Sons, Hoboken, NJ, pp. 143-158.

[8] G.A. Gazonas, J.W. McCauley, R.H. Kraft, B.M. Love, J.D. Clayton, D. Casem, B. Rice, I. Batyrev, N.S. Weingarten, B. Schuster, 2010. Multiscale modeling of armor ceramics: focus on AlON. In: *Proceedings of 27th Army Science Conference*, Orlando, FL, Nov. 29-Dec. 2.

[9] J.W. McCauley, P. Patel, M. Chen, G. Gilde, E. Strassburger, B. Paliwal, K.T. Ramesh, D.P. Dandekar, 2009. AlON: a brief history of its emergence and evolution. *Journal of the European Ceramic Society* 29: 223-236.

[10] M.Y. Lee, R.M. Brannon, D.R. Bronowski, 2005. Uniaxial and triaxial compression tests of silicon carbide ceramics under quasi-static loading condition. *SAND 2004-6005*, Sandia National Laboratories, NM.

[11] D. Zhang, M.S. Wu, R. Feng, 2005. Micromechanical investigation of heterogeneous microplasticity in ceramics deformed under high confining stresses. *Mechanics of Materials* 37: 95-112.

[12] J.D. Clayton, 2010. Modeling nonlinear electromechanical behavior of shocked silicon carbide. *Journal of Applied Physics* 107: 013520.

[13] B. Paliwal, K.T. Ramesh, J.W. McCauley, M. Chen, 2008. Dynamic compressive failure of AlON under controlled planar confinement. *Journal of the American Ceramic Society* 91: 3619-3629.

[14] J.D. Clayton, 2011. A nonlinear thermomechanical model of spinel ceramics applied to aluminum oxynitride (AlON). *Journal of Applied Mechanics* 78: 011013.

[15] R.M. Brannon, A.F. Fossum, O.E. Strack, 2009. Kayenta: theory and user's guide. *SAND 2009-2282*, Sandia National Laboratories, NM.

[16] R.B. Leavy, R.M. Brannon, O.E. Strack, 2010. The use of sphere indentation experiments to characterize ceramic damage models. *International Journal of Applied Ceramic Technology* 7: 606-615.

[17] H.D. Espinosa, P.D. Zavattieri, 2003. A grain level model for the study of failure initiation and evolution in polycrystalline brittle materials. Part I: theory and numerical implementation. *Mechanics of Materials* 35: 333-364.

[18] A.C. Eringen, 1962. *Nonlinear Theory of Continuous Media*, McGraw-Hill, New York.

[19] J.D. Clayton, 2009. A continuum description of nonlinear elasticity, slip and twinning, with application to sapphire. *Proceedings of the Royal Society of London A* 465, 307-334.

[20] J.D. Clayton, 2010. Modeling finite deformations in trigonal ceramic crystals with lattice defects. *International Journal of Plasticity* 26: 1357-1386.

[21] I.G. Batyrev, B.M. Rice, J.W. McCauley, 2009. First principles calculations of nitrogen atomic position effects on elastic properties of aluminum oxynitride (AlON) spinel. In: *Materials Research Society Fall Meeting*, Boston, MA, Nov. 30-Dec. 1, paper LL5.3.

[22] E.K. Graham, W.C. Munly, J.W. McCauley, N.D. Corbin, 1988. Elastic properties of polycrystalline aluminum oxynitride spinel and their dependence on pressure, temperature, and composition. *Journal of the American Ceramic Society* 71, 807-812.

[23] D.P. Dandekar, B.A.M. Vaughan, W.G. Proud, 2007. Shear strength of aluminum oxynitride. In: M. Elert, M.D. Furnish, R. Chau, N. Holmes, J. Nguyen (eds.), *Shock Compression of Condensed Matter 2007*, American Institute of Physics, pp. 505-508.

[24] J.D. Clayton, 2005. Modeling dynamic plasticity and spall fracture in high density polycrystalline alloys. *International Journal of Solids and Structures* 42: 4613-4640.

[25] N.D. Corbin, 1989. Aluminum oxynitride spinel: a review. *Journal of the European Ceramic Society* 5: 143-154.

[26] D.P. Dandekar, P.T. Bartowski, 2001. Spall strengths of silicon carbide under shock loading. In: K.P. Staudhammer, L.E. Murr, M.A. Meyers (eds.), *Fundamental Issues and Applications of Shock-Wave and High-Strain-Rate Phenomena*, Elsevier, New York, pp. 71-77.

[27] J.U. Cazamias, P.S. Fiske, S.J. Bless, 2001. Shock properties of AlON. In: K.P. Staudhammer, L.E. Murr, M.A. Meyers (eds.), *Fundamental Issues and Applications of Shock-Wave and High-Strain-Rate Phenomena*, Elsevier, New York, pp. 181-188.

[28] J. Jung, 2010. Presto 4.16 user's guide. *SAND 2010-3112*, Sandia National Laboratories, NM.

[29] J.D. Clayton, 2006. Plasticity and spall in high density polycrystals: modeling and simulation. In: M.D. Furnish, M. Elert, T.P. Russell, C.P. White (eds.), *Shock Compression of Condensed Matter 2005*, American Institute of Physics, pp. 311-314.

[30] C.J. Shih, V.F. Nesterenko, M.A. Meyers, 1998. High-strain-rate deformation and comminution of silicon carbide. *Journal of Applied Physics* 83: 4660-4671.

[31] A.D. Rollett, P. Manohar, 2004. Monte Carlo modeling of grain growth and recrystallization. In: D. Raabe, F. Roters (eds.), *Continuum Scale Simulation of Engineering Materials*, Wiley, p. 855.

[32] D.R. Curran, L. Seaman, T. Cooper, D.A. Shockey, 1993. Micromechanical model for comminution and granular flow of brittle material under high strain rate application to penetration of ceramic targets. *International Journal of Impact Engineering* 13: 53-83.

[33] A.A. Wereszczak, T.P. Kirkland, K.T. Strong, J. Campbell, J.C. LaSalvia, H.T. Miller, 2010. Size-scaling of tensile failure stress in a hot-pressed silicon carbide. *International Journal of Applied Ceramic Technology* 7: 635-642.

[34] D. Ray, R.M. Flinders, A. Anderson, R.A. Cutler, J. Campbell, J.W. Adams, 2007. Effect of microstructure and mechanical properties on the ballistic performance of SiC-based ceramics. In: L.P. Franks (ed.), *Ceramic Engineering and Science Proceedings: Advances in Ceramic Armor II*, John Wiley and Sons, Hoboken, NJ, pp. 85-96.

[35] K.T. Faber, A.G. Evans, 1983. Intergranular crack-deflection toughening in silicon carbide. *Journal of the American Ceramic Society* 66: C94-C96.

[36] R. Dingreville, C.C. Battaile, L.N. Brewer, E.A. Holm, B.L. Boyce, 2010. The effect of microstructural representation on simulations of microplastic ratcheting. *International Journal of Plasticity* 26: 617-633.

MULTI-SCALE COMPUTATIONAL INVESTIGATIONS OF SiC/B_4C INTERFACES

Christin P. Morrow, Vladislav Domnich, and Richard A. Haber
Ceramic, Composite, and Optical Materials Center, Rutgers University
Piscataway, New Jersey USA

ABSTRACT

Computational tools are used to cross spatial and temporal scales during investigations of ballistic events on armor materials. The ensuing weakening of these materials after such an event is of particular interest, and recent investigations of boron carbide (B_4C) inclusions in silicon carbide (SiC) showed that weaknesses in these materials did not originate at the grain boundaries. In an effort to characterize the development of structural changes, a multi-scale computational approach is employed. Several samples of varying B_4C concentrations in SiC matrices are studied with density functional theory molecular dynamics (DFT-MD) simulations to determine the stability at the SiC/B_4C interface under several high temperature and pressure regimes. DFT calculations employing molecular clusters to represent single sites along the interface are also presented, and this smaller-scale approach aims to isolate the structural changes that occur at each site and their contribution to the stabilities of these SiC/B_4C samples. These two spatial and temporal scales provide insight into how interface dynamics play a role in the development of structural weaknesses. A comparison with experimental observations is also included.

INTRODUCTION

Background Information

Improvement of material properties and ballistic performance of armor materials is an active area of research. Efficient synthesis methods to maximize performance of these materials are paramount to this improvement. Both SiC and B_4C are used as armor ceramics because of their material properties[1-5] and performance under high temperature and pressure conditions.[1-6] Currently, these materials are evaluated separately despite the use of B_4C sintering aids during synthesis of SiC[1, 2, 6-9] and the development of SiC/B_4C composite materials.[10-16] Therefore, investigation of SiC/B_4C interfaces is necessary to improve the synthesis of SiC and to understand the nature of SiC/B_4C composites. However, no investigation of the SiC/B_4C interface has been performed to date. Further, a molecular scale study is warranted to probe the atomic interactions at the interface. Employment of an atomistic computational approach allows one to focus on a single phenomenon within an overall physicochemical process and to calculate experimentally measured quantities. In particular, the advantage is that a computational approach enables insight into experimental observations.

Silicon carbide (SiC) and boron carbide (B_4C) are desirable materials for armor because of their high hardness, low density, high compressive strength,[5] and resistance to degradation under high temperature and pressure conditions. The SiC structure is comprised of tetra-coordinated Si atoms bonded to tetra-coordinated C atoms, and a number of polytypes are possible that differ in packing of the crystal as well as the number of repeating units along the c-axis.[17] Depictions of the 4H and 6H polytypes via GaussView[18] appear in Figure 1, where the 4H polytype has eight atoms and the 6H polytype 12 atoms.[19] The B_4C structure is currently under debate, but common agreement lies with the presence of a 12 atom icosahedron bonded to a side chain comprised of three atoms.[20-24] A representation of the B_4C structure appears in Figure 2.[21] Recently, DFT calculations were used to determine the relative stabilities of 20 different B_4C isomers[22] and several SiC polytypes,[25] and the comparisons of structures appear in Figure 3. The most stable SiC and B_4C species are employed in this work.

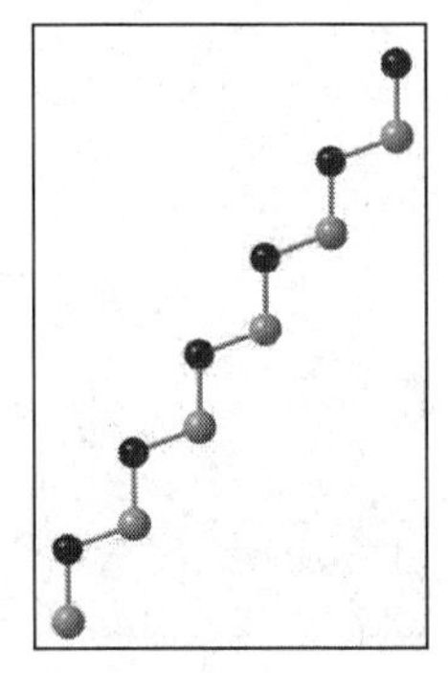

Figure 1: The 4H polytype (left) and the 6H polytype (right) of SiC. The Si atoms are turquoise, and the C atoms are black.

Interfaces of these two materials are found in SiC samples made with B_4C sintering aids[1, 2, 6-9] and SiC/B_4C composites.[10-16] During processing of dense ceramics, sintering aids such as B_4C are added to SiC.[1, 2, 6-9] The presence of B_4C has been shown to influence the SiC polytype distribution.[1, 2, 7] However, the exact cause of this change has not been explained. In addition, sintering aids have been shown to affect the activation energy of SiC synthesis processes.[7] Therefore, one question that remains is how sintering aids affect the energetic stability of SiC polytypes, and an investigation of SiC/B_4C interfaces provides insight into this phenomenon. In SiC/B_4C composite materials, interfaces between constituent components exist throughout a given sample,[10, 11, 13-16] and their prevalence depends upon the synthesis method.[13, 14] The development of localized B_4C and SiC regions within SiC/B_4C composites has been shown,[13-15] and these SiC/B_4C interfaces can be structured such that SiC forms as layers between B_4C grains or individual fragments within a greater B_4C matrix.[13] Thus, investigations into bonding between these two materials as well as their stability over time will provide insight into how SiC and B_4C components interact within composite materials.

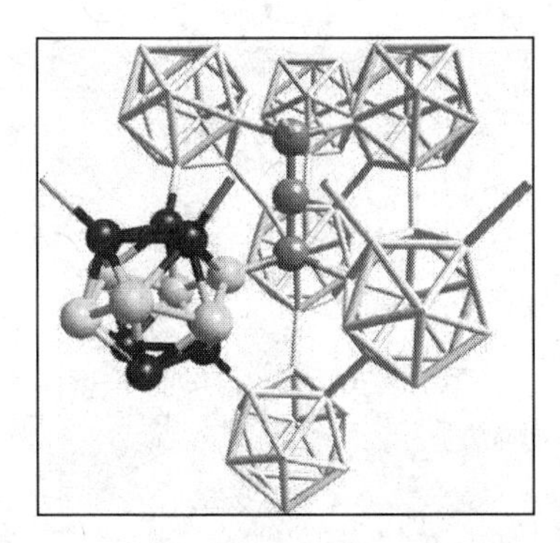

Figure 2: Structure of boron carbide (B_4C). The polar atoms within the icosahedron are black, the equatorial are light gray, and those in the sidechain are dark gray.[21]

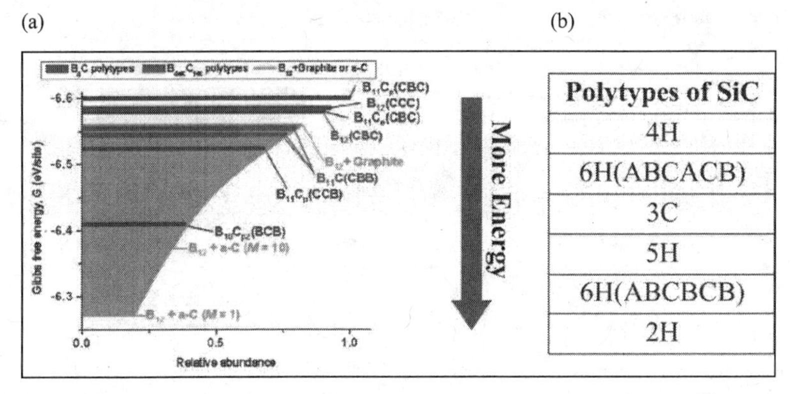

Figure 3: Previously published calculations of the most stable (a) B_4C isomers[22] and (b) SiC polytypes,[25] where energy increases down each series.

Multi-Scale Computational Approach

The strength of atomistic computational methods is the ability to describe a molecular-scale picture of an individual phenomenon within a greater physicochemical scheme. Several temporal and spatial scales are possible using computational tools, and these regimes range from tens of atoms on the femtosecond (fs) scale to millions of atoms without a timescale.[26] The challenge, however, is to include enough chemical information so the calculations provide molecular information while simultaneously representing a system size or timescale that provides meaningful insight. Further, this challenge is compounded by the size of the system that can be studied. Smaller systems are required to include chemical information such as electronic structure,[26] and these calculations are tractable up to ~50 atoms. Larger systems of 10^2 – 10^6 atoms can also be modeled, but the tradeoff is that the electronic structure is not included,[26] which permits these calculations to remain feasible.

A number of methods are available that vary in size of system (both length and number of atoms) and timescale that can be simulated. Figure 4 shows the size and time regimes for a number of computational methods currently employed.[27-32] Starting from the smallest to the largest regime, there are density functional theory (DFT), density functional theory molecular dynamics (DFT-MD), classic molecular dynamics (MD), mesoscale, and continuum theory approaches. DFT models the electronic structure of each atom in a system, while DFT-MD allows for the inclusion of electronic structure as well as the motion of atoms over time. Classical MD also allows for the movement of atoms via Netownian physics, but electrostatic interactions are modeled with fitted potentials, enabling larger system sizes than DFT-MD.[26] In the mesoscale approach, collections of phenomena and molecular regimes are modeled, and lastly, continuum models are used to simulate properties of bulk materials.[33]

Computational investigations across both time and length scales are required for in-depth understanding of armor ceramic materials. As an example, one may consider the impact of a projectile on a target and the ensuing degradation of the material. Simulating such an event would require use of each method in Figure 4, and important information regarding the ballistic event would be given by each. If one traces aspects of a ballistic event from the largest to the smallest regime, the analysis would be as follows. A continuum theory approach would include the simulation of a target with a series of shockwaves moving through it. Moving to the next-smallest regime, a mesoscale approach would also model a shockwave in the target, but now the shockwave is made of individual particles, as opposed to a bulk continuum. In the classical MD approach, the target is made of particles that have a specific identity (SiC). A DFT-MD approach allows for a description of the electronic structure of each Si and C atom while at the same time permitting movement of the Si and C atoms. Lastly, DFT calculations include a quantum mechanical description of the electronic structure of Si and C atoms, and thus, one would have the most accurate structure of polytypes. Throughout the series of methods included in Figure 4, one sacrifices system size for chemical information, but each regime enables prediction of aspects of a ballistic event.

Each of these methods hinges upon describing the system as accurately as possible within the regime simulated. For example, the strength of a classical MD simulation is heavily dependent upon electrostatic potentials that model the atomic interactions effectively. Moreover, computational approaches are also being performed across multiple time and length scales simultaneously. Using multiple tools that overlap in time and size scales allows for identification of which parameters contribute to overall observations. In order for computational methods to have a lasting impact on the armor ceramic materials community, these approaches must be complementary to experiment. Both computational and experimental approaches must be side-by-side endeavors. Each depends upon determining quantities that can be measured both computationally and experimentally. Currently, this is the penultimate approach for accurate and complete description of armor ceramic systems.

To understand SiC/B_4C interfaces, the bonding character of SiC and B_4C as well as the relative stabilities of those interfaces over time and with respect to temperature must first be identified. Molecular scale information of these systems can be delineated via a multi-scale computational approach. This work builds upon that of Fanchini et al. who determined the most stable isomers of B_4C,[22] three of which are used here to model the SiC/B_4C interfaces. The calculations presented are focused on providing insight into SiC/B_4C interfaces that exist at the particle-particle level in SiC samples made with B_4C sintering aids and SiC/B_4C composite materials. As such, they allow for a description of the structure of SiC/B_4C interfaces. These descriptions are important for SiC samples made with B_4C sintering aids, thus having B_4C inclusions and resulting SiC/B_4C interfaces. The B_4C and SiC structures constituting the model interfaces are described, and a discussion of their effects on the interface is given.

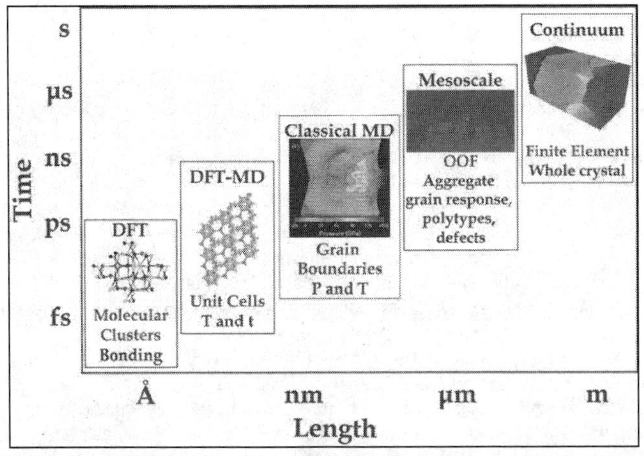

Figure 4: Computational materials science provides tools that cross scales over both length and time.[27-32]

COMPUTATIONAL DETAILS

Density Functional Theory (DFT) Calculations

The main thrusts of the DFT calculations are to optimize SiC unit cells and B_4C isomers as well as to determine how stabilities of interface structures are affected by the presence of various B_4C isomers. The B3LYP functional[34-37] and the 6-31G(d) basis set are coupled as they are included in the Gaussian09 software package.[38] The B3LYP hybrid density functional is comprised of an exchange-correlation functional,[34-37] a generalized gradient approximation (GGA) component,[36, 37] and a contribution of Hartree-Fock exchange.[39] The 6-31G(d) basis set[40, 41] includes *d* orbitals for non-hydrogen elements and enables effective representation of the expanded octets on Si, C, and B atoms. GaussView4W[42] is used to visualize these structures.

Molecular clusters designed to represent SiC unit cells and B_4C isomers are investigated as constituents of SiC/B_4C interfaces. Initially, the three most stable B_4C isomers[20-24] as well as the most stable[25] and commonly observed polytypes[1, 2] of SiC (i. e. 4H and 6H) are energy-minimized. Structures residing at energy minima are characterized by the absence of negative frequencies. These clusters are then compared to previous data to ensure they are accurate representations of B_4C and SiC.

Density Functional Theory Molecular Dynamics (DFT-MD) Simulations

The main goals of the DFT-MD simulations are to transition from single unit cells of SiC and B_4C into bigger systems that enable the investigation of several adsorption sites of B_4C simultaneously and to include experimental parameters such as time and temperature. This computational tool allows for both bond-breaking and bulk scale phenomena to be investigated within one modeling scheme, and such an approach is desirable for the investigation of SiC/B_4C interfaces. Density functional theory molecular dynamics (DFT-MD) simulations are performed with the Vienna Ab-initio Simulation Package (VASP).[43-46] VASP approximates the band structure via a plane wave representation, which is attained by removal of iterative calculations of core electrons from the computation scheme while replicating chemical behaviors of all-electron methods.[47] The Vienna Ab-initio Simulation Package (VASP)[43-46] is employed here because it was designed for materials science applications, particularly in its inclusion of parameter-free functionals.[47, 48] In addition, evaluation of forces on atoms and stresses on the unit cell are possible in this plane-wave approach.[47, 48] Visualization of the DFT-MD simulations is performed with Materials Studio.[49]

The 6H polytype of SiC is used as the substrate in a description of SiC/B_4C interfaces on the nanoscale. The 6H-SiC(ABCACB) substrate is comprised of 9 unit cells for a total of 108 atoms, and this allows for the simulation of multiple surface sites simultaneously. Energy-minimization of the surface enables determination of the surface structure without adsorbates. Thus, any changes that arise once B_4C adsorbates are added can be identified.

RESULTS

Geometry Optimization of B_4C Isomers

Density functional theory (DFT) calculations are employed to energy-minimize the $B_{11}C_pCBC$, $B_{12}CCC$, and $B_{11}C_eCBC$ isomers of B_4C, shown to be the most stable isomers of this stoichiometry.[20-24] Optimized isomers appear in Figure 5, and bond lengths from our calculations are listed in Table I. Previous data from the literature are also included for comparison.[29, 50-52]

Energy-minimization of these structures presents several challenges. Effectively modeling the difference between a polar and an equatorial substitution in the absence of additional icosahedra requires some chemical intuition. The icosahedron is essentially two pentagonal pyramids bonded together,[53] and in the absence of the sidechain, each site within it is chemically equivalent. Defining polar and equatorial sites hinges upon the location of the C substitution with respect to the sidechain. The "polar" substitution is attained by adding the C atom to the three-atom face on the bottom of the icosahedron, as shown in Figure 5a. On the other hand, the "equatorial" substitution is added to the

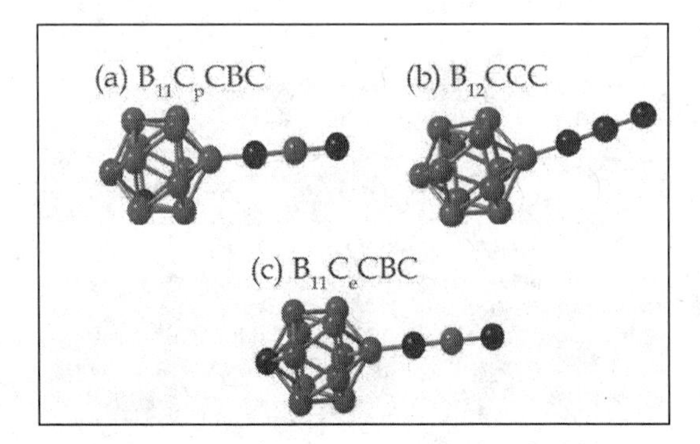

Figure 5: (a) Optimized $B_{11}C_pCBC$, (b) $B_{12}CCC$, and (c) $B_{11}C_eCBC$ structures. The blue atoms are B, and the black are C.

site 180° from the attachment of the sidechain, shown in Figure 5c, and this site is within the hexagonal chair of the icosahedron.[29]

In addition, the $B_{11}C_pCBC$ isomer continually rearranges during optimization. The C substitution within the icosahedron prefers an exo-icosahedral site[52] and migrates to a position where it bonds to four B atoms within the cage. As such, the B–C bond connecting this C atom in its initial position at the base of a pentagonal pyramid to the B atom at the apex breaks each time this structure is optimized without constraints. Therefore, the $B_{11}C_pCBC$ isomer is optimized with constraints on this particular bond. However, the resulting geometry is characterized as an energy minimum because negative frequencies are absent.

Table I: B–B, B–C, and C–C bond lengths (Å) from our B3LYP/6-31G(d) calculations as well as previous works.[29, 50-52]

	Bond Type				
Isomer	B*–B*†	B–B	$B–C_{side}$	$B–C_{subst}$	C–C
$B_{12}CCC$	1.72 – 1.87	1.62 – 1.86	1.49		1.28 – 1.31
$B_{11}C_pCBC$	1.76 – 1.80	1.62 – 1.72	1.35 – 1.52	1.67 – 1.72	
$B_{11}C_eCBC$	1.78	1.65 – 1.68	1.35 – 1.52	1.75 – 1.76	
‡			1.30 – 1.81[a]		
		1.75 – 1.80[b]			
		1.714 – 1.813[c]			
		1.836[d]			
			1.43, 1.59[b]		
			1.409 – 1.788[c]		
					1.328 – 1.336[c]

†B*–B* bonds are between the B bonded to the sidechain and the surrounding B atoms in the icosahedron.
‡Bond lengths are from several possible B_4C isomers. See references for details.
[a]Lazzari et al. 1999.
[b]Balakrishnarajan et al. 2007.
[c]Aydin et al. 2009.
[d]Konovalikhin et al. 2010.

The bond lengths from the B_4C structures in this work fall within the range of those from previous calculations, as listed in Table I. In our calculations, however, we further categorize the B–B bonds as those adjacent to the sidechain (B*–B*) and those remaining throughout the icosahedron (B–B). This is particularly important because those bonds adjacent to the sidechain are more elongated. Shorter B–B bonds are present on those B atoms directly opposite the equatorial B atom to which the sidechain is bonded, likely an effect of excess electron density donated by the sidechain, while those B atoms farther away maintain shorter bond lengths. Moreover, the structure of the icosahedra are distorted such that the positions of the B atom to which the sidechain is attached as well as the atom diametrically opposed from it, either B or C, are not entirely symmetrical.

Geometry Optimization of SiC Unit Cells

As with the energy-minimization of the B_4C isomers, DFT calculations are employed to energy-minimize the unit cells of 4H-SiC and 6H-SiC, and the optimized structures appear in Figure 6. The bond lengths calculated from our calculations are listed in Table II, and previous data from the literature are also included.[54] Initial model clusters comprised solely of the four Si and four C atoms (4H) or six Si and six C atoms (6H) minimize to structures devoid of tetrahedral structural integrity. To overcome this issue, H atoms are added to each Si and C. A vacant site is left on the terminal Si atom in each SiC unit cell for bonding to B_4C.

Although the tetrahedral geometry is maintained, one must consider whether H atoms have a chemical effect. This is particularly important because our goal is to understand the relative energetic stability of each interface modeled. The bond lengths listed in Table II are identical to those from the literature, and thus the presence of these H atoms is not altering the structure. This shows that the H atoms contribute minimally to the energy of each system, which is expected because H atoms have previously been used to terminate SiC clusters.[55]

Addition of H atoms throughout these unit cells change the electronic state from singlet to doublet, which is often challenging from a DFT perspective. The terminal C atoms at the opposite end of the unit cell from where the Si–B bond will occur, shown as the topmost atoms in Figure 6, have three H atoms bonded to them, while each other Si and C atom has two. As such, this leads to an odd number of electrons in the system. However, these model clusters of the SiC unit cells with H atoms in them were successfully energy-minimized with the B3LYP/6-31G(d) methodology. These SiC unit cells will be bonded to the optimized B_4C isomers to form model clusters of SiC/B_4C interfaces.

Nanoscale SiC/B_4C Interfaces

The B_4C and SiC unit cells modeling surface sites will be scaled up in order to fully develop a multi-scale computational approach. Interfaces comprised of multiple surface sites will be simulated to identify long-range, temperature, and pressure effects. As such, a SiC/B_4C interface comprised of multiple B_4C and SiC unit cells will be designed. Molecular level insight will be provided via comparison between these two size and timescales.

Energy-minimization of a 3x3x1 supercell of the 6H-SiC(ABCACB) substrate is currently underway. Experimental coordinates[56] are used to create the substrate, which is visualized in Figure 7. A 3x3x1 sample constitutes a desirable size because there are multiple sites to which B_4C structures can adsorb and because a substrate of ~100 atoms maintains computational feasibility. Once the 6H-SiC(ABCACB) substrate is energy-minimized, isomers of B_4C will be adsorbed to the SiC substrate. These isomers in combination with the 6H-SiC(ABCACB) polytype will be used as a model interface to investigate the bonding and orientation characteristics of B_4C on SiC. Possibilities for the bonding configuration will be evaluated in turn, and the lowest energy configuration will be employed to model the SiC/B_4C interfaces at temperatures that are experimentally relevant.

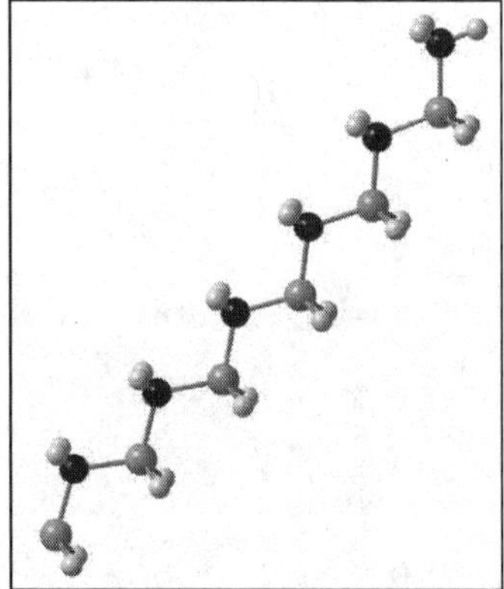

Figure 6: The 4H polytype (left) and the 6H polytype (right) of SiC with H atoms added for structural integrity.[55] The Si atoms are turquoise, the C atoms are black, and the H atoms are light gray.

DISCUSSION

The DFT calculations discussed are centered on understanding bonding between B_4C isomers and Si atoms within SiC unit cells. Those structures examined via DFT calculations will be bonded via one Si–B bond, varying the Si atom within the unit cell. Thus the dependence of the interface stability with respect to bonding within the SiC unit cell will be isolated. In SiC samples sintered with B_4C, SiC is the majority component, and therefore, the current approach has been adopted in these calculations to probe the significance of attachment of B_4C isomers to various Si atoms within the SiC unit cell. Moreover, Si–B bonds are formed in these calculations because our hypothesis is that they would be more stable than Si–C bonds as the C atoms are present as electron donors in these substances; further expanding their octets would be energetically unfavorable.

The expected implications of these DFT calculations pertain to providing insight into the structure and stability of possible SiC/B_4C interfaces. Using these molecular clusters allows us to investigate the stability of bonding B_4C isomers to SiC unit cells. In addition, varying the Si atom to which the B_4C is bonded enables the determination of how the location of the B_4C within the unit cell affects the stability of the interface. From these calculations, we can predict which SiC polytypes and B_4C isomers form the most stable interfaces and thus which of these interfaces are more likely to form overall.

The expected implications of these DFT-MD simulations are primarily to understand what

Table II: Si–C, Si–H, and C–H bond lengths (Å) for the 4H and 6H polytypes of SiC from our B3LYP/6-31G(d) calculations as well as those from Ching et al.[54]

	Bond Type		
Polytype	Si–C	Si–H	C–H
4H	1.89 – 1.90	1.49	1.09 – 1.10
6H	1.89	1.49	1.10
4H	1.890,1.887[a]		
6H	1.890,1.887[a]		

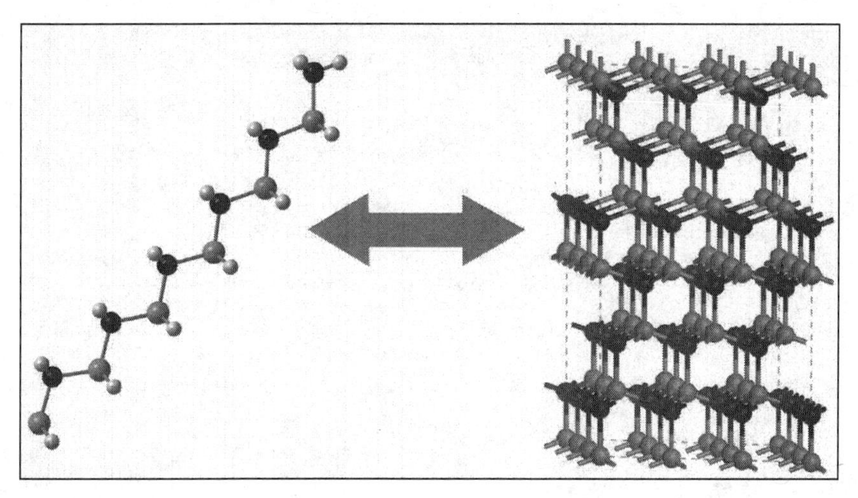

Figure 7: Single unit cell of 6H-SiC (left) is scaled up to a 3x3x1 supercell of 6H-SiC(ABCACB) (right). The color scheme is the same as Figure 6.

[a]Ching et al. 2006.

effects system size, time, and temperature have on SiC/B_4C interfaces. Scaling up the DFT calculations to move from one single Si–B bonding site to several throughout an interface permits the determination of whether separate surface sites behave similarly. In addition, the MD component of DFT-MD simulations allows for the investigation of systems with respect to time and temperature, and therefore, the effects of each of these parameters in turn may also be ascertained relative to structural or energy changes throughout the simulation.

Several future directions are planned upon the successful completion of the research outlined in this manuscript. In addition to varying bonding sites within the SiC unit cell, the energetic dependence of SiC unit cell attachment to polar, equatorial, or sidechain sites within the B_4C structure will be examined. This portion of the study aims to complete the picture of possible bonding sites within model interface structures. Future simulations of the SiC/B_4C interface using 6H-SiC(ABCBCB) and 4H-SiC substrates are also planned. These simulations in conjunction with those outlined in the present proposal represent the most energetically stable constituents possible for SiC/B_4C interfaces, and this systematic approach enables a well-described molecular scale picture of these systems.

CONCLUSIONS

The expected implications of this study relate to the development of a molecular scale description of SiC/B_4C interfaces via employment of multi-scale computational tools. A molecular scale understanding will be gained through use of cluster and unit cell sized models of SiC/B_4C interfaces. In particular, the effect of bonding B_4C structures to various Si atoms within the SiC unit cell on both the geometrical configuration of atoms and the relative stability will be identified. From here, a prediction can be made about which interfaces are likely to form. Increasing the system size from molecular clusters to several unit cells with multiple binding sites provides insight into the long-range effects of Si–B bonding at the interface. Moreover, inclusion of time and temperature parameters allows for the incorporation of more experimental effects into these computational studies.

Computational methodologies facilitate a molecular scale understanding by focusing on a specific phenomenon within a physicochemical process and through the simulation of size- and timescales difficult to attain experimentally.

ACKNOWLEDGEMENTS

The authors gratefully acknowledge support from the NSF IUCRC Ceramic, Composite, and Optical Materials Center and the Army Research Laboratory Materials Center of Excellence via grant DOD-W911NF-06-2-0007 as well as computational resources provided by the National Center for Supercomputing Applications (NCSA) and the TeraGrid clusters via Startup Allocation DMR100115. In addition, Dr. Steven Miller, Dr. Steven Mercurio, and Mr. Daniel Maiorano of the Department of Materials Science and Engineering at Rutgers University are gratefully acknowledged for stimulating discussions, and Dr. Karsten Krogh-Jespersen of the Department of Chemistry at Rutgers University is gratefully acknowledged for assistance with computational strategies. Dr. Alexander Goldberg of Accelrys, Inc. is gratefully acknowledged for assistance with the preparation of the SiC supercell in Materials Studio.

REFERENCES

[1]Ray, D.; Flinders, R. M.; Anderson, A.; Cutler, R. A.; Campbell, J.; Adams, J. W., Effect of Microstructure and Mechanical Properties on the Ballistic Performance of SiC-Based Ceramics. In *Advances in Ceramic Armor II: Ceramic Engineering and Science Proceedings*, Prokurat, L.; Wereszczak, A.; Lara-Curzio, E., Eds. John Wiley & Sons, Inc.: Hoboken, NJ, USA, 2007; Vol. 27, pp 85-96.

[2]Ray, D.; Flinders, R. M.; Anderson, A.; Cutler, R. A.; Rafaniello, W., Effect of Room-Temperature Hardness and Toughness on the Ballistic Performance of SiC-Based Ceramics. In *Advances in Ceramic Armor: Ceramic Engineering and Science Proceedings*, Swab, J. J.; Zhu, D.; Kriven, W. M., Eds. John Wiley & Sons, Inc.: Hoboken, NJ, USA, 2005; Vol. 26, pp 131-142.

[3]Bakas, M.; Greenhut, V. A.; Niesz, D. E.; Adams, J.; McCauley, J., Relationship Between Defects and Dynamic Failure in Silicon Carbide. In *27th Annual Cocoa Beach Conference on Advanced Ceramics and Composites: A: Ceramic Engineering and Science Proceedings*, Kriven, W. M.; Lin, H.-T., Eds. John Wiley & Sons, Inc.: Hoboken, NJ, USA, 2003; Vol. 24, p doi: 10.1002/9780470294802.ch52.

[4]Chen, M. W.; McCauley, J. W.; Hemker, K. J., Shock-induced localized amorphization in boron carbide. *Science* **2003,** 299, (5612), 1563-1566.

[5]David, N. V.; Gao, X. L.; Zheng, J. Q., Ballistic Resistant Body Armor: Contemporary and Prospective Materials and Related Protection Mechanisms. *Appl. Mech. Rev.* **2009,** 62, (5), 050802.

[6]Zhang, X. F.; Yang, Q.; De Jonghe, L. C., Microstructure development in hot-pressed silicon carbide: effects of aluminum, boron, and carbon additives. *Acta Mater* **2003,** 51, (13), 3849-3860.

[7]Ray, D. A.; Kaur, S.; Cutler, R. A.; Shetty, D. K., Effect of additives on the activation energy for sintering of silicon carbide. *J. Am. Ceram. Soc.* **2008,** 91, (4), 1135-1140.

[8]Stobierski, L.; Gubernat, A., Sintering of silicon carbide I. Effect of carbon. *Ceram. Int.* **2003,** 29, (3), 287-292.

[9]Stobierski, L.; Gubernat, A., Sintering of silicon carbide - II. Effect of boron. *Ceram. Int.* **2003,** 29, (4), 355-361.

[10]Magnani, G.; Beltrami, G.; Minoccari, G. L.; Pilotti, L., Pressureless sintering and properties of alpha SiC-B^4C composite. *J. Eur. Ceram. Soc.* **2001,** 21, (5), 633-638.

[11]Shipilova, L. A.; Petrovskii, V. Y., Structure formation, electrophysical and mechanical properties of an electrically conducting ceramic composite based on silicon and boron carbides. *Powder Metall. Met. C+* **2002,** 41, (3-4), 147-149.

[12]Lee, K. S.; Han, I. S.; Chung, Y. H.; Woo, S. K.; Lee, S. W., Hardness and wear resistance of reaction bonded SiC-B4C composite. *Mater. Sci. Forum* **2005,** 486-487, 245-248.

[13]Tkachenko, Y. G.; Britun, V. F.; Prilutskii, E. V.; Yurchenko, D. Z.; Bovkun, G. A., Structure and properties of B_4C-SiC composites. *Powder Metall. Met. C+* **2005,** 44, (3-4), 196-201.
[14]Hayun, S.; Frage, N.; Dariel, M. P., The morphology of ceramic phases in BxC-SiC-Si infiltrated composites. *J. Solid State Chem.* **2006,** 179, (9), 2875-2879.
[15]Hayun, S.; Dariel, M. P.; Frage, N.; Zaretsky, E., The high-strain-rate dynamic response of boron carbide-based composites: The effect of microstructure. *Acta Mater.* **2010,** 58, (5), 1721-1731.
[16]Hayun, S.; Weizmann, A.; Dariel, M. P.; Frage, N., Microstructural evolution during the infiltration of boron carbide with molten silicon. *J. Eur. Ceram. Soc.* **2010,** 30, (4), 1007-1014.
[17]Umeno, Y.; Kinoshita, Y.; Kitamura, T., Ab Initio DFT Study of Ideal Shear Strength of Polytypes of Silicon Carbide. *Strength Mater+* **2008,** 40, (1), 2-6.
[18]Dennington, R., II; Keith, T.; Millam, J. *GaussView, Version 4.1*, Semichem, Inc.: Shawnee Mission, KS, 2007.
[19]Konstantinova, E.; Bell, M. J. V.; Anjos, V., Ab initio calculations of some electronic and elastic properties for SiC polytypes. *Intermetallics* **2008,** 16, (8), 1040-1042.
[20]Vast, N.; Besson, J. M.; Baroni, S.; Dal Corso, A., Atomic structure and vibrational properties of icosahedral alpha-boron and B4C boron carbide. *Comp. Mater. Sci.* **2000,** 17, (2-4), 127-132.
[21]Mauri, F.; Vast, N.; Pickard, C. J., Atomic structure of icosahedral B4C boron carbide from a first principles analysis of NMR spectra. *Phys. Rev. Lett.* **2001,** 87, (8), 085506.
[22]Fanchini, G.; McCauley, J. W.; Chhowalla, M., Behavior of disordered boron carbide under stress. *Phys. Rev. Lett.* **2006,** 97, (3), 35502.
[23]Saal, J. E.; Shang, S.; Liu, Z. K., The structural evolution of boron carbide via ab initio calculations. *Appl. Phys. Lett.* **2007,** 91, (23).
[24]Vast, N.; Sjakste, J.; Betranhandy, E., Boron carbides from first principles. *J. Phys. Conf. Ser.* **2009,** 176, 012002.
[25]Kobayashi, K.; Komatsu, S., First-principles study of BN, SiC, and AlN polytypes. *J. Phys. Soc. Jpn.* **2008,** 77, (8), 084703.
[26]Leach, A. R., *Molecular Modeling: Principles and Applications*. Pearson Prentice Hall: New York, 2001.
[27]FlexFEM Software - Sandia National Laboratory.
[28]Chawla, N.; Patel, B. V.; Koopman, M.; Chawla, K. K.; Saha, R.; Patterson, B. R.; Fuller, E. R.; Langer, S. A., Micro structure-based simulation of thermomechanical behavior of composite materials by object-oriented finite element analysis. *Mater. Charact* **2003,** 49, (5), 395-407.
[29]Balakrishnarajan, M. M.; Pancharatna, P. D.; Hoffmann, R., Structure and bonding in boron carbide: The invincibility of imperfections. *New J. Chem.* **2007,** 31, (4), 473-485.
[30]Becker, R., High level view of multi-scale modeling. In *JOWOG32M*, June 18 2007.
[31]Eker, S.; Durandurdu, M., Pressure-induced phase transformation of 4H-SiC: An ab initio constant-pressure study. *Europhys. Lett.* **2009,** 87, (3), 36001.
[32]Branicio, P. S.; Kalia, R. K.; Nakano, A.; Vashishta, P., Nanoductility induced brittle fracture in shocked high performance ceramics. *Appl. Phys. Lett.* **2010,** 97, 111903.
[33]Buehler, M. J.; Ackbarow, T., Fracture mechanics of protein materials. *Mater. Today* **2007,** 10, (9), 46-58.
[34]Vosko, S. H.; Wilk, L.; Nusair, M., Accurate spin-dependent electron liquid correlation energies for local spin density calculations: a critical analysis. *Can. J. Phys.* **1980,** 58, 1200-1211.
[35]Lee, C.; Yang, W.; Parr, R. G., Development of the Colle-Salvetti correlation-energy formula into a functional of the electron density. *Phys. Rev. B* **1988,** 37, (2), 785-789.
[36]Becke, A. D., A new mixing of Hartree-Fock and local density-functional theories. *J. Chem. Phys.* **1993,** 98, (2), 1372-1377.
[37]Becke, A. D., Density-functional thermochemistry. III. The role of exact exchange. *J. Chem. Phys.* **1993,** 98, (7), 5648-5652.

[38]Frisch, M. J.; Trucks, G. W.; Schlegel, H. B.; Scuseria, G. E.; Robb, M. A.; Cheeseman, J. R.; Scalmani, G.; Barone, V.; Mennucci, B.; Petersson, G. A.; Nakatsuji, H.; Caricato, M.; Li, X.; Hratchian, H. P.; Izmaylov, A. F.; Bloino, J.; Zheng, G.; Sonnenberg, J. L.; Hada, M.; Ehara, M.; Toyota, K.; Fukuda, R.; Hasegawa, J.; Ishida, M.; Nakajima, T.; Honda, Y.; Kitao, O.; Nakai, H.; Vreven, T.; Montgomery, J. A., Jr.; ; Peralta, J. E.; Ogliaro, F.; Bearpark, M.; Heyd, J. J.; Brothers, E.; Kudin, K. N.; Staroverov, V. N.; Kobayashi, R.; Normand, J.; Raghavachari, K.; Rendell, A.; Burant, J. C.; Iyengar, S. S.; Tomasi, J.; Cossi, M.; Rega, N.; Millam, N. J.; Klene, M.; Knox, J. E.; Cross, J. B.; Bakken, V.; Adamo, C.; Jaramillo, J.; Gomperts, R.; Stratmann, R. E.; Yazyev, O.; Austin, A. J.; Cammi, R.; Pomelli, C.; Ochterski, J. W.; Martin, R. L.; Morokuma, K.; Zakrzewski, V. G.; Voth, G. A.; Salvador, P.; Dannenberg, J. J.; Dapprich, S.; Daniels, A. D.; Farkas, Ö.; Foresman, J. B.; Ortiz, J. V.; Cioslowski, J.; Fox, D. J. *Gaussian09, Revision A.01*, Gaussian, Inc.: Wallingford, CT, 2009.
[39]Sousa, S. F.; Fernandes, P. A.; Ramos, M. J., General performance of density functionals. *J. Phys. Chem. A* **2007,** 111, (42), 10439-10452.
[40]Krishnan, R.; Binkley, J. S.; Seeger, R.; Pople, J. A., Self-consistent molecular orbital methods. XX. A basis set for correlated wave functions. *J. Chem. Phys.* **1980,** 72, (1), 650-654.
[41]Clark, T.; Chandrasekhar, J.; Spitznagel, G. W.; von Rague Schleyer, P., Efficient Diffuse Funcion-Augmented Basis Sets for Anion Calculations. III.* The 3-21+G Basis Set for First-Row Elements, Li-F. *J. Comp. Chem.* **1983,** 4, (3), 294-301.
[42]Dennington, R., II; Keith, T.; Millam, J. *GaussView, Version 4.1.2*, Semichem, Inc.: Shawnee Mission, KS, 2007.
[43]Vienna Ab-initio Simulation Package (VASP) Group Page. http://cms.mpi.univie.ac.at/vasp/ .
[44]Kresse, G.; Furthmuller, J., Efficiency of ab-initio total energy calculations for metals and semiconductors using a plane-wave basis set. *Comp. Mater. Sci.* **1996,** 6, 15-50.
[45]Kresse, G.; Furthmuller, J., Efficient iterative schemes for ab initio total-energy calculations using a plane-wave basis set. *Phys. Rev. B* **1996,** 54, 11169-11186.
[46]Kresse, G.; Joubert, D., From ultrasoft pseudopotentials to the projector augmented-wave method. *Phys. Rev. B* **1999,** 59, (3), 1758-1775.
[47]Hafner, J.; Wolverton, C.; Ceder, G., Toward computational materials design: The impact of density functional theory on materials research. *MRS Bull.* **2006,** 31, (9), 659-665.
[48]Hafner, J., Materials simulations using VASP - a quantum perspective to materials science. *Comput. Phys. Commun.* **2007,** 177, (1-2), 6-13.
[49]*Materials Studio*, Accelrys: San Diego, CA.
[50]Lazzari, R.; Vast, N.; Besson, J. M.; Baroni, S.; Dal Corso, A., Atomic structure and vibrational properties of icosahedral B4C boron carbide. *Phys. Rev. Lett.* **1999,** 83, (16), 3230-3233.
[51]Aydin, S.; Simsek, M., Hypothetically superhard boron carbide structures with a B11C icosahedron and three-atom chain. *Phys. Status Solidi B* **2009,** 246, (1), 62-70.
[52]Konovalikhin, S. V.; Ponomarev, V. I., Estimation of the upper limit of carbon concentration in boron carbide crystals. *Russ. J. Phys. Chem. A* **2010,** 84, (8), 1445-1448.
[53]Jemmis, E. D.; Balakrishnarajan, M. M., The ubiquitous icosahedral B-12 in boron chemistry. *B. Mater. Sci.* **1999,** 22, (5), 863-867.
[54]Ching, W. Y.; Xu, Y. N.; Rulis, P.; Ouyang, L. Z., The electronic structure and spectroscopic properties of 3C, 2H, 4H, 6H, 15R and 21R polymorphs of SiC. *Mat. Sci. A-Struct* **2006,** 422, (1-2), 147-156.
[55]Hara, H.; Morikawa, Y.; Sano, Y.; Yamauchi, K., Termination dependence of surface stacking at 4H-SiC(0001)-1x1: Density functional theory calculations. *Phys. Rev. B* **2009,** 79, (15), 153306.
[56]Shaffer, P. T. B., A Review of the Structure of Silicon Carbide. *Acta Crystallogr.* **1969,** B25, 477-488.

SIMULATION OF THE BALLISTIC IMPACT OF TUNGSTEN-BASED PENETRATORS ON CONFINED HOT-PRESSED BORON CARBIDE TARGETS

C.G Fountzoulas[1], J.C. LaSalvia[2]
U.S. Army Research Laboratory,
[1]RDRL-WMM-B, and [2]RDRL-WMM-E, APG, MD 21005-5069, USA

The rapid advancement of the computational power and the recent advances in the numerical techniques and materials models have resulted in improved simulation tools for ballistic impact into single and multi-layer armor configurations. However, the ability of a numerical model to realistically predict the response of ceramic armor to ballistic impact depends mainly on the selection of appropriate material models and availability of appropriate data.
An initial study of the ability of the existing material models to predict the observed damage induced by 93% tungsten heavy alloy (WHA) cylindrical projectiles striking confined cylinders of hot-pressed boron carbide targets (B_4C) at velocities between 819 m/s to 1205 m/s was performed. It was determined that the damage patterns were highly dependent on the properties of the confined ceramic and the impacting cylinder, whose failure behavior was difficult to model, and the strength and failure material models used for the modeling. The current paper will detail the results of parametric studies conducted of various model parameters in an attempt to simulate the observed damage and will detail ongoing efforts to improve the numerical results in order to accurately simulate the ballistic response of confined hot-pressed boron carbide targets.

INTRODUCTION

Ceramics are materials that possess characteristics such as low density, high hardness and high compressive strength that make them ideal for use in light armor; however, ceramics are also brittle and have a low tensile strength, which complicates the design of such systems. Numerous experimental investigations have been performed since the 1960s to develop an understanding of the behavior of ceramics under high velocity impact, the phenomenology of ceramic failure, and the behavior of the failed ceramic under large pressures. These studies have been crucial in the development of ceramic material models [1].

The pioneering experimental and numerical work of Wilkins and co-workers gave a great deal of insight into the penetration process and failure phenomenology of ceramics utilized in composite armor [2]. Their observations of ceramic failure were further expounded upon by the subsequent work of Shockey et al [3] who performed a series of experiments with confined ceramic targets impacted by tungsten-nickel-iron rods at velocities from 0.8-1.4 km/s and from the recovered targets were able to detail the failure sequence in the ceramic. It was determined that if the projectile velocity is sufficient or if the ceramic is not adequately confined, the projectile loading begins to introduce tensile cracks near the impactor periphery. Ceramics possess low tensile strength and when large tensile stress fields are introduced in the radial direction, they may form ring cracks, concentric around the impact point. Initially shallow, upon continued loading these ring cracks propagate in the direction of maximum principal tensile stress, typically on angles from 25°-75° normal to the surface. They progress to the back surface of the ceramic and form large Hertzian cone cracks. As the loading increases further, the compressive strength beneath the penetrator is exceeded and microcracking begins. After the cone cracks form, the principal stresses redistribute to a circumferential direction, which result in tensile radial cracks forming like spokes on a hub from the impact site. Lateral cracks also form below the impact surface and intersect the radial and cone cracks forming fragments, which can break away from the impact site leaving a crater [3].

Although fractured and fragmented, the ceramic still resists penetration. If confined, the penetrator can only proceed by moving the ceramic out of the way. To do so, the projectile must pulverize it and force the pulverized material to flow laterally and then rearward [3]. This zone of pulverized ceramic ahead of the projectile is known as the comminuted zone or the Mescall zone, after the work of Mescall et al [4]. Ceramic penetration resistance is a function of its compressive strength, hardness, pulverization, ceramic powder flow properties and abrasiveness, which serves to erode and consume the lead surface of the projectile [3].

Subsequent studies by Bless et al. [5] and Hauver et al [5] also observed that if the ceramic is sufficiently confined, the projectile will flow laterally on the surface of the ceramic and not penetrate. These phenomena have become known as interface defeat or infinite dwell [6]. Recent x-ray experiments by Lundberg et al. [7] have been able to capture the transition from the projectile dwelling on the ceramic surface to penetration.

However, perhaps the most widely utilized and well known of the material models developed for ceramics are the ones developed by Johnson and Holmquist [8, 9], which have become known as JH-1 and JH-2. Johnson, Holmquist and Beissel have also extended the ceramic model to account for phase transformation and this version is known as the JHB model [10]. One significant difference between the models is that the JH-2 allows for a gradual degradation of the ceramic strength from the intact surface as the damage accumulates, while the ceramic strength in the JH-1 and JHB models does not degrade from the intact strength curve until the material is fully damage (D=1.0).

This work investigates the ability of the existing ceramic material models to predict the observed damage, such as the lack of comminuted region, and short and long cracking introduced in confined cylinders of B_4C struck by cylinder of tungsten carbide at high (819 m/s-1205 m/s) velocities [11]. The effect of specific strength and failure model parameters as well as the inclusion of tensile crack softening on the damage development are described and the contribution and sensitivity of the these parameters on the accuracy of the solution when compared to the experimentally determined damage patterns are discussed.

EXPERIMENTAL DETAILS

Details of the impact experiments have been reported by LaSalvia et al [11] and the experimental setup is shown in Figure 1. The targets were center impacted at velocities ranging from 819 – 1205 m/s. Following the cylinder impact experiments, photomicrographs were taken of the impacted surfaces. To preserve the impact surface integrity, the cylinders were impregnated with a cold mount epoxy and then extracted from their Ti-6Al- 4V cups by carefully sectioning the cups (lengthwise) into several pieces. The final cross-sections were 0.1 – 0.3 mm from the apparent center of impact and were subsequently examined by optical and scanning electron microscopy. A cross section showing the damage of the B_4C at impact velocity at 1205 m/s is shown in Figure 2. The target was confined at bottom of the Ti-6Al-4V "cup" (V = 0 m/s)

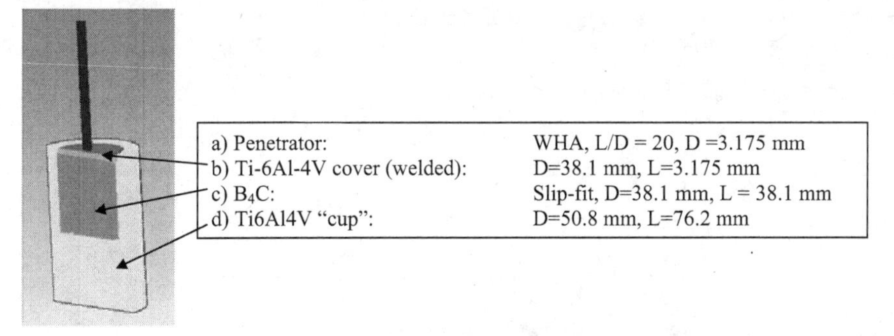

Figure 1. Target configuration.

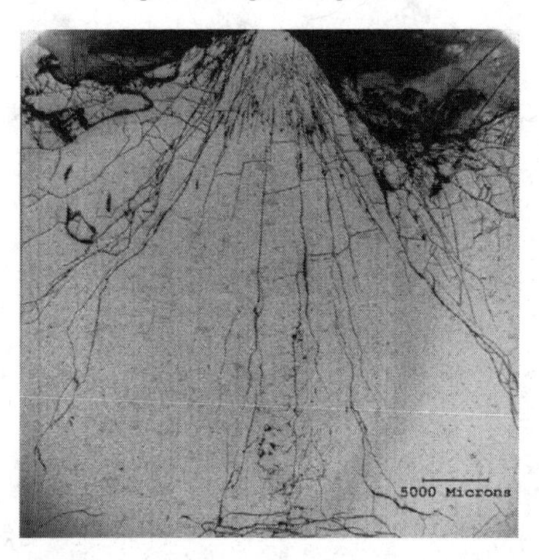

Figure 2. Photomicrograph of cross section of impacted B_4C target at 1205 m/s

NUMERICAL SIMULATIONS

The ballistic behavior of all the targets (Fig. 1) was studied by 2D and 3D models and simulated using the non-linear ANSYS/AUTODYN commercial package [5]. The material models used were obtained from the AUTODYN library [12]. Using the geometry detailed above, the Ti-6Al-4V "cup" and cover plate were modeled either by using Lagrange and/or SPH solvers. The SPH particle size and the mesh of the Ti-6Al-4V "cup" and cover plate were set to 0.25 mm and 0.5 mm for 2D and 3D modeling respectively. The projectile and B_4C were discretized using Smooth Particle Hydrodynamic (SPH) with a particle size of 0.25 mm and 0.5 mm for the 2D and 3D modeling respectively. However, the Ti-6Al-4V sleeve was modeled using a Puff equation of state (EOS), Von Mises strength and a Grady spall failure models taken from the Autodyn Material Library. The WHA projectile was modeled using a shock EOS, and a Johnson-Cook strength model [5]. The B_4C was modeled using a polynomial EOS, JH-2 strength and failure models. In addition, the tensile failure criterion for the B_4C was set to either the hydro-tensile or minimum pressure, or principal stress with or without crack softening. The symmetry of the 2D modeling was set to axisymmetric or planar. The 3D modeling half symmetry was used.

As a first attempt, all materials were modeled using their corresponding material EOS, strength and failure models as they appear at the AUTODYN library without any modification. In particular, the tensile failure of the JH-2 failure material model of B_4C was set to hydro-tensile (P_{min}). The results for all simulations did not correlate well with the experimental observations. According to LaSalvia [11] the B_4C for all impact velocities, 819, 1052, 1063, 1198 and 1205 m/s was cracked and the projectile was fractured (Figure 3). While the simulated extent of damage induced to the B_4C as

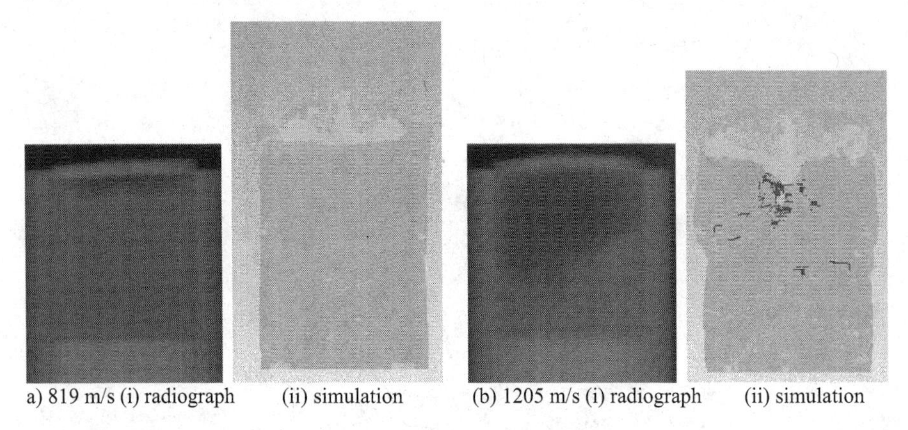

a) 819 m/s (i) radiograph (ii) simulation (b) 1205 m/s (i) radiograph (ii) simulation

Figure 3. Static x-ray radiographs of B_4C and simulated damaged status (40 μs) using hydro-tensile failure model

shown in Figure 3 resembles qualitatively the extent of damage shown on the x-ray radiographs, little cracking of the B_4C appears only at 1205 m/s impact velocity after 40 μs. However, the cracking shown on Figure 2 has not been reproduced by the simulation of the impact. Moreover, while the x-ray radiograph does not show target penetration at 1205 m/s, the simulation show that the impactor has already penetrated the B_4C target. In addition, while the X-ray radiographs show no deformation of the Ti-6Al-4V "cup", the simulations show that at 1205 m/s the "cup" has been heavily deformed. Continuation of the simulation at 1205 m/s impact showed complete WHA impactor fracture and heavy "cup" deformation with little B_4C cracking.

Subsequent studies were performed by utilizing changing the tensile failure of the JH-2 failure model of the B_4C to principal stress without crack softening, and crack softening with stochastic failure with stochastic variance $\gamma = 16$. The resulting simulation predictions of damage induced in the B_4C also did not correlate well with the experimental observations beyond 30 μs of simulation time. Figures 4 and 5, and Figures 6 and 7, show the damage evolution of the target for the cases of principal stress criterion without crack softening, and with crack softening with stochastic variance at impact velocities 819 m/s and 1205 m/s respectively. Below 30 μs the replication of the B_4C cracking show major cracks and minor ones, emanating from the major cracks, similar to the pattern observed experimentally. Simulations for times beyond 40 μs for impact velocity 819 m/s and 25μs for impact velocity 1205 m/s showed complete B_4C cracking. Figure 8 show the 3D modeling damage evolution after 25 μs with crack softening with stochastic variance at 1205 m/s impact velocity. Figure 9 shows the damage status for 2D axisymmetric modeling after 25μs for impact velocity 1205 m/s using hydro-tensile criterion (a), principal stress (b), and principal stress with crack softening and stochastic variance (c) and comparing them with the planar modeling damage status after 25 μs.

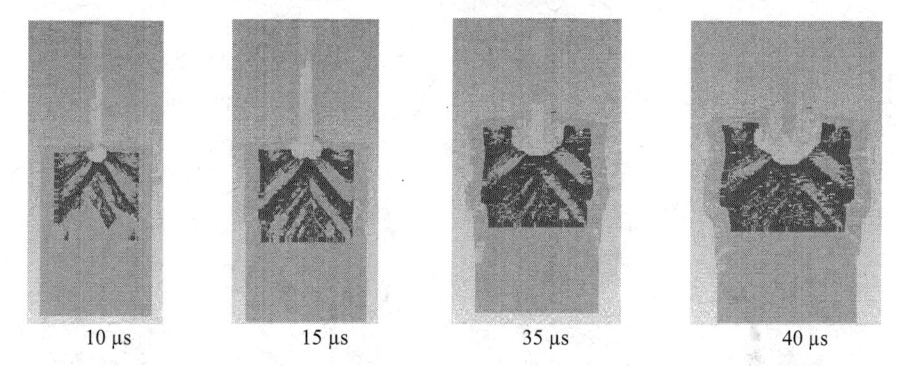

Figure 4. Damage evolution at 819 m/s for principal failure criterion without crack softening

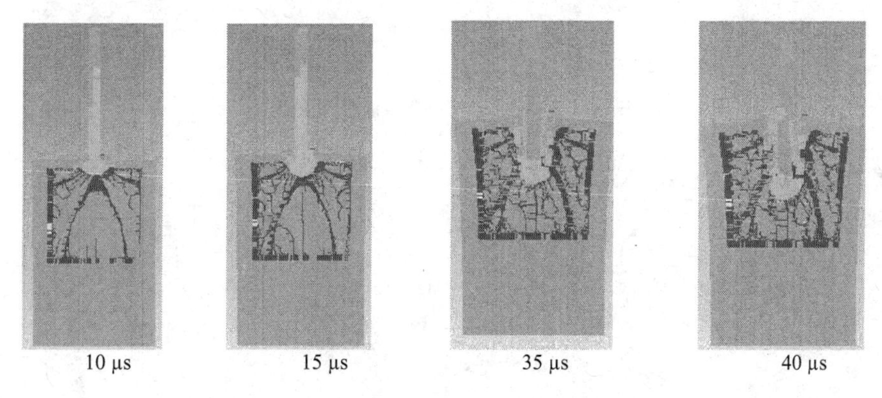

Figure 5. Damage evolution at 819 m/s for principal failure criterion with crack softening and stochastic variance $\gamma = 16$

With the JH models, the damage is a function of the deviatoric strain of the material, which is representative of material in the high compression region ahead of an impacting projectile. In areas away from the impact region, the tensile and shear stresses may be of equivalent magnitude, which for brittle materials, may lead to tensile cracking. The ability of the JH models to model this tensile cracking depends on the maximum tensile hydrostatic pressure, T. The tensile crack softening algorithm has been developed and implemented specifically for enhanced capability of predicting the tensile cracking in brittle materials, such as ceramics and concrete [13] and was utilized in conjunction with the Johnson-Holmquist failure models.

As it becomes apparent from Figures 3-9 no failure applied were able to replicate the observed damage in the B_4C for velocities between 819 m/s to 1205 m/s. Depending on the failure model studied, some models replicated the damage to a satisfactory level up to a certain simulation time but none until the projectile defeat.

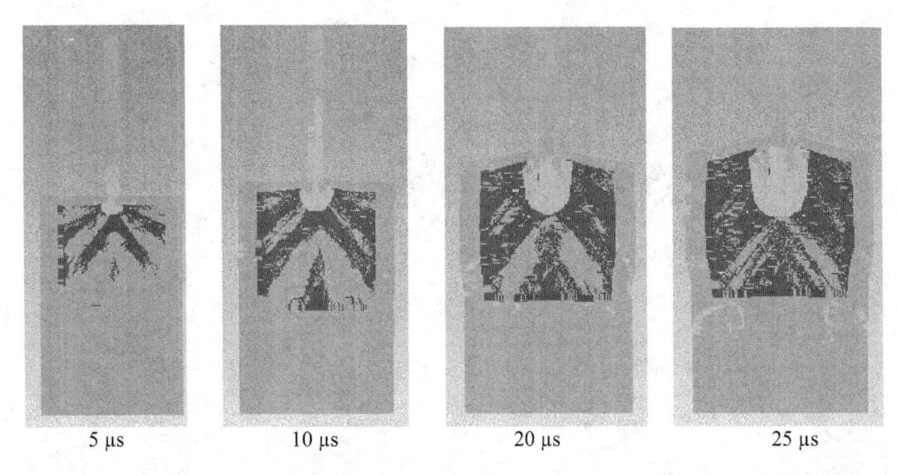

Figure 6. Damage evolution at 1205 m/s for principal failure criterion without crack softening

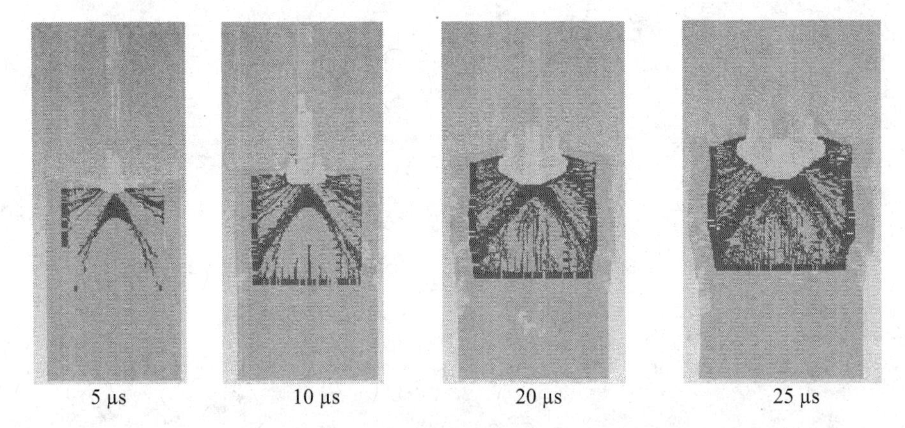

Figure 7. Damage evolution at 1205 m/s for principal failure criterion with crack softening and stochastic variance $\gamma = 15$

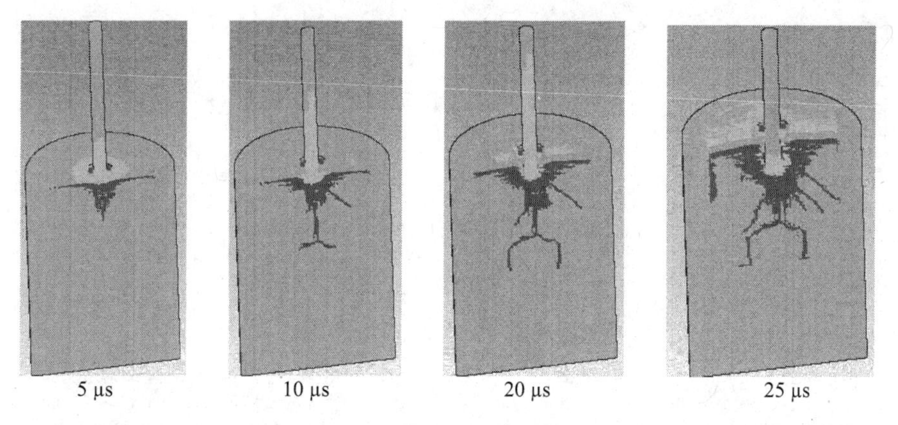

Figure 8. Damage evolution at 1205 m/s for principal failure criterion with crack softening and stochastic variance $\gamma = 15$ (3D modeling)

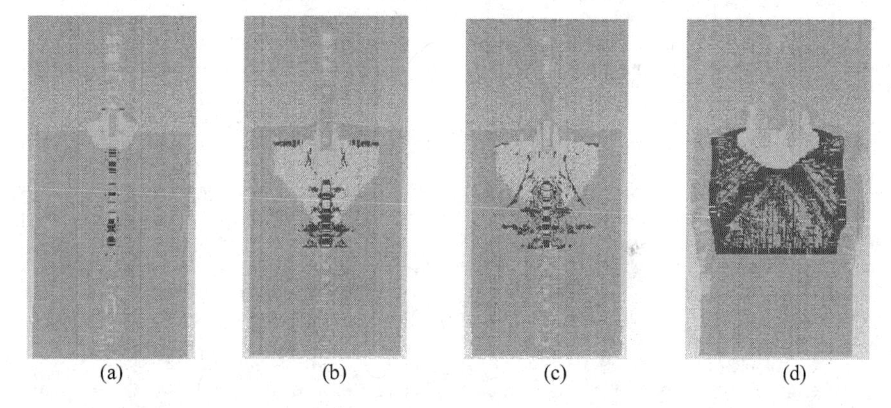

Figure 9. V_{imp} =1205 m/s after 25 μs: a-c: Axisymmetric modeling, hydro-tensile, principal stress, and principal stress with crack softening and stochastic variance $\gamma = 15$; and (d) Planar symmetry with principal stress with crack softening and stochastic variance $\gamma = 15$

DISCUSSION

Available confined hot pressed B_4C targets, impacted by WHA cylindrical projectiles, show significant ceramic cracking. The hydro-tensile failure criterion was not able to introduce cracking to the B_4C target even when it was decreased from 7300 MPa [12] to 2000 MPa (Fig. 3). For planar symmetry for the 2D modeling the cracking was replicated better up to 35 μs by using the principal stress failure model with crack softening and stochastic variance of the ceramic. When using the principal stress failure model with crack softening, only the ceramic crack is excessive for both, low and high impact velocity. For the case of the 2D axisymmetric simulation, also the principal stress failure model with crack softening and stochastic variance induced some cracking to the ceramic target, but it failed to capture most of it. Moreover, 2D axisymmetric modeling introduces an artificial target failure along the axis of symmetry (Fig. 9). Similarly, for the 3D modeling, the latter failure model replicated poorly the ceramic failure up to 40 μs, but replicated the experimental data better than any other failure model used. After 40 μs of simulation, the ceramic target was completely fractured and no individual cracks were able to be observed, opposite to what is observed experimentally.

However, no B_4C failure model managed to stop the projectile from penetrating the target while introducing cracking. Few initial attempts to vary the parameters of the strength and failure models of both WHA and Ti-6Al-4V in conjunction with the B_4C material models were not successful in preventing the target penetration and cracking the target at the same time. The target penetration resulted in bulging of the Ti-6Al-4V "cup", which was not observed experimentally for the impact velocity range used for the testing.

The failure of the existing strength and failure material models of all the parts of the target architecture to capture the ceramic failure without target penetration, may be attributed to the inability of these models to account for the continuously developing compressive stresses caused by the ceramic confinement during the impact.

The simulations provide a significant amount of insight into the problem and could provide another measure that could be quantified through experiments. The simulations also showed that the higher the target cracking the more the retardation of the penetration process. The authors believe that due to the ceramic confinement, the strength of the failed ceramic is higher than the predicted by the B_4C existing failure model, thus resulting in more effective defeat of the impacting WHA cylindrical impactor.

CONCLUSIONS

The Johnson-Holmquist series of ceramic material models have been developed for the high velocity impact and penetration of ceramics; their applicability to the confined hot-pressed boron carbide target hit by cylindrical WHA impactor is of interest. The ability of the phenomenological Johnson-Holmquist ceramic model to predict the observed damage patterns induced by the WHA impactor striking confined cylinders of hot-pressed boron carbide at velocities from 819 m/s to 1205 m/s was investigated. Due to difficulties modeling not only the boron carbide, but the Ti-6Al-4V "cover and "cup" and WHA impactor for this target architecture, the B_4C damage was not able to be replicated until the stoppage and or fracture of the impactor satisfactorily. The authors believe that additional characterization of the hot-pressed boron carbide and WHA are needed for more effective material models. Additional investigations are ongoing towards understanding the effect of the ceramic confinement on the impact resistance, its quantification and subsequent introduction to the existing material models.

REFERENCES

[1] C. G. Fountzoulas, M. J. Normandia, J. C. LaSalvia, B. A. Cheeseman , "Numerical Simulation of the Cracking of Silicon Carbide Tiles Impacted by Tungsten Carbide Spheres", 22nd IBS (2005), Vancouver, Canada

[2] Wilkins M.L., Cline C.F. and Hondol C.A., Tech.Rep. UCRL-50694 (Lawrence Rad.Lab. 1969)

[3] Shockey D.A., Marchand A.H., Skaggs S. R., Cort G.E. Burkett M.W., Parker M.W. 1990, "Failure Phenomenology of Confined Ceramic Targets and Impacting Rods," Int. J. Imp. Eng., 9 (3): 263-275

[4] Mescall J, Weiss V. "Materials behavior under high stress and ultrahigh loading rates" *Part II. Proceeding of the 29th Sagamore Army Conference, Army Materials and Mechanics Research Center: Watertown, MA, U.S.A.,* 1984.

[5] Bless S. J., Brar N. S. Kanel G and Rosenberg Z., "Failure waves in glass", J. Am. Cer. Soc." 75 (4):1002-1004, Aprl 1992

[6] Hauver G.E., Netherwood P. H., Benck R. F. and Kecskes L. J.. "Enhanced ballistic performance of ceramic targets" 1994 *19th. Army Science Conference*, USA

[7] Lundberg P., Renström R., Holmberg L. "An experimental investigation of interface defeat at extended interaction times" *Proc. 19th. Int. Symp. Ballistics*, Vol.3, 1463-1469 (2001).

[8] Johnson G. R. and Holmquist T. J., "A computational constitutive model for brittle materials subjected to large strains, high strain rates and high pressure", *Proc. EXPLOMET Conference*, San Diego, (1990).

[9] Johnson G. R. and Holmquist T. J. "An improved computational constitutive model for brittle materials" In *High pressure science and technology*—1993, Vol. 2, 981-984, Edited by S. C. Schmidt, J. W. Shaner, G. A. Samara and M. Ross, AIP Press, New York (1994).

[10] Holmquist, T.J., Johnson, G.R., and Beissel, S.R., J. Applied Physics, 94, 2003.

[11] LaSalvia presentation, "Ballistic Impact Damage in Hot- Pressed Boron Carbide" 34 ICACC, Daytona Beach, FL, January 25-29, 2010

[12] ANSYS/AUTODYN Vol. 12.1, Manual, Century Dynamics Inc., Concord, CA

[13] Clegg, R.A., Hayhurst, C.J., Robertson, I., "Development and application of a Rankine plasticity model for improved prediction of tensile cracking in ceramic and concrete materials under impact," in the *Proceedings of the 14th DYMAT Technical Meeting*, Sevilla, Spain, 14-15 November 2002.

Author Index